D1693577

Integral Materials Modeling

Edited by
Günter Gottstein

1807–2007 Knowledge for Generations

Each generation has its unique needs and aspirations. When Charles Wiley first opened his small printing shop in lower Manhattan in 1807, it was a generation of boundless potential searching for an identity. And we were there, helping to define a new American literary tradition. Over half a century later, in the midst of the Second Industrial Revolution, it was a generation focused on building the future. Once again, we were there, supplying the critical scientific, technical, and engineering knowledge that helped frame the world. Throughout the 20th Century, and into the new millennium, nations began to reach out beyond their own borders and a new international community was born. Wiley was there, expanding its operations around the world to enable a global exchange of ideas, opinions, and know-how.

For 200 years, Wiley has been an integral part of each generation's journey, enabling the flow of information and understanding necessary to meet their needs and fulfill their aspirations. Today, bold new technologies are changing the way we live and learn. Wiley will be there, providing you the must-have knowledge you need to imagine new worlds, new possibilities, and new opportunities.

Generations come and go, but you can always count on Wiley to provide you the knowledge you need, when and where you need it!

William J. Pesce
President and Chief Executive Officer

Peter Booth Wiley
Chairman of the Board

Integral Materials Modeling

Towards Physics-Based Through-Process Models

Edited by
Günter Gottstein

WILEY-VCH Verlag GmbH & Co. KGaA

The Editor

Prof. Dr. Günter Gottstein
Institut für Metallkunde und
Metallphysik der RWTH Aachen
Kopernikusstrasse 14
52074 Aachen
Germany

All books published by Wiley-VCH are carefully produced. Nevertheless, authors, editors, and publisher do not warrant the information contained in these books, including this book, to be free of errors. Readers are advised to keep in mind that statements, data, illustrations, procedural details or other items may inadvertently be inaccurate.

Library of Congress Card No.: applied for
British Library Cataloguing-in-Publication Data
A catalogue record for this book is available from the British Library

Bibliographic information published by the Deutsche Nationalbibliothek
Die Deutsche Nationalbibliothek lists this publication in the Deutsche Nationalbibliografie; detailed bibliographic data are available in the Internet at http://dnb.d-nb.de

© 2007 WILEY-VCH Verlag GmbH & Co. KGaA, Weinheim

All rights reserved (including those of translation into other languages). No part of this book may be reproduced in any form – by photoprinting, microfilm, or any other means – nor transmitted or translated into a machine language without written permission from the publishers. Registered names, trademarks, etc. used in this book, even when not specifically marked as such, are not to be considered unprotected by law.

Printed in the Federal Republic of Germany
Printed on acid-free paper

Typesetting Asco Typesetters, Hong Kong
Printing betz-druck GmbH, Darmstadt
Bookbinding Litges & Dopf Buchbinderei GmbH, Heppenheim
Wiley Bicentennial Logo Richard J. Pacifico

ISBN 978-3-527-31711-0

Contents

Integral Materials Modeling: Towards Physics-Based Through-Process Models
Edited by Günter Gottstein
Copyright © 2007 WILEY-VCH Verlag GmbH & Co. KGaA, Weinheim
ISBN: 978-3-527-31711-0

List of Contributors

G. Gottstein, E. Jannot, C. Schäfer, V. Mohles, M. Schneider
Institut für Metallkunde und Metallphysik
Rheinisch-Westfälische Technische Hochschule Aachen
52056 Aachen
Germany

A. Bührig-Polaczek, B. Pustal
Gießerei-Institut
RWTH Aachen
52056 Aachen
Germany

W. Michaeli, E. Schmachtenberg, M. Brinkmann, M. Bussmann, B. Renner
Institut für Kunststoffverarbeitung
Rheinisch-Westfälische Technische Hochschule Aachen
52056 Aachen
Germany

J. M. Schneider, B. Hallstedt, C. Walter, J. Müller, D. Hajas, E. Münstermann
Institut für Werkstoffchemie
Rheinisch-Westfälische Technische Hochschule Aachen
52056 Aachen
Germany

G. Hirt, L. Neumann, R. Kopp, X. Li
Institut für Bildsame Formgebung
Rheinisch-Westfälische Technische Hochschule Aachen
52056 Aachen
Germany

W. Bleck, U. Prahl, A.-P. Hollands, D. Senk, B. Zeislmair, F. Gerdemann
Institut für Eisenhüttenkunde
Rheinisch-Westfälische Technische Hochschule Aachen
52056 Aachen
Germany

V. Pavlyk, U. Dilthey, O. Mokrov
Institut für Schweißtechnik und Fügetechnik
Rheinisch-Westfälische Technische Hochschule Aachen
52062 Aachen
Germany

W. Bleck
Institut für Eisenhüttenkunde
Rheinisch-Westfälische Technische Hochschule Aachen
52056 Aachen
Germany

Integral Materials Modeling: Towards Physics-Based Through-Process Models
Edited by Günter Gottstein
Copyright © 2007 WILEY-VCH Verlag GmbH & Co. KGaA, Weinheim
ISBN: 978-3-527-31711-0

H. Emmerich, R. Siquieri
Institut für Gesteinshüttenkunde
Lehr- und Forschungsgebiet
Modellbildung in der
Werkstofftechnik
Rheinisch-Westfälische
Technische Hochschule Aachen
52064 Aachen
Germany

R. Nickel, D. Parkot, K. Bobzin, E. Lugscheider
Institut für Oberflächentechnik
Rheinisch-Westfälische
Technische Hochschule Aachen
52062 Aachen
Germany

L. Singheiser, R. Herzog, P. Bednarz, O. Trunova
Institut für Werkstoffe und
Verfahren der Energietechnik 2
Forschungszentrum Jülich GmbH
Leo-Brandt-Strasse
52428 Jülich
Germany

K. Albe, P. Erhart, M. Müller
Institut für Materialwissenschaft
Technische Universität Darmstadt
Petersenstrasse 23
64287 Darmstadt
Germany

P. S. Bate
School of Materials
University of Manchester
Grosvenor Street
Manchester M1 7HS
United Kingdom

Y. Estrin
Institut für Werkstoffkunde und
Werkstofftechnik
Technische Universität Clausthal
Agricolastrasse 6
38678 Clausthal-Zellerfeld
Germany

M. E. Glicksman
Materials Science & Engineering
Department
Rensselaer Polytechnic Institute
Troy, NY 12180-3590
USA

D. Raabe
Max-Planck-Institut für Eisenforschung
Max-Planck-Strasse 1
40237 Düsseldorf
Germany

M. Schneider, W. Schaefer, G. Mazourkevitch
MAGMA Gießereitechnologie GmbH
Kackertstrasse 11
52072 Aachen
Germany

Y. Wang, C. Shen, J. Li, M. J. Mills
Department of Materials Science and
Engineering
Ohio State University
2041 College Road
Columbus, Ohio 43210
USA

M. Thornagel
SIGMA Engineering GmbH
Kackertstrasse 11
52072 Aachen
Germany

I. Steinbach, N. Warnken
ACCESS e. V.
Intzestrasse 5
52072 Aachen
Germany

M. Crumbach
Novelis Technology AG
Badische Bahnhofstrasse 16
8212 Neuhausen
Switzerland

M. Wessen, I. L. Svensson, S. Seifeddine, J. Olsson
Division Component Technology
Jököping University
55111 Jönköping
Sweden

C. Beckermann, K. Carlson
University of Iowa
Department of Mechanical and Industrial Engineering
Iowa City, IA 52242-1527
USA

P. R. Rios
Universidade Federal Fluminense
Escola de Engenharia Industrial Metalúrgica
de Volta Redonda
Avenida dos Trabalhadores, 420
Volta Redonda, 27255-125
Brazil

1
Introduction

This book comprises the proceedings of the final symposium of the Collaborative Research Center (SFB 370) of the Deutsche Forschungsgemeinschaft on "Integral Materials Modeling" which took place in Aachen, Germany, on December 1–2, 2005. It is composed of the final reports of the projects and complementary manuscripts of renowned scientists in the field of materials modeling, covering a broad range of current simulation activities.

The projects are identified by their project numbers in their title. The manuscripts are organized such that after a list of persons involved in the SFB 370 the final through-process modeling exercises (group C) are introduced by the reports on supporting process and materials models (groups A and B) and complemented by the invited contributions. The first article on "Integral Materials Modeling" gives an introduction into the philosophy, history, and structure of the collaborative research center.

With the final symposium the SFB officially ended but its core topic was continued in a transfer program of the Deutsche Forschungsgemeinschaft (TFB 63) on "Industrially Relevant Modeling Tools".

As a chairman of the collaborative research center on "Integral Materials Modeling" (SFB 370) I would like to express my sincere gratitude to my colleagues for their continuous support and encouragement. As a university professor it was my great pleasure to see the interest and engagement of the young doctoral students in the research program, their fascination by the scientific challenge, and their natural openness to interdisciplinary cooperation, discussion, and information exchange. Last but not least my thanks go to the review panels for their valuable advice and the Deutsche Forschungsgemeinschaft which not only funded the collaborative research center for 12 years but also offered unbureaucratic support.

Günter Gottstein
Chairman, SFB 370
Aachen, December 2006

Integral Materials Modeling: Towards Physics-Based Through-Process Models
Edited by Günter Gottstein
Copyright © 2007 WILEY-VCH Verlag GmbH & Co. KGaA, Weinheim
ISBN: 978-3-527-31711-0

Collaborative Research Center (SFB 370) of the Deutsche Forschungsgemeinschaft on "Integral Materials Modeling" (1994–2005)

Board

Prof. Dr. G. Gottstein, IMM
Prof. Dr.-Ing. U. Dilthey, ISF
Prof. Dr.-Ing. R. Kopp, IBF
Prof. Dr.-Ing. G. Hirt, IBF

Members

Prof. Dr.-Ing. W. Bleck, IEHK
Prof. Dr.-Ing. K. Bobzin, IOT
Prof. Dr.-Ing. A. Bührig-Polaczek, GI
Prof. Dr.-Ing. W. Dahl, IEHK
Prof. Dr.-Ing. E. El-Magd, IWK
Prof. Dr.-Ing. S. Engler, GI
Prof. Dr.-Ing. O. Knotek, IWK
Prof. Dr.-Ing. E. Lugscheider, LNWW
Prof. Dr.-Ing. W. Michaeli, IWK
Prof. Dr.-Ing. D. Neuschütz, LTH
Prof. Dr.-Ing. H. Nickel, FZ Jülich
Dr. S. Rex, ACCESS
Prof. Dr.-Ing. P.R. Sahm, GI
Prof. Dr.-Ing. E. Schmachtenberg, IKV
Prof. Dr.-Ing. J. Schneider, MCh
Prof. Dr.-Ing. D.G. Senk, IEHK
Prof. Dr.-Ing. L. Singheiser, FZ Jülich

SFB-Managing Director at RWTH Aachen

Dr. P. van den Brincken

Guest Members

Dr.-Ing. N. Aretz, Stuttgart
Dr.-Ing. M. Keller, Duisburg
Dr.-Ing. U. Lotter, Duisburg
Dr. A. Ludwig, Leoben
Dr.-Ing. L. Löchte, Bonn
Prof. Dr. J. Vehoff, Saarbrücken

Reviewers

Prof. Dr.-Ing. D. Aurich, Berlin
Prof. Dr.-Ing. H. Biermann, Freiberg
Prof. Dr.-Ing. U. Draugelates, Clausthal
Prof. Dr. J. Estrin, Clausthal
Dr. F.J. Floßdorf, Düsseldorf
Dr. A. Kamp, Dortmund
Prof. Dr.-Ing. R. Kawalla, Freiberg
Prof. Dr.-Ing. D. Munz, Karlsruhe
Prof. Dr. P. Neumann, Düsseldorf
Dr. H. Pircher, Duisburg
Prof. Dr.-Ing. H. Potente, Paderborn
Dr. H. Riedel, Freiburg
Prof. Dr. S. Schmauder, Stuttgart
Prof. Dr.-Ing. R.F. Singer, Erlangen
Prof. Dr.-Ing. H.D. Steffens, Dortmund
Prof. Dr.-Ing. M. Wagner, Stuttgart
Prof. Dr. E. Werner, München

DFG-Correspondents (DFG-Senate Committee Members)

Prof. Dr. H. Eschrig, Dresden
Prof. Dr. W. Schulze, München
Prof. Dr.-Ing. E. Ramm, Stuttgart
Prof. Dr. H. Vahrenkamp, Freiburg
Prof. Dr.-Ing. D. Löhe, Karlsruhe
Prof. Dr. M. Reddehase, Mainz
Prof. Dr. S. Müller, Leipzig

DFG-Program Monitors

Mr. T. Leppien
Mrs. Dr. G. Retz-Schmidt
Mrs. Dr. H. Hildebrandt
Dr. J. Kunze

DFG-Expert Advisors

Dr.-Ing. J. Tobolski
Dr.-Ing. F. Fischer
Dr.-Ing. B. Jahnen

2
Integral Materials Modeling

G. Gottstein

Abstract

This chapter reviews the historical background of computational materials science and introduces the scientific concept of "integral materials modeling". The objectives of the collaborative research center on this topic are formulated and an overview is given on the structure development and achievements of the research program.

2.1
Introduction

One of the ultimate dreams of materials science is the theoretical design of new materials. It would save tremendous costs that are currently invested in alloy development, e.g. for operating expensive pilot plants and conducting comprehensive materials testing, and in view of the fact that even today it takes more than 15 years before a new material eventually sees the market. That there is still need for research to develop new metallic materials although they have be successfully processed for more than 5000 years is not due to the large number of potential alloys that can be produced in multicomponent alloys out of 92 elements or even 70 elements with metallic character. On the contrary, if the properties of these alloys would only reflect the property mix of their components, it would be easy with current computer power to predict the properties of virtually any potential alloy system. However, the properties of a material do not reflect the properties of the constituent elements, rather the properties of a material are controlled by the spatial distribution of elements and crystal defects, which is also referred to as microstructure. The microstructure comprises phase distribution, elemental distribution, orientation distribution, as well as crystal defects like grain boundaries, dislocations, and point defects. What is more, the microstructure is seriously affected by materials processing. In essence, the properties of a material are not given by a superposition of elemental properties, but by a complex function of

Integral Materials Modeling: Towards Physics-Based Through-Process Models
Edited by Günter Gottstein
Copyright © 2007 WILEY-VCH Verlag GmbH & Co. KGaA, Weinheim
ISBN: 978-3-527-31711-0

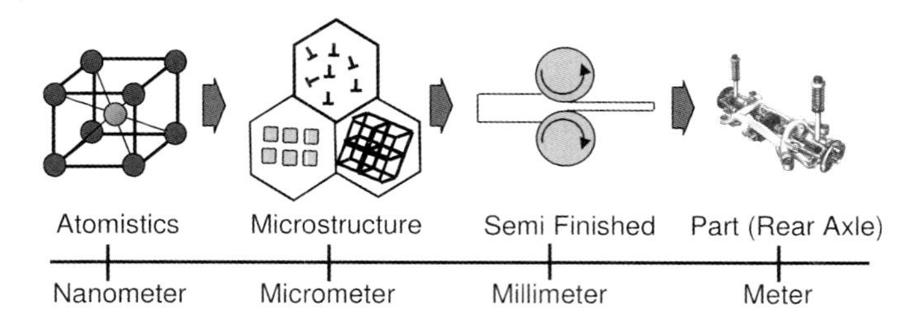

Fig. 2.1 The scale problem of materials modeling: the macroscopic properties are defined by the microstructure which develops by atomic mechanisms.

its chemical composition and its processing history, and there is a virtually infinite number of possible microstructures and, therefore, of material properties. This is good news for alloy development since material properties can be changed by processing at constant chemistry in a wide range but it is a nightmare for materials modeling that aims at predicting material properties from the knowledge of materials chemistry and processing, and it appears hopeless to design computational strategies for optimization of materials chemistry and processing for a given desired property spectrum.

As a result, it is practically impossible to establish relations between materials properties and both processing parameters and the overall chemistry. This is because of the fact that the processing conditions are not the state variables of materials properties even though that would be desirable for the materials engineer who is used to formulate the properties of a product in terms of the engineering control parameters. Processing–property relationship may be most desirable for engineering practice, but unfortunately this approach cannot be successful. It is common in materials engineering to establish correlations between material properties and processing conditions. These correlations, however, are not equations of state for the material, and thus liable to fail when the chemistry or processing parameters are changed.

The only correct way to formulate equations of state for the terminal properties of a processed material is in terms of microstructure development during processing because the microstructure constitutes the state variable of the material. In fact, the microstructure is a fingerprint of the processing history of a material and determines the current properties of a material. That is why geologists hope to derive the history of rocks from today's microstructure to understand the formation of the Earth's crust. Consequently, the very problem of property predictions is to adequately quantify a microstructure and, in particular, to establish microstructure–property relationships. This requires an understanding of the controlling microstructural elements and the microscopic processes that determine a specific property.

An adequate characterization of the microstructure requires knowledge of the atomistic arrangement in a material. Therefore, a deeper physical understanding

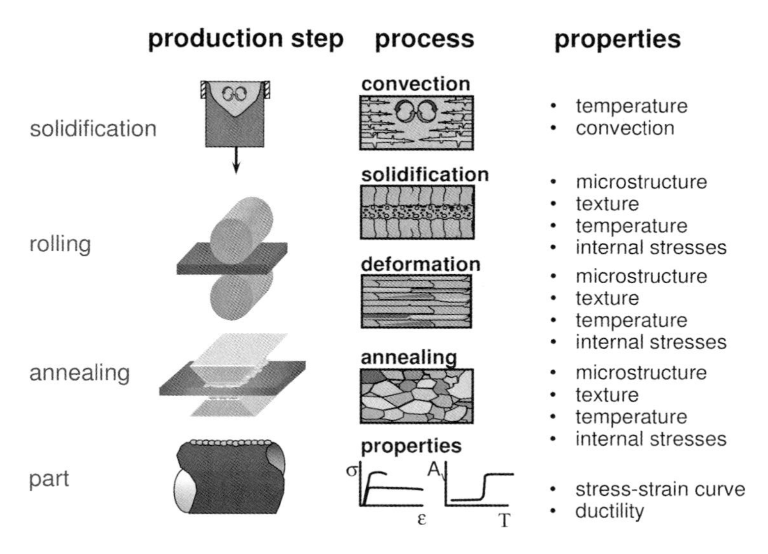

Fig. 2.2 During conventional processing of a metallic material to the final product (here a welded tube) the microstructure changes during each processing step.

of material properties and phenomena could only be developed after the discovery of X-rays and their application to crystallography in the beginning of the 20th century. This engendered an understanding of material behavior on the basis of its atomistic arrangements and atomistic transport processes. Application of X-ray diffraction and spectroscopy complemented much later by electron and neutron diffraction and spectroscopy revealed the importance of nanoscale configurations, notably the crystal structure, crystal defects, and nanoscale chemistry, in terms of solute distribution and dispersion of second phases, for an interpretation of material properties. In fact, materials science in the 20th century was essentially dedicated to an understanding of microstructure evolution and a formulation of microstructure–property relationships.

From this research it became clear, however, that there are only a few although complex physical processes that impact the microstructure of a bulk metallic material. The mechanisms of these processes have been the subject of investigations for many years, and the outcome of this research constitutes the foundations and concepts of modern materials science.

In essence, there are three major processes that affect the microstructure of a material, namely phase transformations, plastic deformation, and restoration processes like recovery, recrystallization, and grain growth. These processes and their impact on microstructure are complex in detail and interdependent in a highly nonlinear fashion. Their thermodynamics, kinetics, and atomic mechanisms have been subject of numerous investigations over the past 50 years, which have substantiated that microstructural evolution is strongly related to the properties

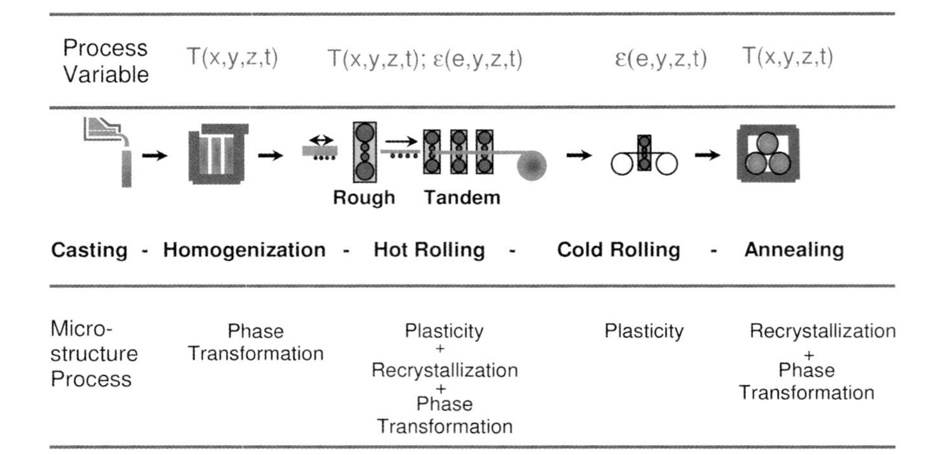

Fig. 2.3 For simulation of microstructure evolution the microstructure processes have to be interfaced to the process variables; here for sheet fabrication of an aluminum alloy.

and behavior of crystal defects: deformation of crystals proceeds by the generation, motion, interaction, and storage of dislocations; phase transformations are controlled by diffusion, the fundamental atomic transport mechanism; and recrystallization and grain growth involve the motion of grain boundaries.

These three microstructural processes affect each other, e.g. precipitation of a second phase hinders dislocation and grain boundary motion and, therefore, influences hardening during crystal plasticity and softening by recrystallization and grain growth. A major complication for the mathematical treatment of these processes is the local inhomogeneity that is introduced or even imposed by processing, e.g. fluctuation of composition, segregation of elements, deformation inhomogeneities, etc.

The theoretical foundations that govern the thermodynamics and kinetics of these processes have been developed during the past 50 years [1–4]. The respective equations of motion and equations of state for these processes are generally formulated in terms of partial differential equations. An analytical solution of these equations was in the vast majority of cases beyond reach, even impossible. Even 20 years ago such mathematical problems could be solved only by experts for very special cases. This situation changed dramatically with the advent of powerful computers, which could be utilized to solve numerically these difficult equations. With the increasingly powerful computers at hand to everybody nowadays, virtually every scientist can tackle these mathematical problems by utilizing sophisticated software. Moreover, the availability of high-performance computers has engendered novel computational techniques to address microstructural changes and thus the option to simulate microstructural evolution on the computer [5].

While the simulation of microstructure evolution constitutes a remarkable progress in computational materials science it does not yet solve the engineering problem to predict the properties of a material. Since the microstructure controls materials properties but is affected by each processing step, the prediction of terminal material properties or the behavior of a part under service conditions requires one to follow microstructural evolution along the entire processing chain, i.e. conventionally from the liquid state to the final product. To simulate this on the computer, it is necessary to connect the microstructural evolution to the processing parameters, which essentially means to subject each microstructural volume element to a temporal change of strain and temperature. In engineering applications this temperature and strain history of a material is typically computed by finite element (FE) approaches. On the other hand, the results of an FE simulation depend on the current properties of a material; hence microstructural evolution along the processing history and the local processing parameters, strain and temperature, are interdependent. In essence, both approaches, processing in terms of FE codes and microstructure in terms of physics-based microstructure evolution codes, have to be connected and interfaced in space and time. This means accounting for local and temporal changes of chemical composition, segregation, defect densities, etc., under changing boundary conditions. Therefore, it also requires advanced interface tools and sophisticated numerical techniques to solve the respective sets of mathematical equations and places a substantial demand on computing power.

2.2 The Collaborative Research Center on "Integral Materials Modeling"

In 1994 the collaborative research center on "Integral Materials Modeling" (SFB 370) of the Deutsche Forschungsgemeinschaft set out to tackle this problem and to develop strategies and techniques to predict terminal material properties from the knowledge of material chemistry and processing conditions. Since each processing step affects the microstructure, the microstructure evolution through the entire processing chain had to be traced to determine the final microstructure at the end of the processing chain in order to predict the properties associated with it. This though-process modeling approach on a physics-based microstructure definition was subject of research for a variety of materials and their processing. Two fundamental processing stages (A and B) were distinguished, which characterized virtually all materials fabrication procedures (Fig. 2.4):

(A) Generation of the solid state from the liquid phase (solidification) or gas phase (condensation).
(B) Processing of the solid by mechanical and thermal treatment.

They determined the properties of the product which constituted stage C:

(C) Determination of the specific properties of interest for a processed product.

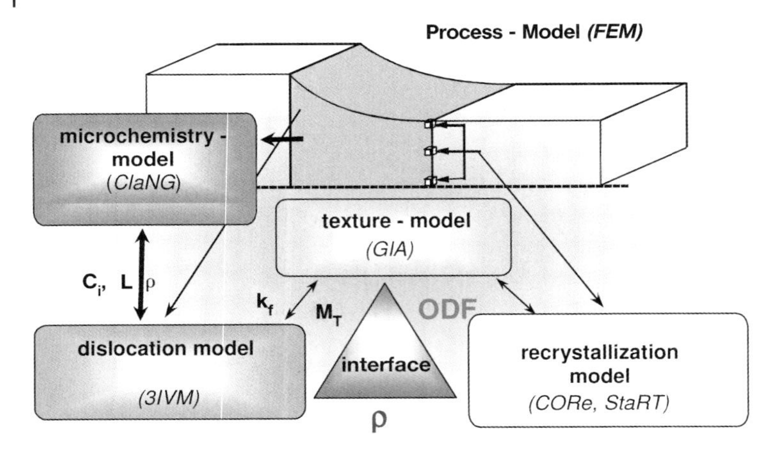

Fig. 2.4 Coupling of microstructure and processing by interfacing FEM and physics-based microstructure models. For each volume element the microstructural information has to be updated in every time step.

Depending on the specific product considered these processing stages can be connected in various ways (Fig. 2.5). For a net shape cast metallic part or an injection molded polymer, stage A directly connects to stage C. Stage A itself may be subdivided in to several steps, e.g. if a cast part is coated. A homogenization anneal

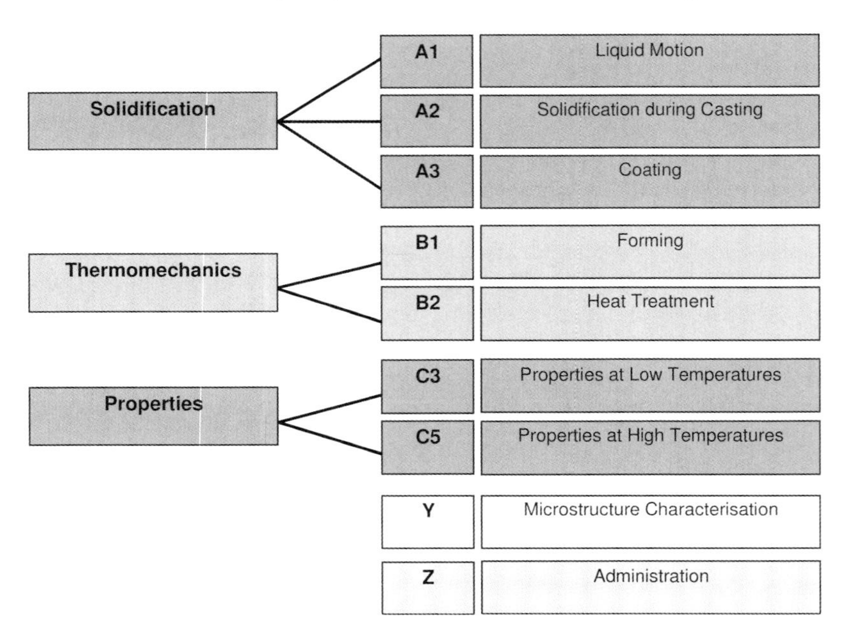

Fig. 2.5 Topical structure of the collaborative research center (SFB 370) "Integral Materials Modeling" of the Deutsche Forschungsgemeinschaft.

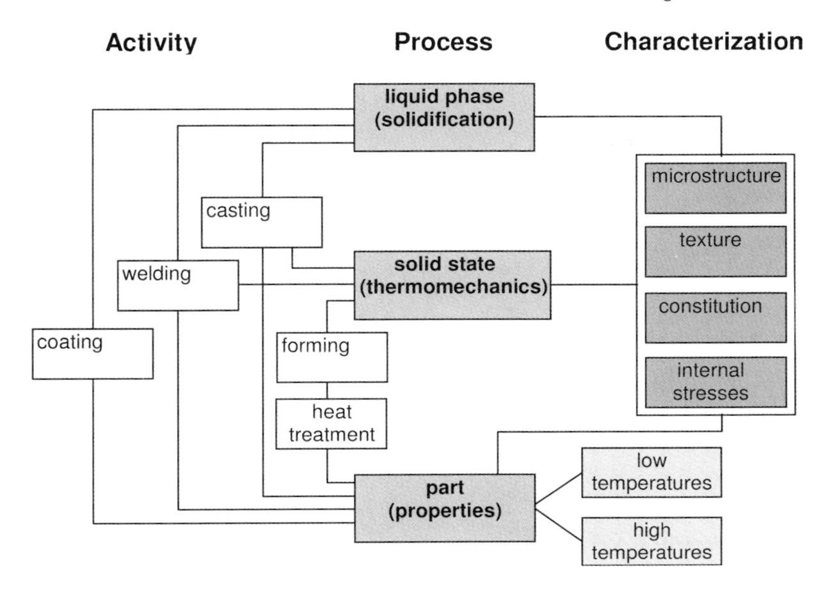

Fig. 2.6 A wide variety of process combinations leads to the final part with specific properties.

of the cast part is usually considered as an integral part of the casting process. The typical processing route for wrought alloys is continuous casting, multistep thermomechanical (or thermal and mechanical) processing, and, finally, shaping (e.g. sheet forming, bending, etc.).

A special and very complicated sequence of processing stages occurs during welding, where local solidification and thermal treatment of a premanufactured microstructure is considered. The heat input leads to a heat-affected zone besides a solidified weld pool.

The processing stages were subdivided into processing steps which constituted the research projects (Fig. 2.6). Twelve institutes of the RWTH Aachen University and the Research Center Jülich were engaged in the program and worked jointly on specific projects.

Processing stages A and B and properties C comprised seven projects:

A1: Fluxes in the melt
A2: Solidification of the liquid state
A3: Condensation from the vapor phase

B1: Thermomechanical processing (hot and cold forming)
B2: Thermal processing at nonstationary temperature

C1: Mechanical properties at low temperatures (strength, toughness)
C2: Mechanical properties at high temperatures (strength, creep resistance)

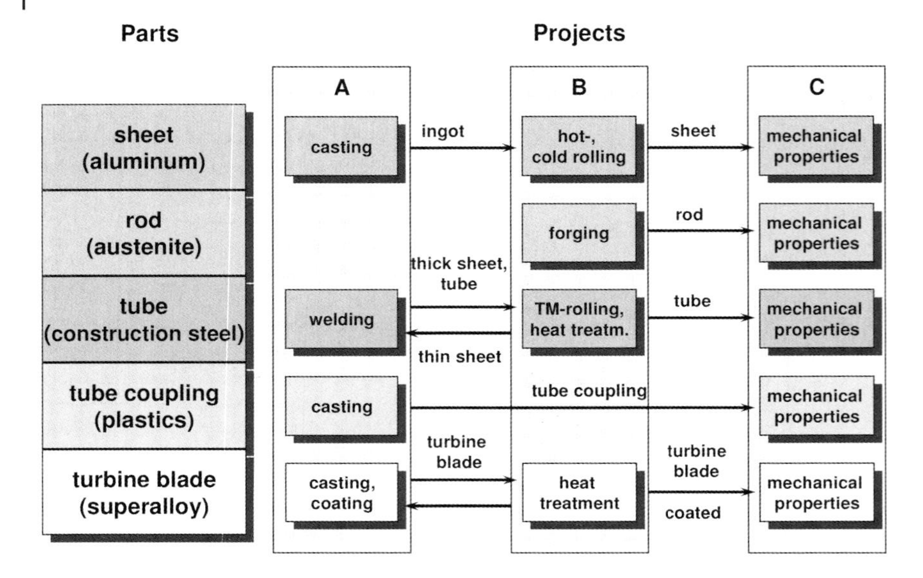

Fig. 2.7 The simulation addressed specific processing schedules and materials. The materials were actually fabricated and characterized along the process chain to validate the models.

The research activities were not limited to the computer modeling but were accompanied by an experimental program, to provide input data to the simulations and to validate the predictions. For this purpose five different materials and parts were processed (Fig. 2.7).

The program ran in four periods of three years each. In the first phase the basic physical approach and computational strategies were formulated. In the second phase respective modules were developed to reflect microstructural evolution and the processing windows. The interfacing and coupling of the modules was subject of research in the third phase.

The status of research was communicated in symposia on "Integral Materials Modeling" and published in [6–8].

2.3 Through-Process Modeling

The final phase (Fig. 2.8) was devoted to specific through-process modeling exercises with the developed tools, i.e.

C6 Production of a sheet of aluminum alloy AA3104 and subsequent cup forming. Target prediction: anisotropy and earing profile.

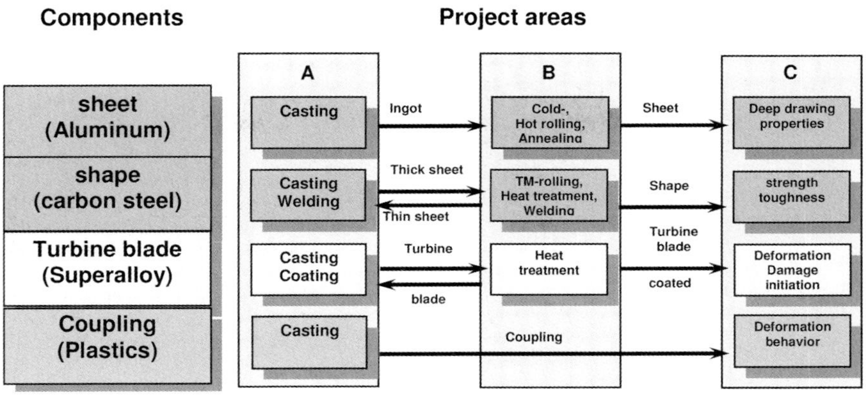

Fig. 2.8 Final through-process fabrication and modeling exercise with defined materials and their terminal part and properties. In project area C specific terminal properties were predicted by microstructure modeling along the whole process chain and compared to experimental results.

C7 Production of an angle profile made of carbon steel C40. Comparison of a welded and a bent structure. Target prediction: material strength and failure.
C8 Production of a coated turbine blade made of a nickel base superalloy. Target prediction: creep and damage of the bond coat.
C9 Production of a plastics tube coupling by injection molding. Target prediction: compressive radial strength.

All simulated processing chains and the resulting property predictions were compared to experimental results. For this the materials were processed exactly as modeled by the participating partners, and the target properties were measured. Concurrently, microstructure evolution during processing was determined by light optical microscopy as well as scanning and transmission electron microscopy.

2.4 Outlook

The results of these through-process exercises were encouraging. They have demonstrated that through-process modeling based on microstructural variables is indeed possible and renders predictions in reasonable agreement with the experimental results. Since several processing chains involve 10 and more processing steps, this is a remarkable success. Of course, the predictions are not perfect, and

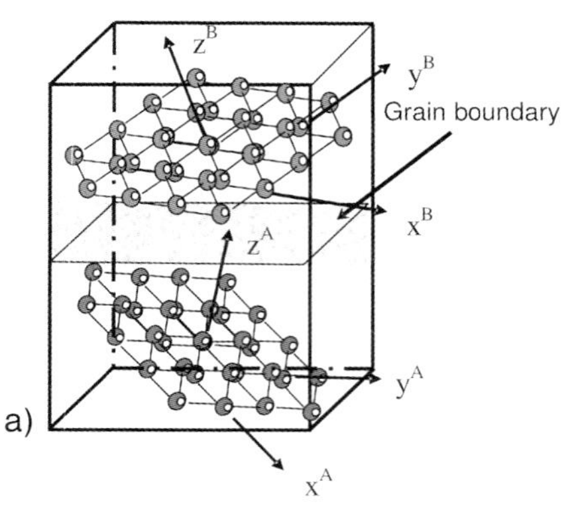

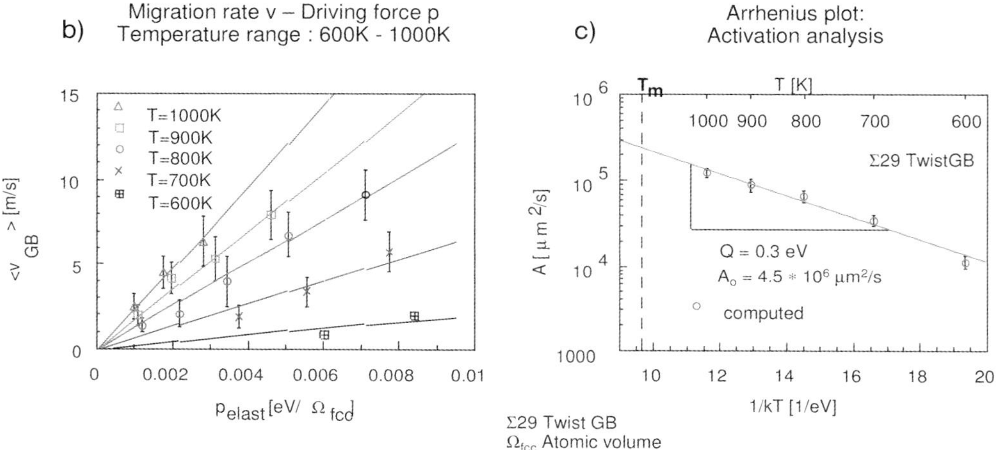

Fig. 2.9 Molecular dynamics simulation of grain boundary motion to determine grain boundary mobility. (a) Setup of modeling cell; (b) grain boundary velocity versus driving force; (c) Arrhenius plot of grain boundary mobility [9–11].

there is room for substantial improvement by more sophisticated approaches, optimized computer codes for shorter computing time and larger systems, as well as improved microstructural models. A major deficiency is the lack of reliable data to be used in the microstructural models like diffusivities, mobilities, interface energies, etc. It is obvious that measurements of such data for all kinds of material will not readily become available. Therefore, such data will have to be provided from atomistic simulations and quantum mechanical calculations. Molecular dynamics simulations were already employed in the program (Fig. 2.9). Ab initio quantum mechanical approaches by electron density functional theory in conjunction with molecular dynamics, cluster expansion methods, etc., have already provided valuable information and are designed to become common practice in the near future [12, 13].

The ultimate dream of a theoretical materials design is not a dream anymore; there remain problems and deficiencies, and the development of virtual materials on the computer is still a long way ahead, but the material modeling community makes big strides to get there. Steadily growing computer power and sophistication of simulation tools promise eventually to allow true virtual materials design by integral materials modeling.

Acknowledgments

The authors gratefully acknowledge the financial support of the Deutsche Forschungsgemeinschaft (DFG) within the Collaborative Research Center (SFB) 370 "Integral Materials Modeling".

As chairman of the collaborative research center SFB 370 I thank all colleagues for their continuous engagement in the program, all industrial partners for their valuable advice, and all scientists and doctoral students for their conspicuous cooperation and dedication to their projects.

References

1 J.W. Christian, *The Theory of Transformations in Metals and Alloys*, 2nd edn, Pergamon Press, Oxford, 1981.
2 J.P. Hirth, J. Lothe, *Theory of Dislocations*, 2nd edn, John Wiley, New York, 1982.
3 R.E. Smallman, *Modern Physical Metallurgy*, 4th edn, Butterworths, London, 1985.
4 G. Gottstein, *Physical Foundations of Materials Science*, Springer-Verlag, Berlin, 2004.
5 D. Raabe, *Computational Materials Science*, Wiley, Berlin, 1998.
6 G. Gottstein, R. Sebald, *Integral Materials Modelling*, Shaker-Verlag, Aachen, 2000.
7 Multiple authors and articles, *Mod. Sim. Mater. Sci. Eng.* 8, 2000, 871–957
8 Multiple authors and articles, *Mod. Sim. Mater. Sci. Eng.* 12, 2004, S1–S8.
9 B. Schönfelder, D. Wolf, S.R. Phillpot, M. Furtkamp, *Interface Sci.* 5, 1997, 245.
10 B. Schönfelder, G. Gottstein, L.S. Shvindlerman, *Acta Mater.* 53, 2005, 1597.
11 B. Schönfelder, G. Gottstein, L.S. Shvindlerman, *Met. Trans. A* 37A, 2006, 1757.
12 G. Kresse, J. Furthmüller, *Phys. Rev. B.* 54, 1996, 11169.
13 M.P. Allen, D.J. Tildesley, *Computer Simulation of Liquids*, Clarendon Press, Oxford, 1999.

3
Aluminum Through-Process Modeling: From Casting to Cup Drawing (TP C6)

L. Neumann, R. Kopp, G. Hirt, E. Jannot, G. Gottstein,
B. Hallstedt, J. M. Schneider, B. Pustal, and A. Bührig-Polaczek

Abstract

The aims and achievements of through-process modeling of a laboratory production of aluminum sheet are presented. The production steps are casting, homogenization, hot rolling, cold rolling, final annealing, and cup drawing. The aim is to predict the anisotropic yielding of the final sheet which is reflected by the cup's earing profile. This prediction was based on through-process simulation rather than on experimental measurements. The latter were carried out for model validation. In order to predict the earing profile of the cup a common yield function is implemented into the finite element method. The yield function was calibrated using texture and dislocation substructure data, which both were calculated in the preceding multipass hot and cold rolling and annealing steps. These steps influence texture and dislocation substructure through deformation and static recrystallization. The respective models are briefly presented. The mechanical properties of the homogenized ingot are predicted by taking its microstructure into account for the thermomechanical treatment. In the preceding processing steps of casting and homogenization, the main microstructural values predicted by the models were grain size, volume fraction of phase, and their composition which characterize the initial state of the material in hot rolling. Special emphasis was put on the interfacing of the models and the micro–macro coupling strategies that were developed to achieve a maximum precision while keeping the computational efforts within reasonable limits.

3.1
Introduction

The goal of this study is to deliver an experimentally validated through-process simulation of the laboratory production of an aluminum cup. The production stages to be addressed with physics-based models that are coupled to process

Integral Materials Modeling: Towards Physics-Based Through-Process Models
Edited by Günter Gottstein
Copyright © 2007 WILEY-VCH Verlag GmbH & Co. KGaA, Weinheim
ISBN: 978-3-527-31711-0

models range from the liquid phase to the final part. Microstructural variables play a vital role as they are the state variables, and they represent the "processing history" of the material. In order to achieve the goals the development of physics-based material models was necessary, in conjunction with the development of efficient micro–macro coupling strategies as well as experimental validation.

A material property that sensitively characterizes the fabricated aluminum sheet and that is of great importance is its anisotropic flow behavior. Anisotropy causes different flow stresses in different directions of the aluminum sheet. The main reason for plastic anisotropy is the sheet texture which is caused by deformation and recrystallization during hot and cold forming (rolling in the present case) as well as during thermal treatment such as inter-pass times at elevated temperatures and annealing treatments.

3.2
Casting and Solidification

3.2.1
The Casting Alloys

Two alloys were used to carry out the experiments. The first alloy was a commercial wrought alloy AA3104, which will be denoted by the letters "KL" in this paper. The second one was a model alloy AlMg1Mn1Fe0.45Si0.2 derived from the commercial alloy AA3104 by using only the five major elements (given in weight fractions), denoted by "ML". It was produced from technically pure components to avoid influences from other elements. Both alloys were provided by Hydro Aluminum Deutschland GmbH for this study. ML was used to validate the various models used during each processing step.

3.2.1.1 Casting

Typically, wrought alloys are produced using the direct chill (DC) casting process. However, DC casting is too time-consuming and too expensive to validate the presented models though it provides the best quality for ingots. Therefore, a die-casting chill was designed to produce a sound and homogenous ingot for the subsequent homogenization and hot rolling process.

The ingot was of rectangular shape. Details of the casting technique are given in Chapter 8. The requirements for the ingot were: no porosity, oxides, or other inclusions, and low microstructure gradient from the bottom to the top of the casting (direction of hot rolling). The homogeneity of the microstructure can be characterized by the dendrite arm spacing (DAS) and the grain size. At positions P1 through P3 (Fig. 3.1) metallographic samples were taken to measure the DAS variation over the length of the ingot. It increases slightly from 38 μm at P1 to 43 μm at P3. An x-ray analysis showed that pores only arose at the bottom of the casting, which was discarded before hot rolling. The front and the back surfaces

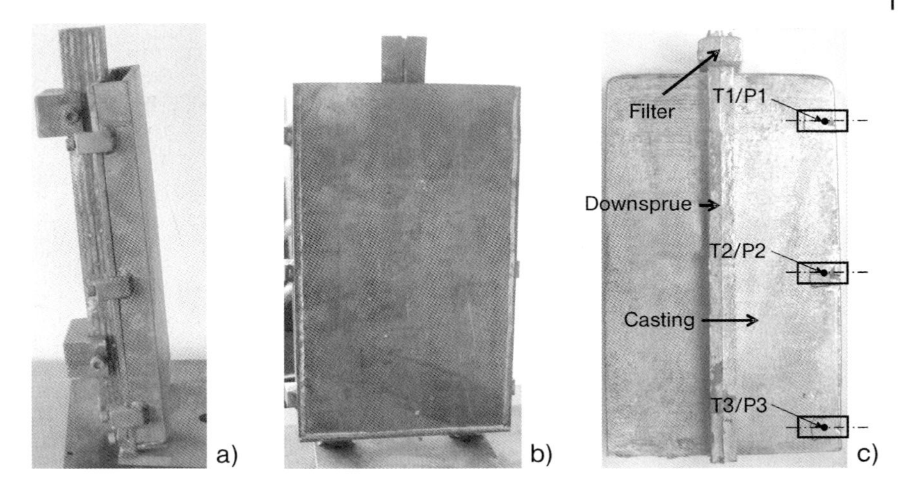

Fig. 3.1 The actual chill used for the casting: (a) side view, (b) front view, (c) casting.

were milled after casting to remove the oxide film at the surface. The final thickness of the ingot was 15 mm.

3.2.2
Simulation of the Casting Process

3.2.2.1 Thermodynamic Description of the Model Alloy

Simplified solidification simulations using thermodynamic equilibrium, Scheil–Gulliver approach by Thermo-Calc [1], and one-dimensional diffusion (by DICTRA [1]) were performed for the ML. For the thermodynamic calculations a modified version of the COST II database [2–4] covering the elements Al, Fe, Mg, Mn, and Si was used. The diffusion simulations utilized a mobility database developed by Prikhodovsky for aluminum alloys [3], a dendrite arm spacing of 38 μm, and an experimental cooling curve corresponding to position P1/T1 (Fig. 3.1). In the diffusion simulation only one solid phase, $Al_6(Mn,Fe)$, could be considered besides the face-centered cubic (fcc) matrix. According to these calculations primary fcc starts to form at 651 °C, followed by $Al_6(Mn,Fe)$ already at 650 °C. The equilibrium solidus is at 626 °C. At equilibrium α-Al(Mn,Fe)Si is predicted to form below 595 °C, and Mg_2Si to form below 493 °C. The diffusion simulation predicts that solidification is complete at 611 °C. According to the Scheil–Gulliver calculation, the solidification is not entirely completed even at 450 °C, but using a criterion of 1% liquid left leads to a temperature of 593 °C for complete solidification. In addition to $Al_6(Mn,Fe)$, $Al_{13}Fe_4$ is predicted to form below 635 °C, and Mg_2Si to form below 594 °C. The calculated fraction of liquid phase using the three methods is shown in Fig. 3.2. Obviously none of the three methods is ideal. The amount of back-diffusion is considerable, as shown by the

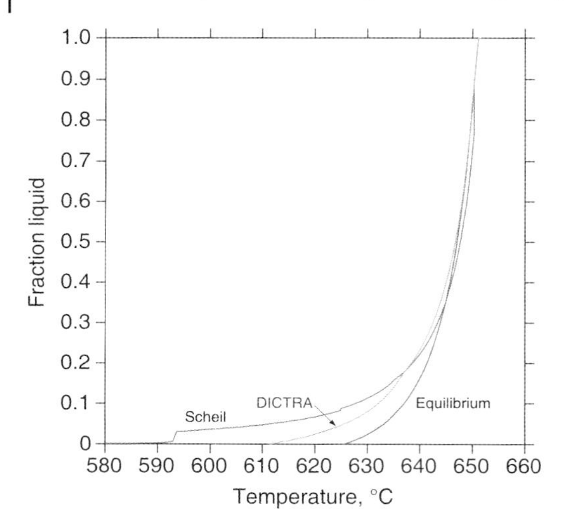

Fig. 3.2 Fraction of liquid phase using the three methods: equilibrium, Scheil, DICTRA (with back-diffusion in solid and liquid phases).

difference between the Scheil–Gulliver and diffusion calculations. On the other hand, additional intermetallic phases cannot be taken into account in the diffusion calculation using DICTRA. Possibly, the best solution is a combination of diffusion simulations followed by equilibrium solidification of the remaining liquid when the second intermetallic phase appears.

3.2.2.2 **Simulation of Grain Nucleation and Growth Using a Multiphase Flow and Solidification Model**

The model describes grain nucleation and growth kinetics under the influence of multiphase flow for globular equiaxed solidification structures. Conservation equations for mass, momentum, species, and energy are solved by using a fully implicit and control-volume-based finite difference method. The computational fluid dynamics (CFD) software FLUENT Rev. 6.1 is used here. The solidification model was realized by adding a conservation equation for grain transport and by quantifying source terms, appearing in the conservation equations, which were modeled by means of a user defined scalar (UDS) and a user defined memory (UDM). There are three phases involved: the liquid melt, the solidifying grains, and air. A volume averaging approach to formulate the conservation equations for each of the three phases was employed. More details about the models of nucleation, growth kinetics, mass transfer rate, solute partitioning at the liquid–solid interface, momentum, and enthalpy transfer among the phases liquid, solid, and air can be found in [5–9].

The orientation of the casting had a 15° deviation from the gravity direction. A low pouring velocity of 0.135 m s^{-1} was used. The connection between downsprue and casting was a thin horizontal ingate. All these measures aimed at casting the melt slowly and smoothly into the permanent mould. Input parameters for the simulation besides boundary and starting conditions for each conservation equation were the geometry and its discretization as well as thermophysical material properties and thermochemical flow parameters. Nucleation parameters were evaluated by undercooling experiments. A thermodynamic database and Thermo-Calc were used to obtain thermodynamic information for the growth model.

Figure 3.3 shows the evolution of solid fraction, composition of phase mixture, and the grain diameter. It is noted that: (a) the flow front is smooth and nearly flat; (b) a high solid fraction occurs near the wall whereas a high liquid fraction exists near the ingate; (c) air is squeezed out gradually; (d) macrosegregation is not significant at all; (e) a sharply defined zone with a large grain size was found near the center line; (f) a high temperature predominates near the gate and flow front, and a low temperature is found near the bottom and side walls; and (g) only the pouring velocity controls the velocity distribution field, because no significant convection is present in the bulk of the casting.

From this simulation the mean grain size could be determined as a characteristic parameter of the microstructure, which is needed as a starting value to study the evolution of the grain size during the hot rolling process.

3.2.2.3 Simulation of Phase Fractions, Dendrite Arm Spacing, and Concentration Profiles Using a Microsegregation Model

MicroPhase is a diffusion-controlled microsegregation model to calculate phase fractions and concentration profiles during solidification for multicomponent and multiphase systems. At the moving solid–liquid interface equilibrium was assumed. The equilibrium information was calculated using the TQ-Interface of the commercial software Thermo-Calc and the thermodynamic database for ML. The representative volume element was half of the dendrite arm spacing, which was expanding with time due to coarsening effects. Furthermore, the model was coupled to the macroscopic FEM software CASTS [10]. At each node of the FEM grid during each time step the phase fractions and the latent heat were calculated using the temperature difference provided by CASTS. The latent heat was passed on to the temperature solver of CASTS to calculate the temperature difference for the following time step.

As input parameters, the boundary and initial conditions for the macroscopic temperature solver and the microsegregation model were needed as well as thermophysical data for the materials used. Furthermore, the geometry of the casting and the discretization had to be specified, as well as diffusion parameters, interface energies, and a thermodynamic database.

As a result, characteristic parameters of the microstructure like the distribution of phase fraction, DAS and virtual EDX line scans can be provided at each node of the FEM grid. The microsegregation model and the principle of coupling were described in detail by Pustal et al. [11]. Figure 3.4 shows a virtual EDX line scan within the representative volume element during solidification. The left ordinate denotes the local relative phase fraction for the FCC_A1**AQ1**, the $Al_6(Mn,Fe)$, and

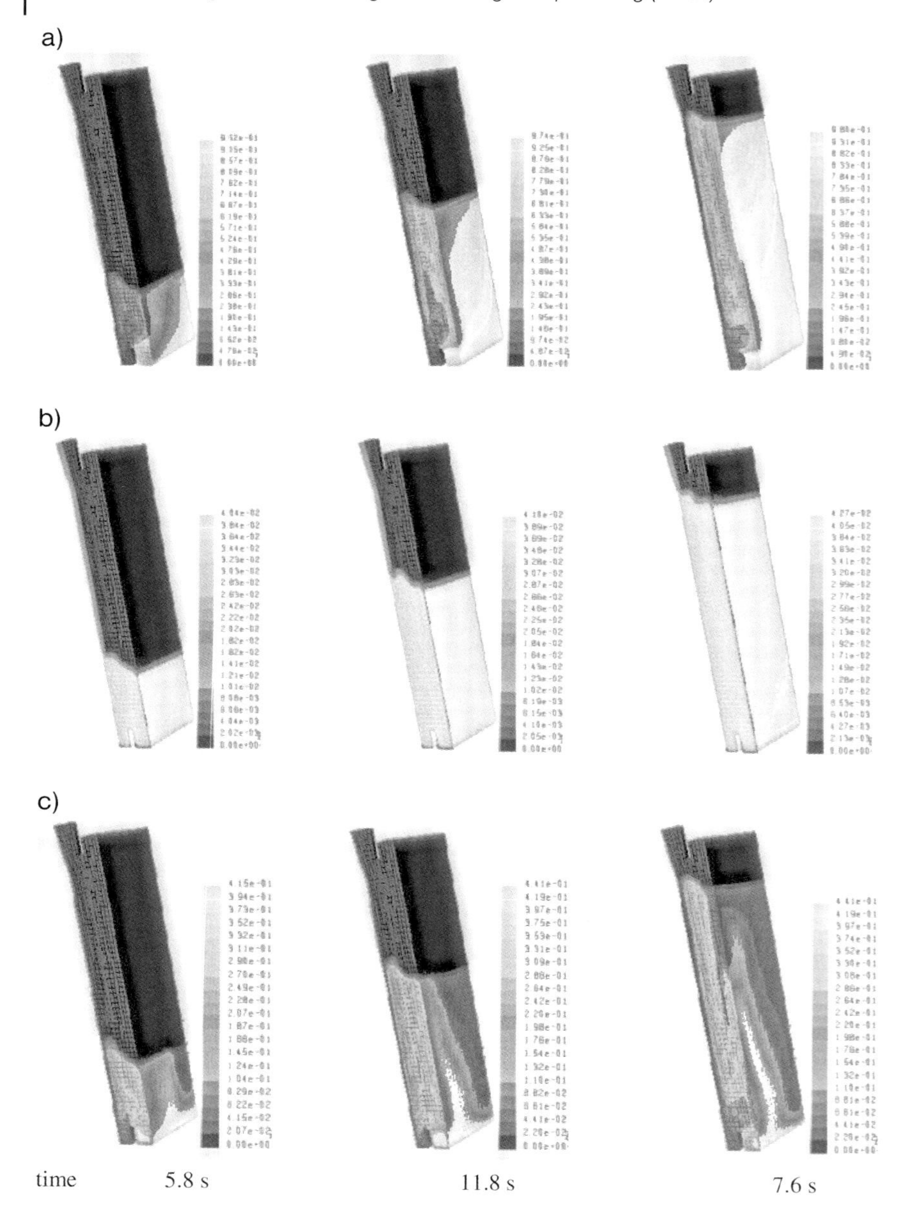

Fig. 3.3 Simulated results of the three-phase volume averaging model showing (a) the solid fraction, (b) the concentration of the phase mixture, and (c) the grain size distribution at different time steps during the casting process.

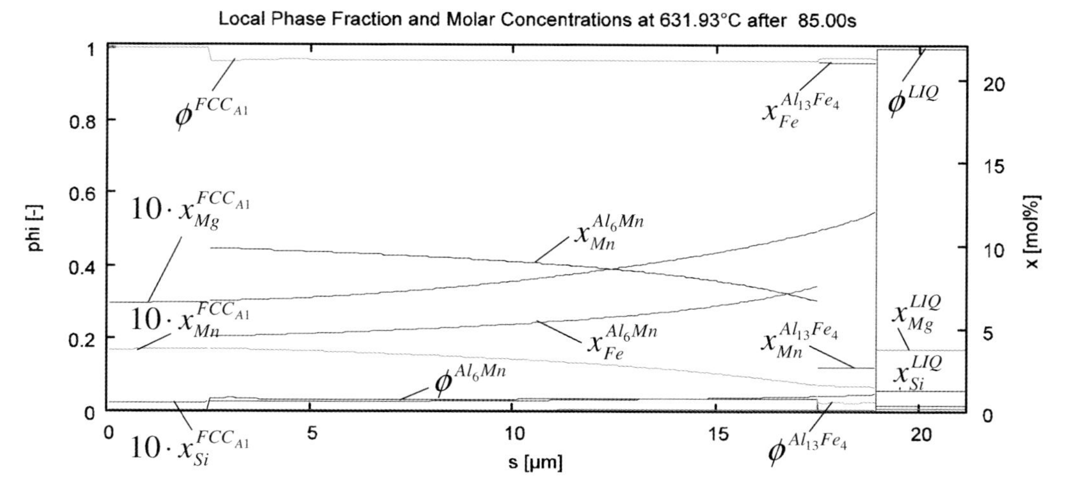

Fig. 3.4 Representative volume element of MicroPhase, the local coordinate s varies within $0 \leq s \leq \lambda/2$, where λ is the dendrite arm spacing, ϕ the local phase fraction, and x the mole fraction of each species.

the $Al_{13}Fe_4$ phase. The right ordinate scales the molar concentrations of the diffusing species within the different phases. The concentrations in the FCC_A1 phase are magnified by a factor of 10. Barely visible concentrations as for Fe in FCC_A1 (0.01 mol%) and in Liquid (0.41 mol%) as well as for Mn in Liquid (0.19 mol%) are not insinuated.

The dendrite arm spacing and the microsegregation profiles are needed to model the subsequent homogenization process using DICTRA.

3.3 Homogenization

3.3.1 Homogenization of Alloy AA3104

After casting, the material has to be heat-treated to reduce the chemical heterogeneities in the material (microsegregations). This thermal treatment performed at high temperatures (see Fig. 3.5) leads to various microstructural changes. On the one hand, by means of diffusion, the composition becomes more homogeneous in the dendrite arms. On the other hand, phase transformations occur. Both phenomena will greatly influence the properties of the material during subsequent industrial processes, which stresses the importance of controlling the microstructure obtained after homogenization. This task is achieved by the design of a suitable annealing program (time-temperature curve).

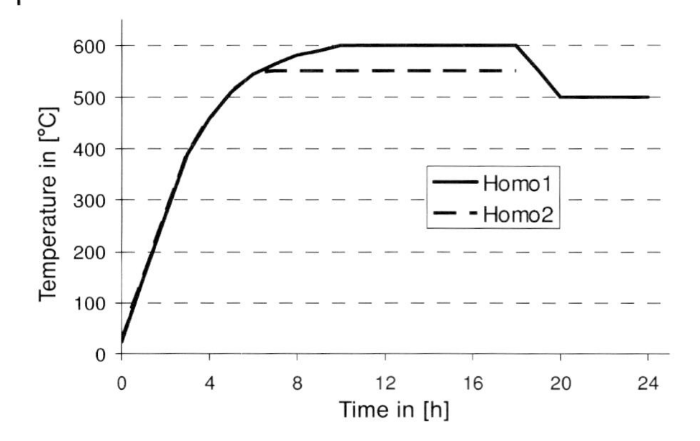

Fig. 3.5 Time–temperature program for the homogenization of AA3104.

In this work, two different homogenization schedules, presented in Fig. 3.5, are considered. The first one corresponds to an industrial two-stage homogenization (curve Homo1), the second to a laboratory one-stage homogenization (curve Homo2). In the following, the two alloys studied in the casting step will be further considered to determine the influence of the alloying elements on the final state. Hence, four different cases are investigated: KL1 (commercial alloy with the homogenization curve Homo1), KL2 (commercial alloy, Homo2), ML1 (model alloy, Homo1), and ML2 (model alloy, Homo2).

In the case of alloy AA3104, the following phase transformations occur. During solidification, primary phases of composition β-$Al_6(Mn,Fe)$ precipitated at the edge of the dendrites. These phases appear at about 600 °C during casting and are thermodynamically unstable when the material is reheated to above 400 °C. During homogenization, the β-phase transforms partially into the stable α-phase with the stoichiometry $Al_{12}(Mn,Fe)_3Si_{1.8}$. Concurrently with this transformation, smaller precipitates, the dispersoids, copiously nucleate in the dendrite arms. A characteristic microstructure after homogenization is shown in Fig. 3.6.

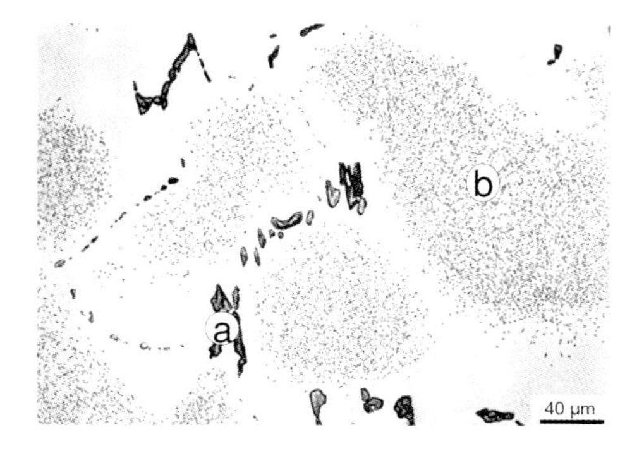

Fig. 3.6 Microstructure after homogenization: (a) primary phases; (b) dispersoids.

In order to assess the quality of the heat treatment, four questions have to be answered:

1. How fast will the microsegregations, introduced during casting, disappear?
2. How will the primary phases "behave"?
3. How many dispersoides and of what size will be present at the end?
4. What is the solute content in the matrix phase in the end?

The aim of this study was to provide answers to these questions with the help of simulations. The quality of the predictions will then be estimated after comparison of the simulated results with values determined experimentally.

3.3.2 Simulation Methods

To answer the previous questions, two simulation techniques were employed. The first one is DICTRA. These calculations help capturing how fast the microsegregations will be removed (question 1) and how fast the primary phases will grow (question 2) during the homogenization. In addition to these spatially resolved diffusion simulations, statistical predictions of the precipitation were performed using the ClaNG model [12] to estimate the population of dispersoids (question 3). This simulation method also delivers the solute concentrations (question 4) in the dendrite arms at the end of the homogenization process.

3.3.2.1 DICTRA Calculations

One-dimensional diffusion simulations were performed using the commercial software DICTRA. The used DICTRA cell, shown in Fig. 3.7, models a dendrite. The concentration profiles in the matrix at the beginning of the simulation were taken from the simulations of the casting step. The same thermodynamic and mobility databases were used as for the solidification simulation described in Section 3.2.1.

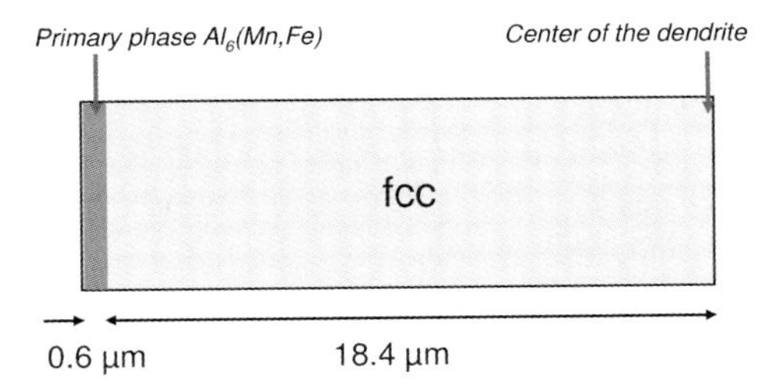

Fig. 3.7 DICTRA cell used to simulate the diffusion during homogenization.

3.3.2.2 ClaNG Model

The ClaNG model is used to predict the precipitation kinetics. This model is based on the classical nucleation and growth theory for precipitation. It follows the Kampmann and Wagner methodology to determine the evolution of the precipitate size distribution during a given heat treatment. A complete description of this model is given in [12].

3.3.3 Experimental Procedure

Electrical resistivity measurements were performed to record the change of solute concentration in solution. These investigations were conducted on a Burster Resistomat type 23074 at the Hydro Aluminium Deutschland GmbH R&D center in Bonn (Germany).

The primary phase changes during homogenization were studied using scanning electron microscopy (SEM; Leo 1530) with field emission gun and equipped with an EBSD detector. The nucleation and growth kinetics of the dispersoids were evaluated from transmission electron microscopy (TEM) investigations with a Jeol JEM-2000-FX2. All measurements were conducted on samples after 0, 3, 6, 10, 18, 20, and 24 h homogenization time.

3.3.4 Comparison between Experimental and Simulation Results

3.3.4.1 Primary Phases

The amount of $Al_6(Mn,Fe)$ as a function of time for the two homogenization schedules was estimated by DICTRA calculations. The results are shown in Fig. 3.8. These calculations indicate that during the homogenization the $Al_6(Fe,Mn)$ phase grows slowly. Starting with a thickness of 0.6 μm, the β-phase layer thickness reached about 0.64 μm at the end of the homogenization treatment (type Homo1). Therefore, no substantial change in the primary phase volume fraction was expected. This result was confirmed by SEM investigations. Because of its complexity, the phase transformation from $Al_6(Mn,Fe)$ to $Al_{12}(Mn,Fe)_3Si_{1.8}$ has not been modeled.

3.3.4.2 Solute Concentrations

The speed of dissolution of microsegregation was estimated by DICTRA calculations. The results indicated that, during the homogenization program Homo1, the alloy becomes completely homogenized after holding for 8 h at 600 °C. On the other hand, the second program, Homo2, with a temperature of 550 °C, is not sufficient to completely remove the microsegregation.

The dispersoid precipitation will strongly modify the solute levels in the dendrite arms. Using ClaNG, this phenomenon was studied, and the resistivity obtained with the simulated solute concentration can be compared with experimen-

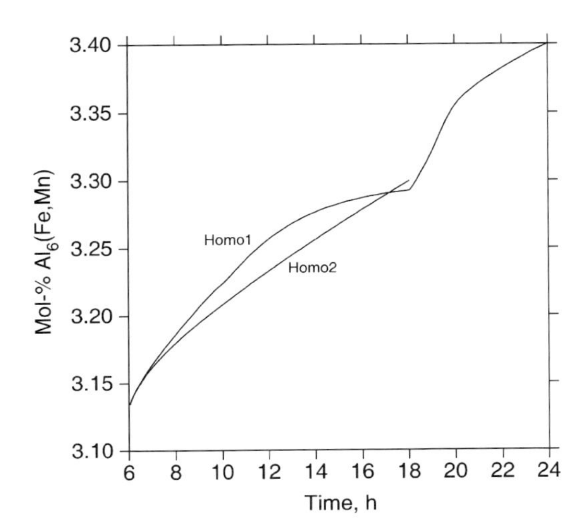

Fig. 3.8 Primary phase mole fraction estimated by DICTRA calculations.

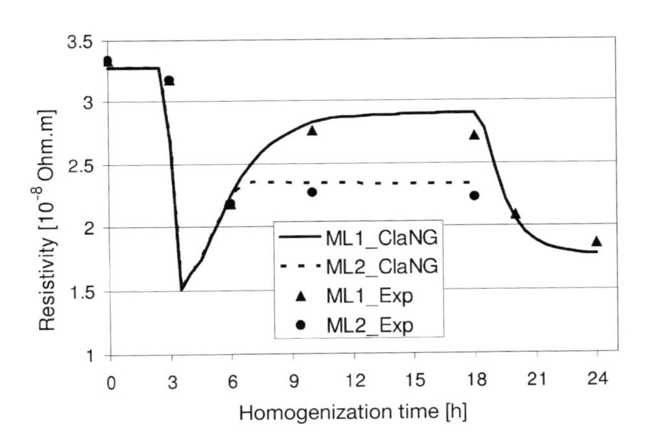

Fig. 3.9 Simulated and experimental resistivity during homogenization.

tal results, as presented in Fig. 3.9 for alloy ML. This exercise shows that the ClaNG model reasonably reproduces the solute level during the heat treatment.

3.3.4.3 **Dispersoid Precipitation**

The dispersoid density and average radius at the end of the heat treatment obtained by ClaNG simulations are summarized in Table 3.1 and compared to experimental values. For both quantities, the simulations yield the correct order of magnitude. However, simulations cannot exactly reproduce the experimental results, especially the precipitate density.

Table 3.1 Precipitate average radius and density after homogenization obtained by simulations and the corresponding experimental values.

Dispersoids		KL1	ML1	KL2	ML2
Simulation	Average radius (μm)	95	97	50	51
	Density (m^{-3})	2.95×10^{18}	2.87×10^{18}	1.14×10^{19}	1.32×10^{19}
Experimental	Average radius (nm)	90	96	62	62
	Density (m^{-3})	3.50×10^{19}	9.73×10^{18}	4.44×10^{19}	1.64×10^{19}

3.4 Hot and Cold Rolling

3.4.1 Flow Stress Modeling

The rolling steps of aluminum sheet production were modeled using the finite element method as a process model. In order to capture the effect of preceding production steps, the effect of micro-segregation of magnesium (Mg) and manganese (Mn) had to be taken into account. This was achieved by calculating the material flow stress during forming with a physics-based model. This model is a statistical dislocation density model using three different dislocation density populations as variables of state: hence the name "three-internal-variables-model" (3IVM).

3.4.2 Texture Simulation

The crystallographic texture that is the result of thermomechanical processing of aluminum sheet plays a central role for subsequent deep-drawing operations of the sheet. The sheet texture will primarily induce anisotropic flow behavior which – in industrial processing – can be a cause for significantly increased scrapping. In the following, "T-Pack", a modeling tool for cold forming is presented. Apart from the process data such as tool geometries, microstructural information – material texture and hardness as expressed in terms of dislocation densities – is input to the simulation.

3.4.3 Recrystallization

For the simulation of the rolling process the FEM package Larstran/Shape was applied. After the finite element simulation of each rolling pass the deformation textures were calculated with the combined deformation texture–hardening model GIA-3IVM in postprocessing, and the recrystallization textures were simulated with ReNuc and StaRT considering the deformation conditions and the microstructure data. The StaRT model and GIA as well as their interaction with the FE model are described in detail in Chapter 10. The final texture of the sheet after hot rolling, cold rolling and the final heat treatment was input to the subsequent cup drawing simulation along with the calculated dislocation density populations.

3.5 Cup Drawing

3.5.1 Anisotropy Update

The model setup developed for hot and cold rolling was extended by a module that took this information into account for prediction of the earing profile in cup drawing that is caused by texture-induced anisotropic flow behavior. To this end, T-Pack offers an interactive update of the yield locus when all modules (full constraints Taylor model for texture, 3IVM for flow stress, and Hill48 for plastic anisotropy) are activated, and the parameters for the chosen yield locus can be calculated directly from the mechanical properties (see Chapter 13 for details). After a preset summed-up strain (usually of the order $\varepsilon_{plast} = 0.02$) for a given element had been reached, texture simulation was invoked for this element. This user setting avoided texture and anisotropy update in finite elements that were not in the forming zone, thus significantly reduced the computational efforts necessary for the simulation. Precision was not compromised up to a value of $\varepsilon_{plast} = 0.02$. Then, the orientation distribution was updated. Next, the finite element and its corresponding orientation distribution were subjected to the testing strain corresponding to the test for Y_0, Y_{45}, Y_{90}, and Y_b, which represent the yield stresses in the rolling direction, in the 45° direction, in the transverse direction, and in the biaxial loading direction, respectively, and used to calculate the yield stresses with the 3IVM. The generated values were used to update the shape of the Hill48 yield locus in the current element. This procedure consisted of the following calculation steps:

1. Appropriate rotation of the orientation distribution for tests Y_0, Y_{45}, Y_{90}, and Y_b.
2. Calculation of the Taylor factor $\overline{M}$ (for each test Y_0, Y_{45}, Y_{90}, and Y_b) after subjecting the finite element to a virtual appropriate displacement gradient tensor.
3. Calculation of Y_0, Y_{45}, Y_{90}, and Y_b with 3IVM and the previously calculated values for the Taylor factor $\overline{M}$.
4. Calibration of the yield locus Hill48 with the calculated values for Y_0, Y_{45}, Y_{90}, and Y_b.

For a more comprehensive overview of the deep-drawing simulation, the reader is referred to Chapter 13.

3.5.2 Results

In the following, results for the deep-drawing simulations of the two variants of AA3104 alloy will be given. The resulting textures were generally very weak – almost random – only showing slightly increased intensity of the Cube orientation. When comparing the commercial alloy AA3104 "KL" to the model alloy "ML" (both underwent homogenization and were hot rolled at 380 °C) the only apparent difference is a somewhat stronger R orientation in the commercial alloy AA3104 (equivalent to the S orientation but slightly shifted after recrystallization). The effect of this difference can be distinguished in the experimentally measured earing profiles (Fig. 3.10). The slightly stronger Cube orientation in the model

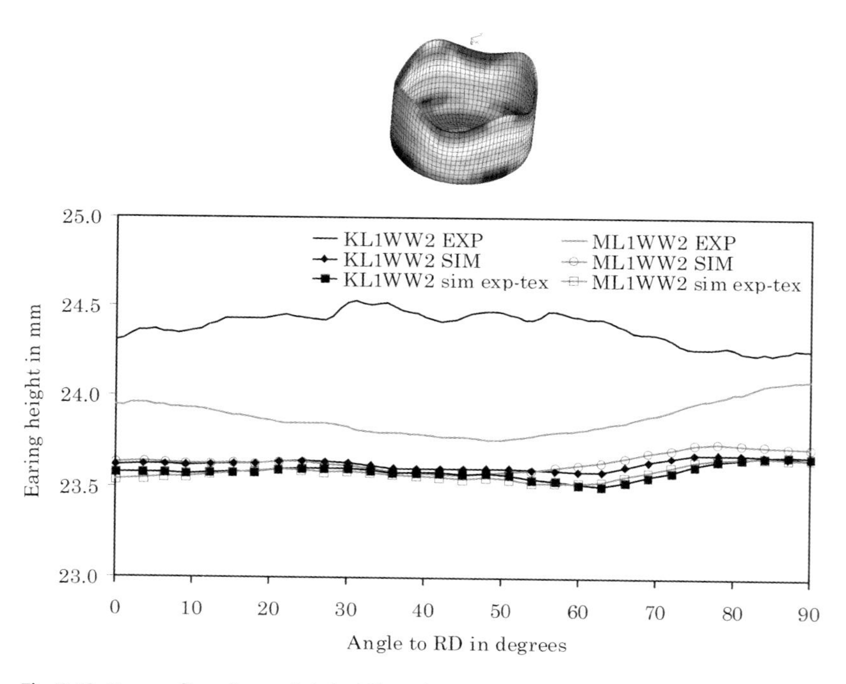

Fig. 3.10 Cup profiles of two slightly differently textured alloys: the commercial alloy AA3104 "KL" and the model alloy "ML". Measurements are marked as "EXP" and simulations (that are results of the simulated through-process texture) as "SIM". Simulations that were carried out using the measured material texture are marked "sim exp-tex".

alloy leads to weak ears in RD and TD with the latter being slightly higher. The commercial alloy AA3104 developed a very low and broad ear around 45° which is typical for textures with a slightly increased intensity around the β-fiber. T-Pack does not capture these differences. Nevertheless, T-Pack does predict extremely flat earing profiles, thus reproducing quite well the quantitative earing height (Fig. 3.10). (Qualitatively, the prediction is more precise for the model alloy than for the commercial alloy AA3104, because in both cases T-Pack predicts a small ear in TD.) This demonstrates an important feature: the quasi-isotropic behavior is excellently reproduced in the simulation while computation times remain within the same order of magnitude as for simulations with the classic phenomenological approach.

3.6 Conclusions and Outlook

This report substantiates that simulations are able to capture the major features arising during the processing steps casting, homogenization, hot rolling, cold rolling, final annealing, and cup drawing. The simulated results are in good agreement with the values determined after experimental investigations. After homogenization, microstructural variables are traced throughout the hot rolling, cold rolling, and annealing steps. Hardening, recovery, and static recrystallization are included in the modeling setup, and their influence on the dislocation substructure and texture evolution is accounted for. For sheet forming operations, T-Pack was extended to account for texture-induced plastic anisotropy. The presented results of a cup drawing operation show good conformance with the experimentally determined earing profile. Further improvement in the modeling setup for through-process modeling can be achieved in the modules themselves as well as in their interaction. For example, a more precise investigation and understanding of recovery and its implementation into 3IVM is desirable. In the finite element simulation, remeshing algorithms and adapted data transfer routines are necessary prerequisites for an application of T-Pack to geometrically more complex processes than flat rolling and deep drawing.

Acknowledgments

The authors gratefully acknowledge the financial support of the Deutsche Forschungsgemeinschaft (DFG) within the Collaborative Research Center (SFB) 370 "Integral Materials Modeling".

Furthermore, the authors thank Dr. H. Aretz, M. Schneider, M. Goerdeler, J. van Santen, M. Nutzmann, Ch. Wiedner, and A. Meyer for assistance and helpful discussions.

References

1 J.-O. Andersson, T. Helander, L. Höglund, P. Shi, B. Sundman, *CALPHAD* 26, 2002, 273.

2 I. Ansara, A.T. Dinsdale, M.H. Rand (eds.), in *Thermochemical Database for Light Metal Alloys*, Vol. 2, COST 507 Final Report Round 2, European Communities, Luxembourg, 1998.

3 A. Prikhodovsky, Dissertation, RWTH, Aachen, 2000.

4 E. Balitchev, T. Jantzen, I. Hurtado, D. Neuschütz, *CALPHAD* 27, 2003, 275.

5 T. Wang, M. Wu, A. Ludwig, M. Abondano, B. Pustal, A. Bührig-Polaczek, *Int. J. Cast Metals Res.* 18, 2005, 221.

6 T. Wang, B. Pustal, M. Abondano, T. Grimmig, A. Bührig-Polaczek, M. Wu, A. Ludwig, *Trans. Nonferrous Metals Soc. China* 15, 2005, 389.

7 A. Ludwig, M. Wu, T. Wang, A. Bührig-Polaczek, Proceedings of the 3rd International Conference on Computational Modeling and Simulation of Materials, Sicily, Italy, 2004.

8 M. Wu, A. Ludwig, *Adv. Eng. Mater.* 5, 2003, 62.

9 M. Wu, A. Ludwig, A. Bührig-Polaczek, M. Fehlbier, P.R. Sahm, *Int. J. Heat Mass Transfer* 46, 2003, 2819.

10 G. Laschet, J. Neises, I. Steinbach, *Lecture Notes of 4iéme Ecole d'été du CNRS*, Porquerolles, C8, 1998, p. 1.

11 B. Pustal, B. Böttger, A. Ludwig, P.R. Sahm, A. Bührig-Polaczek, *Metall. Mater. Trans. B*, 34B, 2003, 411.

12 M. Schneider, Dissertation, RWTH, Aachen, 2006.

13 R. Sebald, G. Gottstein, *Acta Mater.* 50, 2002, 1587.

14 R. Sebald, G. Gottstein, in *Recrystallization and Grain Growth*, Vol. II, ed. G. Gottstein and D.A. Molodov, Springer-Verlag, 2001, p. 1027.

4
From Casting to Product Properties: Modeling the Process Chain of Steels (TP C7)

U. Prahl, W. Bleck, A.-P. Hollands, D. Senk, X. Li, G. Hirt, R. Kopp, V. Pavlyk, and U. Dilthey

Abstract

This chapter deals with modeling of the whole process chain of steel fabrication including continuous casting, hot rolling, cooling, welding, and forming at room temperature. The experimental work was accomplished on a laboratory scale. An integrative model is presented, which combines experimental, semiempirical, and physical approaches for the different processing steps. Internal variables describing the microstructure are handed over between interactive modules. Several simulation tools are implemented and linked in order to establish a multidisciplinary and multiscale approach. Final mechanical properties are predicted for a pipeline and an angle profile.

4.1
Introduction

Steel is the most common construction material used in many transport and energy applications like shipping, off-shore, pipelines, bridges, pressure vessels, etc. The reason is its attractive combination of various mechanical and technological properties, low price, and ecological recyclability. Highly accurate processes and optimized alloying concepts cause a steady increase in the usage of this classic material. Modern computational tools and numerical modeling increasingly play a role in current materials and process development of steel.

Generally, the necessary material properties are determined by the application. The final mechanical properties of steels are mainly determined by its microstructure which is influenced by all processing steps like casting, hot forming, phase transformation, cold forming, and annealing [1]. Welding, as an additional process step, severely modifies the microstructure and redefines the local properties owing to an inhomogeneous rapid thermal treatment including melting, solidification and transformation.

Integral Materials Modeling: Towards Physics-Based Through-Process Models
Edited by Günter Gottstein
Copyright © 2007 WILEY-VCH Verlag GmbH & Co. KGaA, Weinheim
ISBN: 978-3-527-31711-0

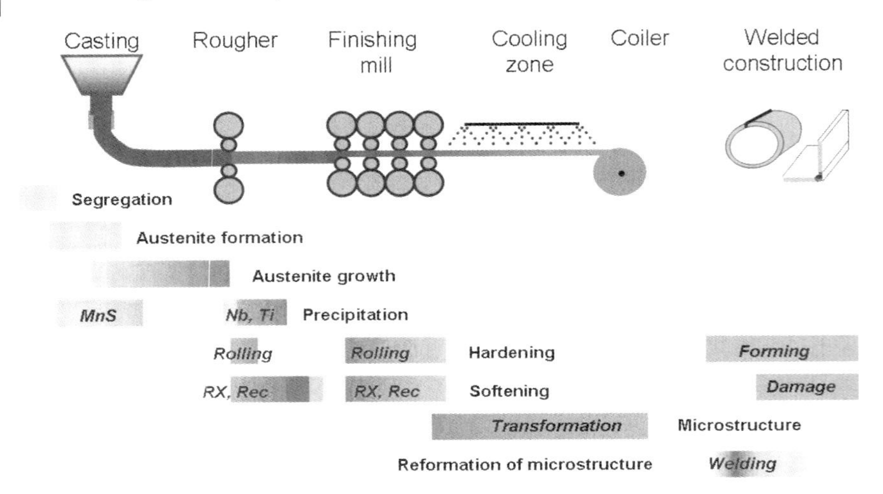

Fig. 4.1 Processing steps and mechanisms during processing which influence the microstructure of steel.

The microstructural parameters of particular importance are phases and precipitates in terms of phase fractions, ferrite grain size, pearlite lamellae spacing, martensite packet size, and void fraction. The processes and mechanisms that influence the final microstructure during different processing steps are shown in Fig. 4.1.

Concerning solidification, the temperature field as well as the primary and secondary dendrite arm spacing are simulated and used for modeling austenite grain formation and MnS precipitation. The hot rolling simulation starts with the primary austenite grain size and includes the metallurgical mechanisms of precipitation, hardening, recovery, and recrystallization. Phase transformation kinetics and microstructure-dependent mechanical properties depend on the austenite condition after hot forming. For applications, welding is investigated, where temperature evolution, weld geometry, local mechanical properties, and residual stresses are to be predicted. At the end of the process chain, failure due to cold forming (bending) is predicted using mechanical properties and inclusions calculated in the former simulation steps.

The work was carried out on a S460 pipeline steel and a C45 construction steel. The chemical compositions are given in Table 4.1. As an example, a four-point bending test of a pipeline and a technological bending test of both a cold formed and a welded construction angle are investigated.

In this work the process chain ranged from casting to the final application. Each individual processing step is briefly described, while a detailed account of the modeling approaches is given in Chapters 7, 11, and 12.

Table 4.1 Chemical composition (mass %) of investigated steels.

Steel	C	P	S	Al	N	Cr	Cu
C45	0.47	0.015	0.028	0.054	0.011	0.22	0.22
S460	0.11	0.010	0.009	0.07	0.005	0.200	n.d.

Steel	Mn	Mo	Ni	Si	Nb	Ti	V
C45	0.74	0.03	0.19	0.32	<0.001	0.022	0.039
S460	1.37	n.d.	0.04	0.28	0.036	<0.001	0.004

4.2
Continuous Casting Simulation

In order to predict the evolution of microstructure during continuous casting, a model has to be formulated for calculating the temperature field as well as the most important metallurgical features as there are primary and secondary dendrite arm spacing, primary austenite grain size and MnS precipitation.

Therefore, semiempirical models were developed for individual microstructural features and combined for a joint approach. The parameters of these models were deduced from experiments. Once a valid set of parameters for a certain material was found, these correlations yielded reliable predictions. For well-known materials such as steels this is a promising and practical approach to couple process and material scale.

As time and time derivative are the most important boundary conditions for the above mentioned metallurgical mechanisms, a special mould was constructed with variable width and a nickel coated *surface on one side*. This special construction allowed a constrained solidification due to different conductivities at opposite mould walls. Using this mold, melting experiments were performed under conditions similar to continuous casting with various solidification constraints.

Based on these experiments a model was formulated describing the secondary dendrite arm spacing depending on the local cooling rate in a way as known from Wolf and Kurz [2] and Liu et al. [3]. For the simulation of the grain sizes a model by Andersen and Grong was used [4], where parameters of the model were varied, e.g. the mobility of the grain boundaries. The MnS precipitation depends on the cooling rate, and it is formulated in a model according to Schwerdtfeger [5].

A more refined approach incorporated the cooling rate, Mn and S fraction and secondary arm spacing [6].

The models were implemented in a continuous casting simulation algorithm that offered a combined thermochemical and thermodynamical simulation of

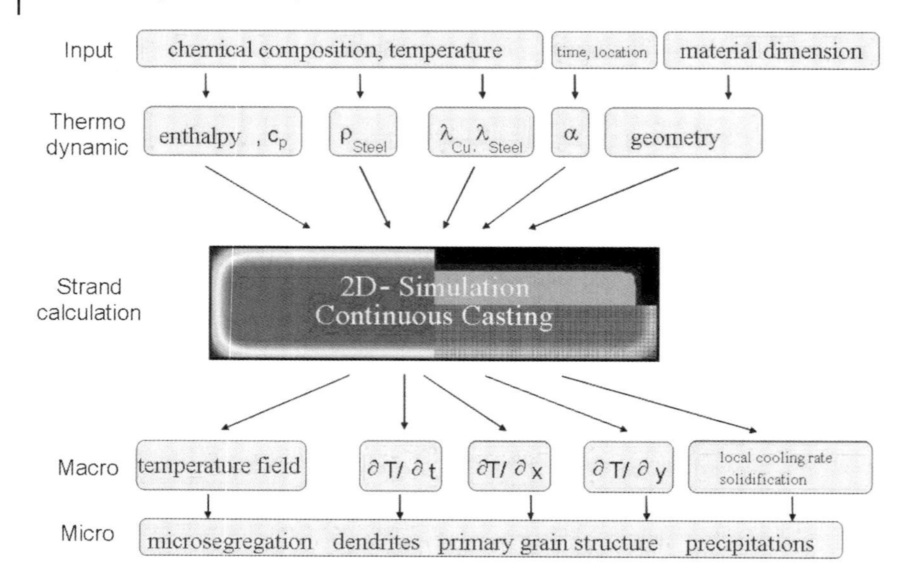

Fig. 4.2 Simulation algorithm for continuous casting [7].

macro- and micro-quantities (Fig. 4.2). A user-friendly windows-based interface was developed and industrial process parameters were possible input parameters. The results can be sketched graphically, and they can be handed over to the following process simulation tools. More detailed information on the continuous casting simulation tool can be found in [7].

4.3 Hot Rolling Simulation

On the macroscopic scale, the most important process parameters to be described during hot rolling are rolling force and temperature. These parameters are strongly coupled with precipitation and with the evolution of the austenite grain size due to recovery and static or dynamic recrystallization.

Single mechanisms were investigated using the hot deformation simulator and the deformation dilatometer at the department of ferrous metallurgy (IEHK). In order to enable a detailed investigation of the complete process chain, casting and rolling in one heat was realized on a laboratory scale (Fig. 4.3). Here, a computer-controlled continuous casting simulation was performed at the IEHK, and the ingot was handed over to the institute of metal forming (IBF) in a hot transport box that ensured temperature stability above 1300 °C during the entire transport. At the IBF reverse rolling was accomplished where the temperature was controlled with up to three furnaces.

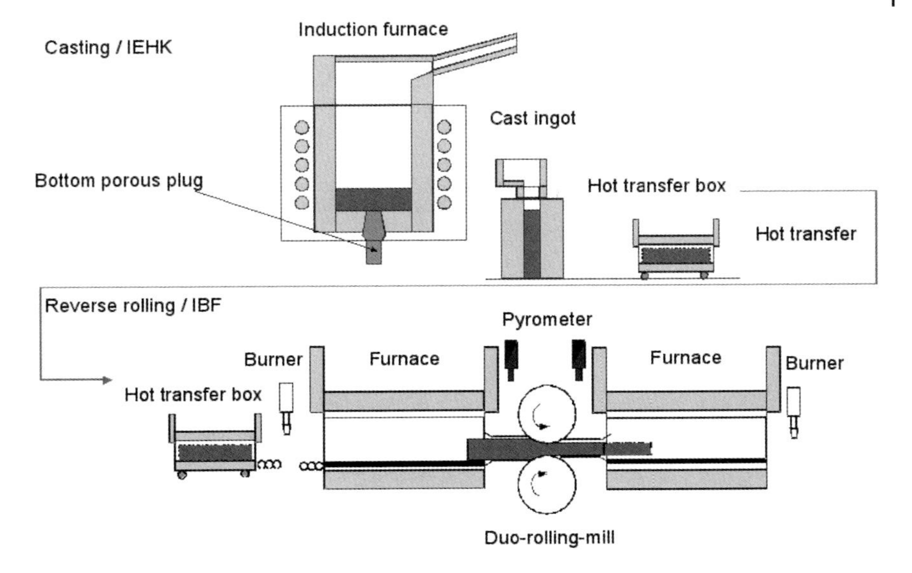

Fig. 4.3 Casting and rolling in one heat on a laboratory scale [8].

For the flow curve description a parameter separation approach was used [9]:

$$k_f(\varphi, \dot{\varphi}, T) = \dot{\varphi}^{c_1(\varphi)T} \dot{\varphi}^{c_2(\varphi)} \exp[c_3\varphi(T)]c_4(\varphi)$$

where c_1, c_2, c_3, and c_4 are empirical functions of the strain φ.

For recovery (Rec) and recrystallization (Rx) a bilinear softening description was formulated and the material-dependent parameters were identified for C45 [10]:

$$\ln\ln(E_{\text{Rec}}) = \ln(42) + 0.24\ln(\dot{\varphi}) - \left(\frac{-50.5}{RT}\right) + 0.34\ln(t_{\text{p}})$$

$$\ln\ln(E_{\text{Rx}}) = \ln(183\,078) + 0.17\ln(\dot{\varphi}) - \left(\frac{-131.9}{RT}\right) + 0.54\ln(t_{\text{p}})$$

E_{Rec} and E_{Rx} are the fractions of the total softening due to recovery and recrystallization, respectively, and t_{p} is the holding time.

The Dutta–Sellars [11] approach was modified for the description of thermally and deformation-induced Nb(C,N) precipitation in the austenitic and in the intercritical range, and also for high-temperature MnS precipitation [6].

For the coupled rolling simulations at IBF, microstructure based models were implemented in the StrucSim software and validated by means of multipass hot rolling schemes on a laboratory scale. As an example the comparison of simulation and experimental results of temperature and grain size during a three pass rolling scheme of C45 can be found in Fig. 4.4.

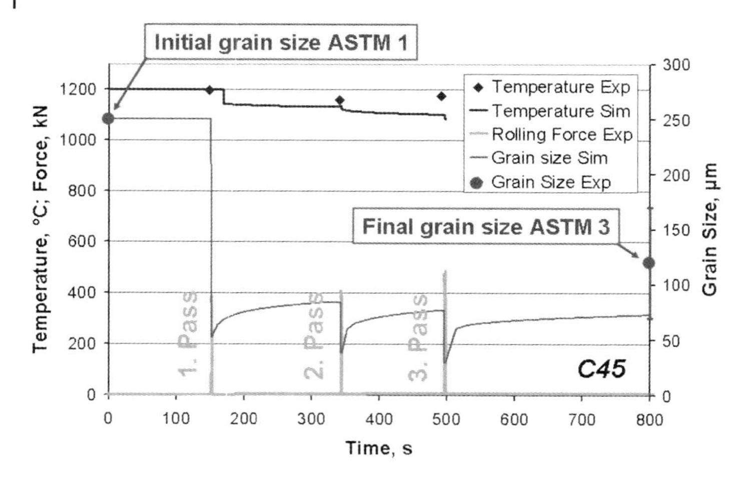

Fig. 4.4 Experimental and simulation results of temperature and grain size during a three-pass rolling scheme.

4.4 Simulation of Phase Transformation

Within the current approach, the γ–α phase transformation was described during isothermal holding, continuous cooling, or influenced by austenite conditioning. Here, physical as well as semiempirical approaches were used. For the simulation of isothermal transformation kinetics, a microstructure based phase field approach was used as it is implemented in the MICRESS code developed by ACCESS [12].

Concerning continuous cooling as well as for austenite conditioned transformation, a Johnson–Mehl–Avrami–Kolmogorov (JMAK)-based approach was developed and implemented in the SimZTU program at the IEHK to compute the TTT and CCT diagrams [13].

4.4.1 Physical Modeling of Isothermal Proeutectoid Ferrite Transformation

The isothermal proeutectoid ferrite transformation evolution is strongly influenced by ferrite nucleation kinetics. Therefore, nucleation densities and relative numbers on triple junctions were identified depending on the undercooling by means of isothermal holding experiments. Real microstructures and measured nucleation densities were used as input for phase field calculations by MICRESS, and the ferritic phase transformation was calculated including partitioning (Fig. 4.5).

By these calculations it was found that the ferrite transformation is a two-step process. First it is controlled by C diffusion where other alloying elements X still remain homogeneously distributed (LENP). For longer times, the transformation

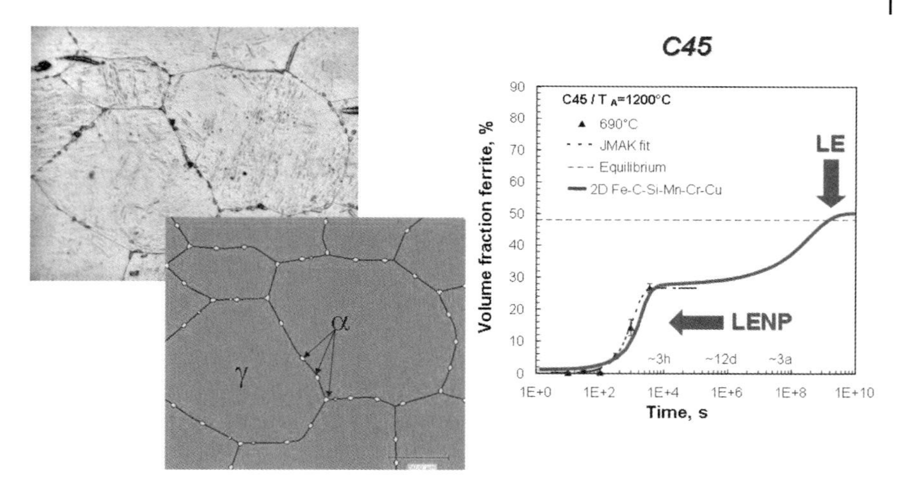

Fig. 4.5 Microstructure-based physical modeling of isothermal proeutectoid ferrite transformation using the phase field method [14].

is diffusion controlled by the other alloying elements until these are adequately partitioned among ferrite and austenite (LE).

4.4.2 Semiempirical Modeling of Phase Transformation

A semiempirical modeling approach was used in order to quantify the effect of different process parameters. The algorithm is based on a computational scheme proposed by Hougardy [15]. This approach uses isothermal TTT diagrams as input described by the 1% and 99% curves of transformed fraction of each microstructure. Here, diffusion-controlled transformation was described by the JMAK equation [16], and martensite transformation was described by using the Koistinen–Marburger equation [17]. In order to calculate CCT diagrams on the basis of the input data, an incremental method proposed by Hougardy was implemented using a stepwise isothermal approximation for continuous cooling paths.

The effect of austenite conditioning can be incorporated by modifying the isothermal transformation kinetics serving as input data for the continuous cooling simulations, where the most important parameters were assumed to be grain size, effective grain boundary area per unit volume, and strain. Here, recrystallized and deformed austenite can be taken into account as well as shear bands [18].

As an example, Fig. 4.6 shows a calculated and an experimentally measured CCT diagram of S460 during continuous cooling after austenitizing at 1250 °C followed by two deformation steps at 970 and 900 °C. The influence of austenite conditioning on transformation start can be predicted quantitatively quite well.

The influence of deformation on the ferritic transformation was tested by assuming a three-pass hot rolling scheme. Here, the last pass is just above the

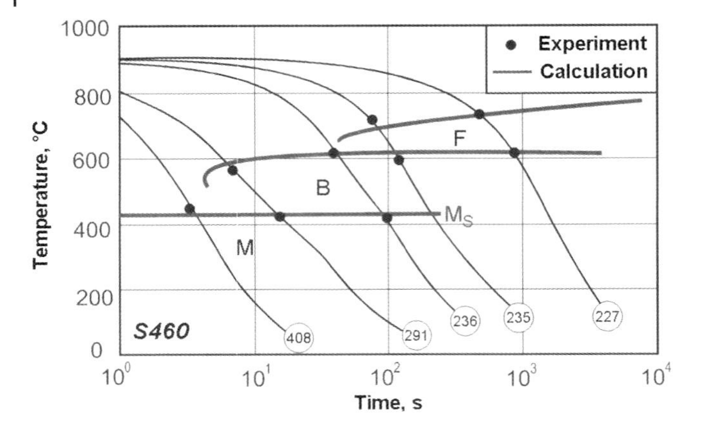

Fig. 4.6 Calculated and measured continuous cooling transformation [19].

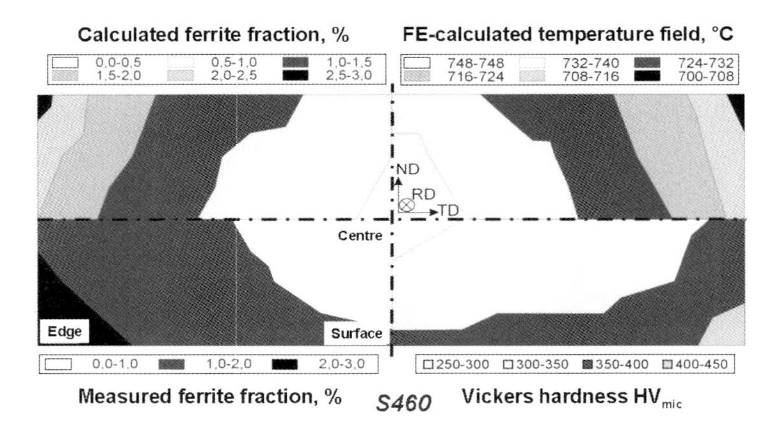

Fig. 4.7 Calculated and measured ferrite formation due to cooled edges [19].

transformation temperature A_{c1} in such a way, that at the cooled edges the transformation is accelerated. The simulation was performed by a FE calculation of the rolling process followed by the simulation of austenite conditioned phase transformation. Results of calculation and experiment are shown in Fig. 4.7.

4.5
Simulation of Mechanical Properties

Concerning the mechanical properties like strength and formability, it is assumed that the main microstructural parameters are phases and precipitates in terms of

$$\sigma = \sigma_0 + \Delta\sigma + \alpha \times \mu \times b \times \frac{1}{\sqrt{b}} \sqrt{\frac{1 - \exp(-M \times k_2 \times \varepsilon)}{k_2 \times L}}.$$

Peierls stress

σ_0 = 77 + 80 x (%Mn) + 750 x (%P) + 60 x (%Si) + 80 x (%Cu) + 45 x (%Ni) + 60 x (%Cr) + 11 x (%Mo) + 5000 x ($\%N_{ss}$)

influence carbon	$\Delta\sigma$ = 5000 x ($\%C_{ss}$)	(Ferrite)
resp.	$\Delta\sigma$ = 3065 x ($\%C_{ss}$) - 161	(Martensite)
resp.	$\Delta\sigma$ = 5000 x ($\%C_{ss}$) - $3\mu b\lambda^{-1}$	(Pearlite)
constant	α = 0.33	
shear modulus	μ = 80.000 MPa	
Burgers vector	$b = 2{,}5 \times 10^{-10m}$	
Taylor factor	M = 3	
recovery rate	$k_2 = 10^{-5} / d_\alpha$	(Ferrite)
resp.	k_2 = 20	(Martensite)
resp.	k_2 = 7	(Pearlite)
free dislocation length	$L = d_\alpha$	(Ferrite)
resp.	$L = 3{,}8 \times 10^{-3}$	(Martensite)
resp.	$L = 2\lambda$	(Pearlite)

Fig. 4.8 Flow curve model [20].

phase fractions, ferrite grain size, pearlite lamellae spacing, martensite packet size, and void fraction. For a flow curve prediction the model of Rodriguez was used, which is available for ferrite, pearlite, and martensite [20]. The essential parameters of this model are displayed in Fig. 4.8.

In order to predict the formability the microstructure-based damage model of Gurson–Tveergard–Needleman (GTN) [21] was used (Fig. 4.9). It describes ductile failure as a sequence of void nucleation due to inclusions, void growth, and void coalescence, which forms the first microcracks.

The initial void nucleation can be linked to measured or calculated inclusion densities. Void growth is caused by incompressibility of the plastic matrix, and void coalescence originates from void interaction and by plastic failure of the ligament. The latter mechanisms were quantified using a homogenization approach of axisymmetrical unit cells with a single spherical void [22].

It is well known that the FE solution of the continuum mechanical problem is not independent of mesh size. As a pragmatic solution, Steglich and Brocks propose a lower bound for the mesh size which depends on the fracture surface analysis [23].

Using the combined flow curve and damage approach, crack propagation was calculated (Fig. 4.10). The prediction is accurate for the first stage of deformation, whereas, in the later stage, lower forces were measured than calculated. This is due to combined ductile and cleavage fracture, evident from the fracture surface (Fig. 4.10). Since the GTN model describes ductile failure, cleavage is not taken into account during the simulations which causes a higher strength prediction of the structure.

Continuum damage yield potential Φ of Gurson-Tvergaard-Needleman:

$$\Phi = \frac{\sigma_v^2}{Y^2} - \left(1 + q_1^2 f^{*2} - 2q_1 f^* \cosh\left(\frac{3}{2} q_2 \frac{\sigma_h}{Y}\right)\right) \quad \text{and } Y = Y(\varepsilon_v^p)$$

void nucleation at yield start f_0 or during loading f_N:

$$f_0 \quad \longleftrightarrow \quad \dot{f}_N = \frac{f_n}{s_n 2\pi} \exp\left(-\frac{1}{2}\left(\frac{(\varepsilon_v^p - \varepsilon_n)}{s_n}\right)^2\right) \dot{\varepsilon}_v^p$$

void growth f_W (volume constance due to plastic slip λ):

$$\dot{f}_W = (1-f)tr(\dot{\varepsilon}_v^p) = (1-f)\dot{\lambda}\frac{\partial\Phi}{\partial\sigma_h}$$

bilinear effective damage f^* (void interaction f_f, crack initiation f_c):

$$f^*(f) = f \qquad (f < f_c)$$

$$f^*(f) = f_c + \frac{f_f^* - f_c}{f_f - f_c}(f - f_c) \qquad (f > f_c)$$

damage f, growth parameters q_1, q_2, q_3, continuous nucleation parameter f_n, s_n, ε_n, equiv. plastic strain ε_v^p, equiv. stress σ_v, yield stress Y, hydrostatic stress σ_h

Fig. 4.9 Damage model based on microstructural features [21].

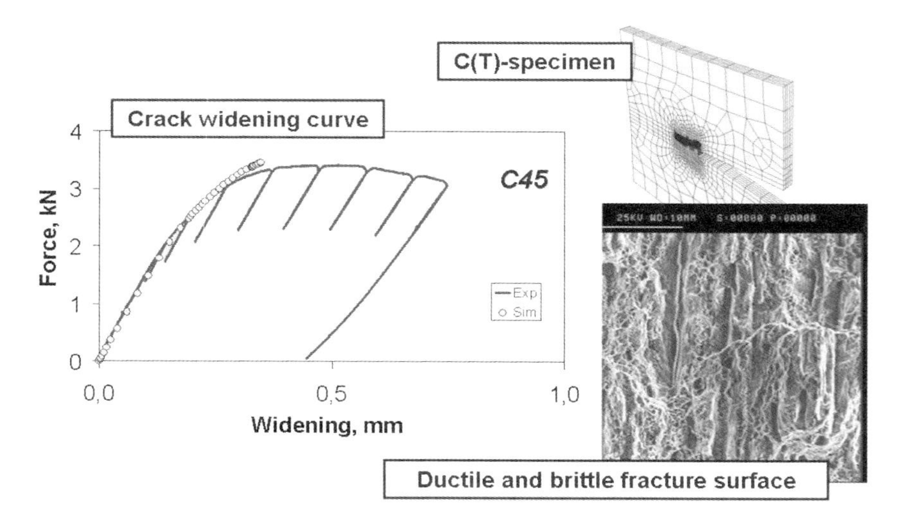

Fig. 4.10 Fracture mechanics prediction using ductile damage simulation.

4.6
Welding Simulation

Concerning welding, temperature distribution, weld geometry, mechanical properties, and residual stresses have to be predicted. For this, submodels for heat

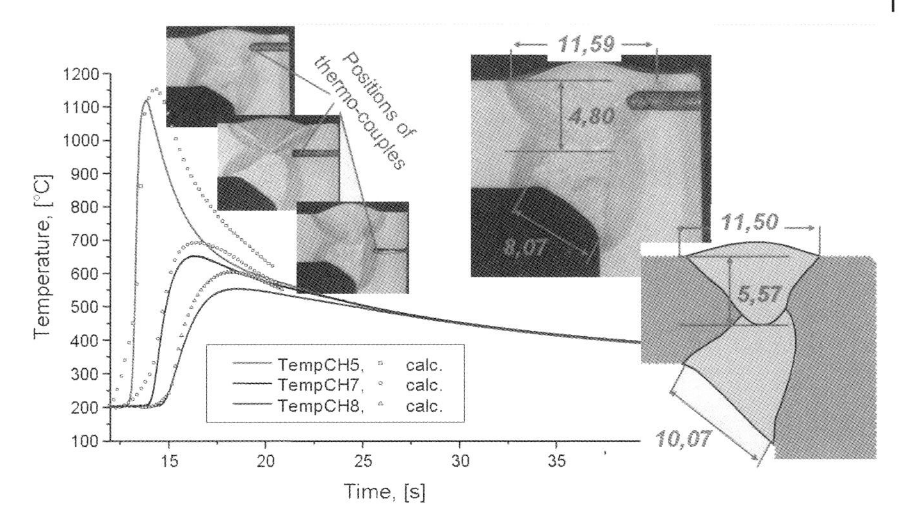

Fig. 4.11 Calculated and measured temperature evolution and weld geometry.

source, heat flow, and free surface formation during gas–metal arc (GMA) welding were combined and implemented in the SimWeld program [24].

Several thermal cycles were measured at different locations during a two pass welding of an angle profile and compared to simulation results (Fig. 4.11). Calculated and measured weld geometry showed good agreement [24].

In order to predict transformation and mechanical properties, rapid thermal cycles were performed using a deformation dilatometer and heat treatment simulation machine. For a rapid heat-treated specimen hardness, strength and toughness were measured and related to the maximum temperature during heating and to $t_{8/5}$, which is the cooling time between 800 and 500 °C.

The results were linked to the temperature computed in the welding simulation which enabled a local prediction of the mechanical properties. As an example, in Fig. 4.12 two hardness scans along the real and the simulated weld geometry are shown.

Simulation and experiment show an increasing hardness level in the heat-affected zone. The difference between measurement and simulation for the location of the maximum is due to an error in the geometry prediction. The low maximum hardness in the second experimental line scan was caused by a hole used for temperature measurement, which had changed the local cooling conditions.

Beside hardness and property simulations, the prediction of residual stresses due to thermal strain and transformation dilatation are of major importance for industrial applications. A simplified approach taking into account thermal expansion was compared with experimental results of neutron diffraction measurements [25]. For a gas–tungsten arc weld, the simulation predicts reasonable results (Fig. 4.13).

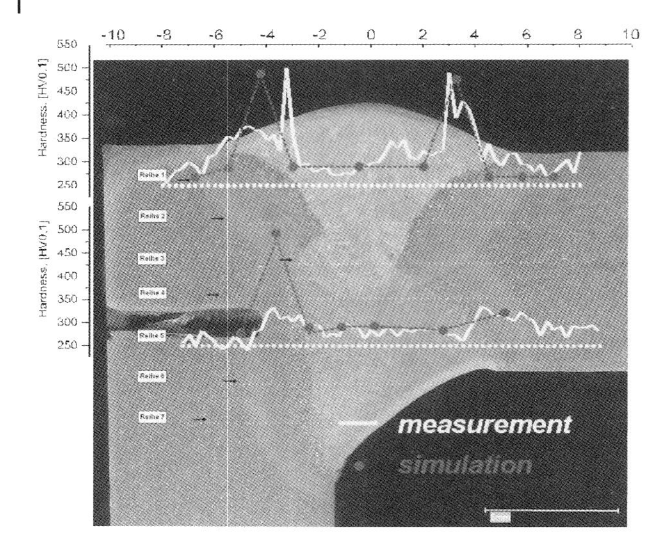

Fig. 4.12 Calculated and measured hardness distribution.

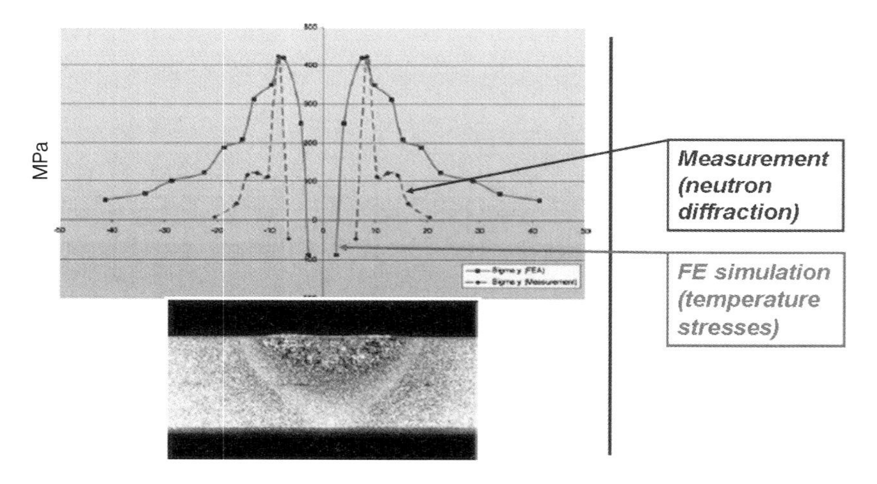

Fig. 4.13 Calculated and measured residual stresses in a single pass weldseam.

4.7 Application

For application, two tests were performed: bending of a S460 pipe and bending of a C45 angle profile. Failure was predicted using the GTN damage model.

The background of the first test is the offshore laying of a welded pipe on the bed of the ocean (Fig. 4.14). During this operation the pipe is bent, and the head

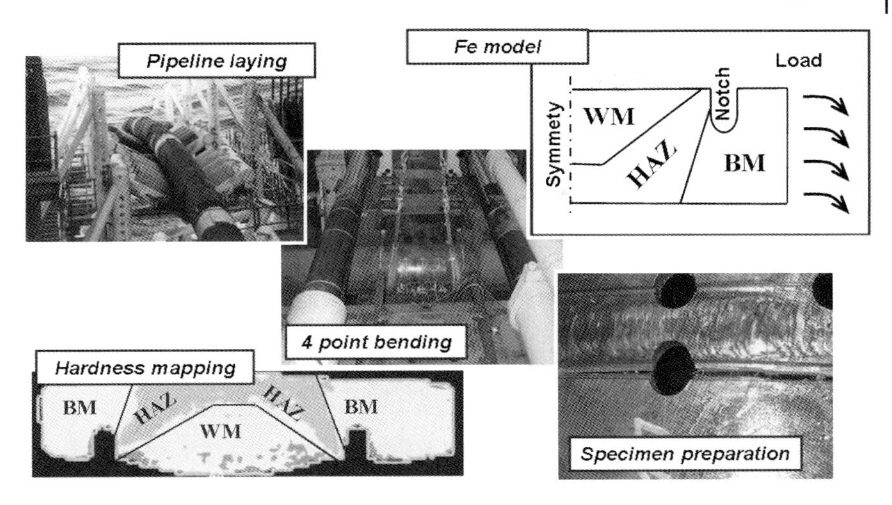

Fig. 4.14 Pipe bending during laying and four-point bending test.

weld at the upper side of the pipe is maximally loaded. A notch-like welding error was defined and artificially introduced by machining. The weld geometry was described by three zones: base material (BM), heat-affected zone (HAZ) and weld material (WM). For each zone specific material properties were defined. The special loading situation was simulated on a laboratory scale by means of a four-point bending test of a 4 m long pipe. In order to quantify the local damage fraction, holes were drilled out of the head weld.

To predict component failure, the concept of failure curves was used which renders the critical strain, that for microcrack formation, as a function of the local triaxiality [26]. The damage curve was directly predicted using the continuum damage approach of GTN with microstructure-based parameters that were determined by electron microscopy combined with automated image analysis and x-ray diffraction [27]. The result of the damage curve prediction and of the failure prediction for the welded and notched pipe under four-point bending load can be found in Fig. 4.15. Here, the connected results are compared to experiments on notched tensile specimens with different notch geometry [28].

An angle profile made of the steel grade C45 was manufactured by 90° cold bending of a hot rolled sheet. During the bending operation failure may occur depending on the bending radius. The damage distribution was calculated and failure was predicted using the GTN approach with the above mentioned parameters.

Simulation and experimental results are shown in Fig. 4.16. The material was not able to sustain a deformation with a bending radius of 5 mm, whereas at a bending radius of 10 mm the angle was formed without cracks. These results were predicted numerically and validated experimentally. Thus, it can be concluded, that the microstructure based damage parameter identification is a suitable tool for a prediction of formability.

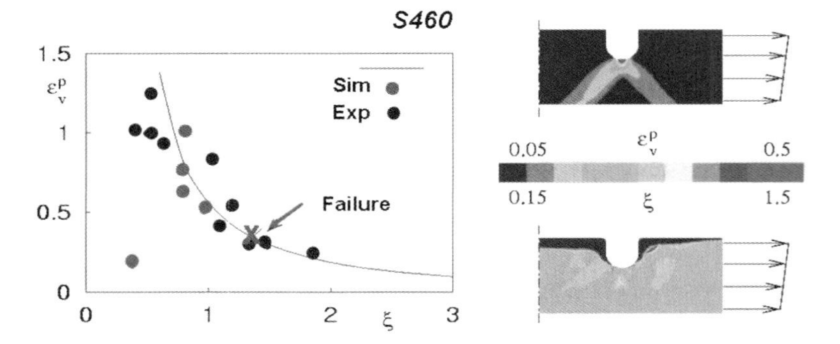

Fig. 4.15 Failure prediction using damage curves predicted by means of GTN modeling and experimentally measured in [27].

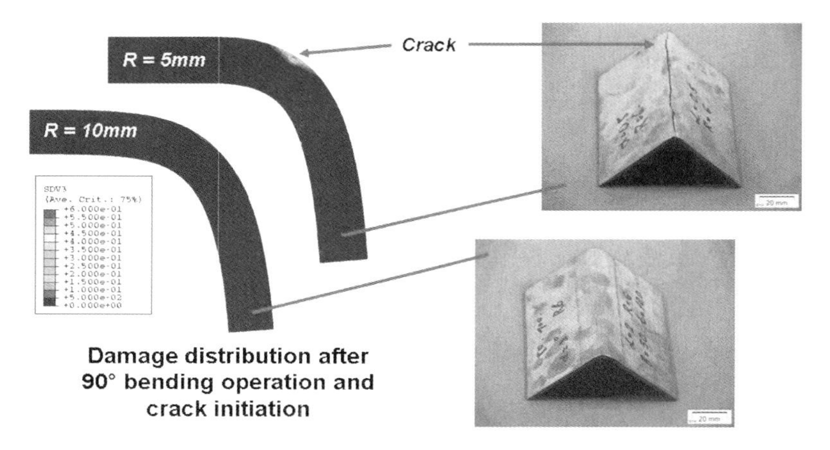

Fig. 4.16 Formability prediction for cold forming.

4.8 Summary

The process chain of steel was modeled from casting to application, where physical and semiempirical approaches were used in combination. By the combination of different models for the individual process steps, a detailed description of the evolution of microstructure was realized. Finally, mechanical properties were predicted using microstructural quantities as input parameters.

It is shown that the complete process chain can be predicted, but many parameters still have to be obtained experimentally. Nevertheless, first models have been formulated and implemented based on physical concepts that are more general.

Acknowledgment

The authors gratefully acknowledge the financial support of the Deutsche Forschungsgemeinschaft (DFG) within the Collaborative Research Center (SFB) 370 "Integral Materials Modeling".

References

1 W. Bleck et al., *Werkstoffkunde Stahl für Studium und Praxis*, 1st edn, Verlag Mainz, Aachen, 2001.
2 M. Wolf, W. Kurz, *Met. Trans. B* 12B, 1981, 85.
3 Z. Liu, J. Wie, K. Cai, *ISIJ Int.* 42, 2002, 958.
4 I. Andersen, Ø. Grong, *Acta Mater.* 43, 1995, 2673.
5 K. Schwerdtfeger, *Archiv Eisenhüttenwesen* 43, 1972, 201.
6 R. Diederichs, Dissertation, RWTH Aachen, Shaker Verlag, 2004.
7 D. Senk, M. Safi, H. Hamadou, Proc. 5th Eur. Continuous Casting Conf., Nizza, 2005 (CD).
8 R. Kopp, M. Nutzmann, O. Ziegelmayer, W. Bleck, R. Diederichs, *Steel Res.* 73, 2002, 321.
9 K. Schotten, Dissertation, RWTH Aachen, Shaker Verlag, 2000.
10 C. Beste, Dissertation RWTH Aachen, in preparation, 2006.
11 B. Dutta, C. M. Sellars, *Mater. Sci. Technol.* 3, 1987, 187.
12 www.MICRESS.de.
13 U. Prahl, U. Lorenz, A. Guindani, W. Bleck, Proc. TMP'04 2nd Int. Conf. on Thermomechanical Processing of Steels, Liege, Belgium, 2004, p. 77.
14 G. Pariser, Proceedings of BHT, 56. Berg- und Hüttenmännischer Tag, Freiberger Forschungsforum, Microstructure Analysis in the Materials Science, Freiberg, Konferenz-Einzelbericht: Freiberger Forschungshefte, Reihe B, Werkstofftechnologie, 2005, p. 87.
15 H.-P. Hougardy, K. Yamazaki, *Steel Res.* 57, 1986, 466.
16 M. Fanfoni, M. Tomellini, *Il Nuovo Cimento D* 020, 1998, 1171.
17 D. P. Koistinen, R. E. Marburger, *Acta Metall.* 7, 1959, 59.
18 Z. Husain, EUR-Report EUR 17907 EN, Luxemburg, 1994.
19 W. Bleck, R. Diederichs, U. Lorenz, R. Kopp, L. Neumann, *Steel Res.* 74, 2003, 631.
20 R. Rodriguez, I. Gutierrez, Proceeding of TMP '04, B-Liege, 2004, p. 356.
21 A. Needleman, V. T. Vergaard, *J. Mech. Phys. Solids* 35, 1987, 151.
22 U. Prahl, W. Rehbach, C. Kuckertz, D. Weichert, W. Bleck, *Key Eng. Mater.* 251–252, 2003, 351.
23 D. Steglich, W. Brocks, *Comp. Mater. Sci.* 9, 1997, 7.
24 U. Dilthey, O. Mokrov, V. Pavlyk, *Paton Welding J.* 4, 2005, 2.
25 U. Dilthey, U. Reisgen, M. Kretschmer, *Mod. Sim. Mater. Sci. Eng.* 8, 2000, 911.
26 A. Schlüter, Dissertation, RWTH Aachen, Shaker Verlag, 1997.
27 W. Rehbach, U. Prahl, C. Kuckertz, W. Bleck, Proc. Microscopy and Microanalysis, San Antonio, 2003, pp. 91–92 (CD).
28 U. Achenbach, Dissertation, RWTH Aachen, Shaker Verlag, 2000.

5
Status of Through-Process Simulation for Coated Gas Turbine Components (TP C8)

R. Herzog, N. Warnken, I. Steinbach, B. Hallstedt, C. Walter, J. Müller, D. Hajas, E. Münstermann, J.M. Schneider, R. Nickel, D. Parkot, K. Bobzin, E. Lugscheider, P. Bednarz, O. Trunova, and L. Singheiser

Abstract

This chapter gives an overview on integrative, through-process modeling and simulation approach for coated turbine blades on different scales. The approach includes the modeling of the production, materials properties, and in-service degradation processes and is accompanied by the actual production and testing of coated CMSX-4 single-crystal turbine blades and laboratory specimens. Especially, solidification of the blade alloy during casting, microstructural changes during homogenization and aging heat treatments, chemical vapor deposition of an Al_2O_3 diffusion barrier coating, physical vapor deposition (sputtering) of a (Ni,Co)CrAlY bond coat, atmospheric plasma spraying of a Y_2O_3 stabilized ZrO_2 thermal barrier coating, microstructural changes, and development of critical stresses under in-service conditions are addressed.

5.1
Introduction

As highly efficient turbine blades do no longer consist of a single metallic alloy, but rather of a complex material composite which has to perform several tasks, the development during the past 20 years was strongly focused on the optimization of the interaction between design and construction, cooling techniques, and the choice of materials for the blade body as well as the applied coatings. A turbine blade's body is commonly manufactured from high-temperature-resistant nickel-based alloys with an adapted chemical composition and microstructure achieved by advanced single-crystal casting technology which guarantees optimized strength and fatigue properties. Complex multilayer coatings, as the one displayed exemplarily in Fig. 5.1, protect the base material. Metallic coatings

Integral Materials Modeling: Towards Physics-Based Through-Process Models
Edited by Günter Gottstein
Copyright © 2007 WILEY-VCH Verlag GmbH & Co. KGaA, Weinheim
ISBN: 978-3-527-31711-0

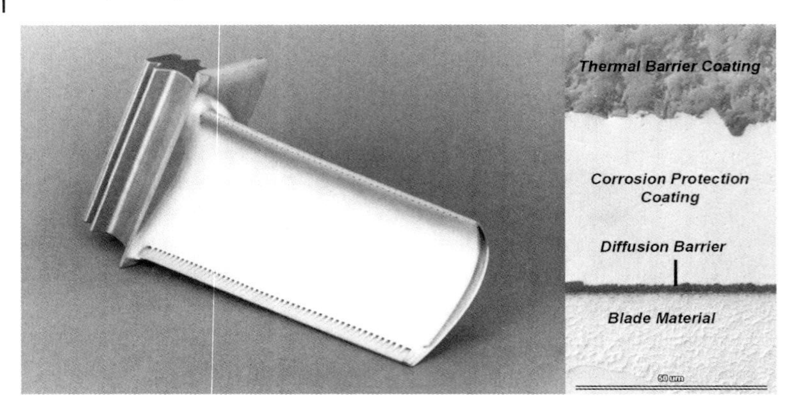

Fig. 5.1 Example of a coated gas turbine blade and a cross-sectional view of a corresponding material composite with an additional diffusion barrier.

with high aluminum and chromium contents improve the high-temperature corrosion and oxidation resistance; ceramic thermal barrier coatings (TBCs) provide thermal insulation and prevent overheating in order to extend their lifetime.

The increasing complexity of the material composite is associated with a growing number of parameters and processes limiting the turbine blade's lifetime. The microstructure and the interaction between blade material and coatings like interdiffusion and oxidation, resulting in phase changes close to the material interfaces, thermally induced strains due to thermal expansion mismatches, and so forth, affect the durability of a coated blade. Accordingly, a better comprehension of the production steps and their effect on singular and integral properties of coated blades and of the degradation processes during service incorporate a high potential for improving the in-service performance of advanced gas turbine blades.

The presented modeling and simulation approaches cover the whole production process of the turbine blade starting with the solidification of the γ'-strengthened single-crystal superalloy CMSX-4, the respective investment casting process, and the resulting dendrite structure and γ'-distribution which affect the high-temperature creep resistance and the phase changes during heat treatment. This is followed by the different coating processes, namely the CVD process for alumina deposition as a diffusion barrier, the sputter PVD process for the deposition of an MCrAlY corrosion protection coating, and the APS process for the ceramic thermal barrier coating including the prediction of macroscopic coating properties such as the elastic modulus, which affects significantly the service life of a coated blade. Finally, the work includes the simulation of degradation processes, such as bond coat oxidation, stress development, and crack formation for service-like loading, in particular for prolonged cyclic thermal and high-temperature loading.

5.2
Solidification and Heat Treatment of the Nickel-Based Superalloy

Single-crystal turbine blades are produced through directional solidification, involving a furnace with separated heating and cooling zones. The melt-containing mold is withdrawn from the heating to the cooling zone to achieve a strong temperature gradient and a directional heat flux. This leads crystals to grow directionally along the temperature gradient. As the velocity of the moving mold can be easily controlled, precise control over the crystal growth is achieved. A helix-shaped grain selector enables the selection of only one grain orientation from the initially large variety of orientations. Typically a $\langle 001 \rangle$ orientation with low Young's modulus is selected to provide a good fatigue resistance. The superior high-temperature creep strength of Ni-based superalloys is achieved by a large fraction of coherent γ' (Ni_3Al) precipitates in a γ matrix acting as effective obstacles against dislocation motion. The alloying elements Al, Ta, and Ti strongly promote the formation of γ'. Other elements are used to increase the matrix strength (W, Cr), corrosion resistance (Cr), and solidus temperature (Co), or to reduce the coarsening kinetics of the γ' precipitates (Re). As a side effect, all elements other than Al, Ta, and Ti decrease the γ' fraction. As single crystals do not contain grain boundaries, grain boundary strengthening elements such as C are kept out of the alloy [1, 2]. During solidification the segregation of the alloying elements strongly promotes the formation of interdendritic γ' precipitates. To avoid this, a homogenization heat treatment is required after the solidification.

The present work package is aimed at the development of a through-process simulation, which reflects the influence of the process parameters on the evolution of the distribution of chemical elements and microsegregations as well as microstructural parameters such as dendrite structure (spacing) and γ' precipitates (volume fraction, size distribution) during solidification and homogenization heat treatment. Thus, the models and simulation methods should provide a tool to predict the microstructural parameters, which in particular control the creep properties of the superalloy. Two superalloys were involved, namely the commercially available alloy CMSX-4 and a model alloy.

The modeling of solidification and heat treatment generally involves different length scales, which differ by up to six orders of magnitude, that is approximately the range from nanometer to millimeter. The respective processes at different length scales, for instance the diffusion at γ/γ'-phase boundaries at nanometer scale as well as the motion of the mold and the resulting macroscopic temperature gradients at the millimeter scale during the Bridgman process were simulated using individual models and simulation approaches. The complete through-process simulation was then achieved by an appropriate linking of the individual models.

The simulation of dendrite formation during the Bridgman process was the starting point. Experimental analysis has shown that directionally growing primary dendrites form regular arrangements with slowly varying interdendritic spacing with respect to the cross-section of single-crystal blades. The interdendritic spacing is basically controlled by the local cooling rate in the blade during solidification. This macroscopic level of the simulation comprising the Bridgman furnace and blade geometry was accomplished by simulating the nonstationary temperature field in the solidifying melt using the FEM package CASTS [3]. The calculations provide the locally resolved cooling rate and the interdendritic spacing. Thermophysical data such as the heat capacity and the amount of solid fraction were calculated from thermodynamic and kinetic databases using Thermo-Calc and DICTRA.

The coupling of the macroscopic process simulation and the detailed microscopic simulation of the dendrite formation was achieved by defining a unit cell with the size of the primary dendrite spacing [4]. For simulating the dendrite formation in the unit cell the multiphase, multicomponent phase field method [5, 6] was applied. It allows one to calculate the evolution of the free interface between transforming phases and thus the time-dependent morphology and composition of growing phases. An equation of motion for the moving interface was derived from free energy functionals using double well or double obstacle potentials.

The local equilibrium conditions at the solid/liquid interface were provided by coupling the phase field method to thermodynamic calculations. Figure 5.2 illustrates the final coupling scheme for the simulations as it was implemented in MICRESS. The central part is the phase field solver, coupled to the solutal and thermal diffusion solver. The thermodynamic calculations were brought in by coupling to Thermo-Calc [7, 8] via the TQ interface.

The presence of microsegregations within interdendritic regions after solidification degrades the mechanical properties, particularly the creep resistance of

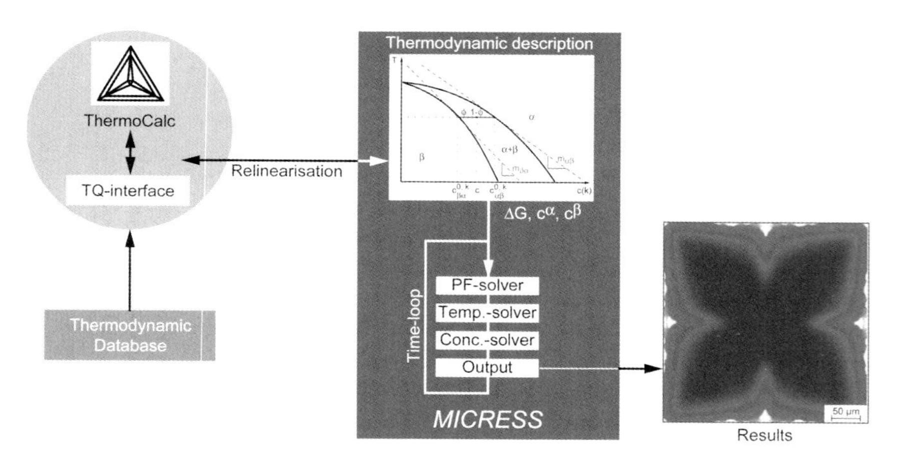

Fig. 5.2 Coupling scheme for the calculation of dendrite formation using the phase field method coupled to thermodynamic databases. The phase field part of the code provides the evolution of the phase boundary, matter, and energy transport through diffusion, while the thermodynamics provides the local equilibrium data.

the superalloy [1, 9]. In order to achieve the desired mechanical properties a homogenization heat treatment is always required to dissolve microsegregations and interdendritic phases. This so-called solution heat treatment is generally followed by an aging treatment to tailor the strengthening γ'-phase, with respect to its size and spatial distribution.

A solution heat treatment typically takes 12 to 20 hours and is conducted by a stepwise temperature increase with dwell times. The kinetics of the homogenization is determined mainly by the diffusion kinetics of the alloying elements, but also by the stability of the interdendritic precipitates. The phase field method coupled to thermodynamic calculations was used to simulate the heat treatment.

The results from the prior solidification simulations were generally used as the starting condition for the simulation of the further heat treatment. The already defined unit cells were used, so the influence of the solidification parameters, interdendritic spacing, and cooling rate, was respected this way. Further details of the simulation of solution heat treatment and locally resolved chemical composition, particularly the prediction of the composition-dependent solidus temperature for avoiding incipient melting during heat treatment are described in [10] and Chapters 8 and 12.

The solution heat treatment is generally followed by an aging treatment, which aims at an optimized microstructure with a uniform γ' size distribution and a tailored mean particle size and spacing of the γ' precipitates. Because the application of a statistical Ostwald ripening model cannot account for the influence of coherency stresses, which determine particularly the shape evolution of γ' precipitates, the phase field method was also applied for the precipitation and ripening processes during ageing treatment. No a priori assumptions about the size and volume fraction of the precipitates had to be made. The phase field equation was coupled to stress and strain calculations via the driving force term. Depending on the sign, the appearing stresses in the matrix promote or hinder the growth of the precipitates. As the molar volume of γ' is larger than the molar volume of γ, compressive stresses hinder the γ' growth. The evolution of size and shape of the γ' precipitates was primarily controlled by the interfacial properties of precipitates and matrix, diffusion in the matrix, thermodynamics, and the coherency stresses at the interface. The development of a cubic shape was in particular due to the anisotropic elastic properties of the γ matrix and the occurrence of coherency stresses. Even if the detailed γ' particle parameters were not quantified as yet, it has been demonstrated that the output of the present simulations comprises the γ' particle parameters, which determine the creep and therefore also the stress relaxation properties of single-crystal turbine blades cast from Ni-based superalloys.

At this point, the developed simulation tools can be coupled to a microstructure-based constitutive creep equation such as that developed by Fleury [11] in order to provide a modeling interface to the macroscopic engineering material properties. However, the simulations done so far were not matching experimentally observed γ' particle distributions satisfactorily, and further work associated with underlying model restrictions is planned.

5.3 CVD Processing of an Alumina Interdiffusion Barrier

Highly loaded blades in advanced gas turbines are almost always applied with corrosion-protective coatings to increase their high-temperature corrosion and oxidation resistance. The high Al, Cr, and Co contents of currently applied MCrAlY (M = Ni, Co) coatings cause large gradients in chemical composition towards the actual Ni-based blade material. The concentration gradients in turn lead to interdiffusion processes at high temperature between the MCrAlY coating and blade and thus to the formation of a broad interdiffusion zone. Resulting undesirable phase changes alongside both sides of the original interface, Al loss from the bond coat, or the formation of Kirkendall porosity cause a deterioration of both the mechanical properties and the oxidation resistance of the coated blade. By applying a thin intermediate Al_2O_3 layer, interdiffusion between the bond coat and substrate can be completely or almost completely inhibited [12–15]. The Al_2O_3 layer should be as thin as possible in order to minimize thermally induced loads from lattice mismatch, but if it is less than about 3 μm thick, it may become unstable during high-temperature exposure and may not block interdiffusion sufficiently [12]. In the present work a two-layered TiN/Al_2O_3 diffusion barrier coating was applied prior to the corrosion protective coating using chemical vapor deposition (CVD). The TiN layer avoids whisker growth [12–14, 16, 17] and thus the formation of open porosity combined with fast diffusion paths. TiN also appears to act as a buffer by adapting the different thermal expansions of blade material and Al_2O_3 scale [12].

The CVD coating process was simulated to predict and thus better control the deposition kinetics and the coating thickness. The PHOENICS-CVD code [18] was applied to simulate the α-Al_2O_3 deposition from an $AlCl_3$–CO_2–H_2–HCl precursor system [14, 19]. The associated CVD experiments were performed in a vertical tubular hot-wall reactor. The simulations, which comprise in general the prediction of local deposition rates in the reaction furnace, were performed using the CFD (Computational Fluid Dynamics) code PHOENICS with a CVD module developed within the European Esprit project 7161 and implemented by Kleijn and Kuijlaars [20]. The mass flow included diffusion according to Wilke [21]. Thermodiffusion was not included since its contribution was found to be negligible. Heat transport by the mass flow, radiation, and conduction in the gas and solids was included. Absorption of radiation by the gas was not considered.

The overall agreement between simulation and experiment is reasonably good, but the dependence on the position in the reactor tube is quite different. In particular when the amount of HCl in the inlet gas was small, the experimental deposition rate decreased strongly from the lower to the upper end of the isothermal zone, whereas the simulated deposition rate shows a maximum followed by a moderate decrease. The first (lowest) substrate in the chain typically shows a comparably high deposition rate. At a higher partial pressure of HCl (20 hPa) the simulations showed better agreement with experimental data. Moreover, at higher p(HCl) values the deposition rate was more uniformly distributed in the

CVD reactor indicating that the partial pressure of HCl acts as a key parameter for obtaining a constant coating thickness. Thus, the reasonable agreement between simulations and experimental data within the studied parameter space demonstrates that the current CVD simulations provide a helpful tool for better controlling the CVD process in regard to a uniformly coating thickness.

5.4 Magnetron Sputter Process of NiCoCrAlY Corrosion-Protective Coating

As mentioned in the previous section corrosion-protective coatings are a prerequisite for highly loaded blades to increase their high-temperature corrosion and oxidation resistance. While the already described diffusion barrier to the blade material was provided by CVD, the corrosion protective coating is frequently applied using physical vapor deposition (PVD). The principle is to evaporate solid material which condensates on the turbine blades' surface.

In the current approach the magnetron sputter process was selected for depositing an MCrAlY coating, which is favorable for land-based gas turbines. Here, the MCrAlY target material is evaporated by transferring the kinetic energy of impacting plasma ions, commonly argon. Due to the different physical effects and the great differences in the time scales during the different subprocesses, the simulation of the sputter PVD is separated into three parts, namely evaporation, particle transport through the plasma, and film growth. The results of each part are used as input for the following simulation step.

In the first step, the simulation of the evaporation is performed using two different ways. The first method is using analytical descriptions of the necessary distributions, the kinetic energy, and the flight direction of the evaporated atoms as well as the sputtering yield, which is the number of atoms leaving the surface per incident argon ion. The calculated quantities mostly depend on the surface binding energy, which is influenced not only by the chemical binding energy of the surface atoms but also by all nonlinear properties of the surface such as surface roughness. A good estimation for the surface binding energy is the heat of sublimation, which is well known for single-element targets. For alloy targets like the present NiCoCrAlY, the surface energy depends on the mole fraction of each component and their binding mechanisms. Several approaches were used to calculate the average surface binding energy for binary and ternary alloys [22, 23]. The experimental data that were used for fitting the analytical functions were taken from the NIFS Database SPUTY [24]. The second method for the simulation was using the simulation package SRIM [25], which calculates the sputtering yield and the energy distribution of each component of the target alloy, respectively.

The evaporated target material looses a part of its kinetic energy during the transport phase due to elastic collisions with the process gas. Assuming that the process gas is sufficiently rarefied and the sputtered material is mostly neutral, which is a good approximation in conventional sputtering processes, the movement of the particles between two collision processes is regarded as force free allowing to simply calculate the energy changes caused by the collision and the new collision position according the mean free path in the argon at given temperature and pressure. Each target atom is traced starting with its evaporation position until it hits the substrate or one of the walls of the process chamber. As interaction potential between the process gas atoms and the target particles, the variable hard-sphere model has been chosen. Both evaporation and scattering can be considered as statistical processes and are therefore simulated with Monte Carlo methods. The results of this simulation step are the spatial distribution of the adsorbed atoms, their energy, and angle of impact at the substrate surface. The simulated film thickness values were scaled to relative units and compared with experimental data showing reasonable agreement.

Additionally an algorithm for meshed three-dimensional substrates was developed and the CAD model of the turbine blade already used in the solidification simulations was converted and implemented in the coating simulation. The simulation was performed in a first step for a single-element target and later for the MCrAlY alloy showing again reasonable agreement with experimental data.

5.5
Atmospheric Plasma Spraying of Ceramic TBC

Atmospheric plasma spraying (APS) is used for the deposition of ceramic TBCs for the application in gas turbine components. Zirconia partially stabilized with about 7% yttria (PYSZ) is mainly used as material for TBCs. A large amount of defects such as pores and microcracks is desired to decrease the thermal conductivity for better thermal insulation as well as to decrease the elastic stiffness in order to achieve low stress levels arising from thermally induced strains. These material properties, but also others such as creep and stress relaxation of the TBC at high temperature, are strongly affected by a number of process parameters like the mass flow, the injection velocity, and the diameter distribution of powder particles, the plasma jet characteristics (enthalpy, velocity, pressure), and the substrate temperature.

The present simulation approach of the APS process has to be divided into the simulation of the subprocesses related to plasma torch, plasma free jet, sprayed powder particle characteristics, and coating formation [26–28]. The simulation tools developed for the subprocesses are linked by interfaces to realize the complete APS process simulation [29]. The gas and plasma flow field variables (velocity, temperature, enthalpy, pressure) inside the torch and inside the plasma jet were numerically calculated by the finite volume method using the CFD tool Fluent [30]. A separate calculation approach is required because of different geometric dimensions which may cause numerical instabilities during the calculations. The nozzle at the end of the torch acts as an interface between both subprocesses. Within the plasma jet the injected powder particles are heated and accelerated. The calculation of the heat and momentum transfer to the particles was conducted using the software PLASMA2000 [31, 32] or alternatively a coupled dis-

Fig. 5.3 Micrograph of a thermal barrier coating made from zirconia partially stabilized with about 7% yttria (PYSZ).

crete phase model approach in the CFD tool Fluent. The process of coating formation on the turbine blade was simulated using the finite element method (FEM), whereby the plasma jet and sprayed particle simulation define the initial and boundary conditions for the FEM approach [33, 34]. The simulation steps are described in more detail in [10] and Chapter 9.

A subsequent homogenization method was applied to obtain macroscopic material properties from the microstructure with defects like pores and cracks (Fig. 5.3).

Two homogenization methods were applied in this project to predict macroscopic and thus continuum material properties depending on this microstructure. One of the two methods developed is based on the physical equivalence homogenization [35]. SEM images such as in Fig. 5.3 were selected and then digitized as a black and white graph and implemented as a meshed structure within an FE code.

The corresponding material properties for the dense ceramic were applied for the "white" phase. The unit cell containing the defects was then virtually loaded by means of an FE simulation. The stress response to constant rate straining was simulated, and the Ramberg–Osgood model was adjusted to the computed stress–strain curve in order to obtain material parameters reflecting the homogenized, macroscopic material behavior [36].

The calculated elastic material parameters can be directly used for subsequent simulations of the stress response under service-load conditions as described in the next section.

5.6
Stress Response and Crack Formation at the Bond Coat/TBC Interface During Cyclic Thermal Loading

Thermal and mechanical exposure during service operation causes crack formation, crack growth, and final spallation of the ceramic thermal barrier coating along the bond coat/TBC interface and thus limits the durability of ceramic-coated gas turbine components (combuster parts, transition pieces, turbine blades). The damage evolution is accompanied and affected by degradation processes such as bond coat oxidation (formation of a thermally grown oxide scale, TGO) and sintering of the ceramic top coat. In the case of plasma-sprayed TBCs, which are increasingly applied in gas turbines for power generation, the remaining component from which a TBC was spalled off appeared frequently as a mixture of black and white colors. This type of fracture surface resulted from a failure crack which propagated partly in the thin TGO (black) and partly in the TBC (white), thus leaving some parts of the TBC on the component. The "black and white" or "gray" failure mode of APS TBCs can be reproduced by furnace cyclic tests or cyclic oxidation tests in the laboratory, which comprise thermal cyclic loading with a dwell time of a few hours at high temperature.

The simulations presented in this chapter were focused particularly on this type of failure. It was envisaged to simulate the long-term stress response to this load profile as well as the formation of microcracks in or near the TGO as well as the influence of different stiffness values of the APS TBC on the stress response.

In order to compare the FE simulations with experimental results, a series of furnace cyclic tests was carried out on the TBC composite which was investigated in the SFB 370. The experiments were conducted up to selected fractions of the time-to-spallation (lifetime). Figure 5.4 shows an example of crack pattern observed at early stages of exposure (~10% of life). It illustrates crack formation at the metal/ceramic interface of TBCs at the micrometer scale. The crack marked by arrows represents a type of microcrack which was frequently observed and indicates a critical site of crack formation at the TGO (the darker gray area marks the TGO). In detail, the microcrack was formed at an internal roughness peak along the bond coat/TGO interface and partly crossed the TGO towards the TBC. The crack tips of this type of cracks were sometimes located within the TBC (right side) or at the interface of TGO and TBC (left side). More detailed

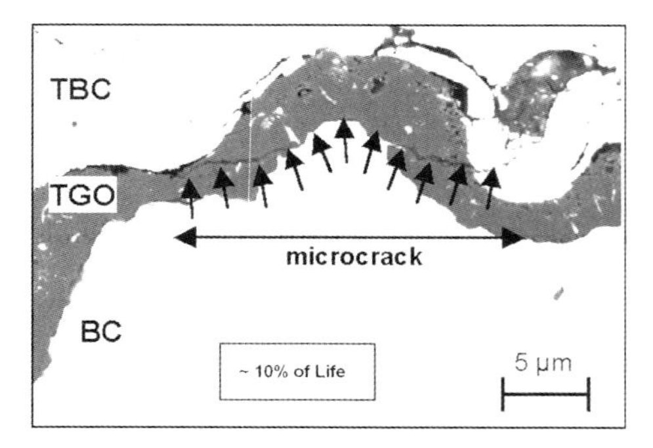

Fig. 5.4 Frequently observed crack pattern at early stages of exposure indicating weak points for crack formation and initial crack growth.

results about the evolution of damage in APS TBCs are described in Chapter 14 and [37].

The model selected for the FE simulations corresponded to the cylindrical specimen geometry used in the experiments. It consisted of four layers, namely base material, bond coat (BC), TGO, and TBC (Chapter 14). The base material (CMSX-4) was treated as an entirely elastic material, because of the low stresses which occurred during pure thermal cycling without additional external creep or fatigue loading.

It was shown in [38] that creep and resulting stress relaxation processes in the base material do not affect the lifetime relevant stresses near the TGO in a decisive manner when thermal cycling with high-temperature dwell time as one of the basic load profiles for TBC degradation is taken into account. Therefore, it is also not reasonable to consider the developed simulation approach for the single-crystal casting process and its impact on the creep properties when the simulation aims at the degradation processes of TBCs. The bond coat was considered as elastic–viscoplastic, the TGO as elastoplastic, and the TBC as viscoelastic. All material properties were temperature dependent, experimentally determined on actual coatings (Chapter 14) [39], and summarized in [40, 41].

Bond coat oxidation was simulated as a continuous volume increase of the alumina scale at high temperature using the swelling option in ABAQUS. It was modeled as an orthotropic swelling strain of the TGO, whereby lateral TGO growth (length increase) was considered as a constant fraction of thickness growth (generally 5%).

It has been shown in [38] that during few thermal cycles the stress response near the TGO develops in such a way that the room temperature stresses became much higher than those at high temperature. High-temperature stresses did not vanish due to stress relaxation after a certain number of thermal cycles and cannot be neglected as it is sometimes mentioned in the literature, but they were considerable smaller than those at ambient temperature. Thus, the following results comprise the stress state at room temperature after each thermal cycle.

It is shown in Chapter 14 that the simulation of crack formation resulted in early cracking at this interface corresponding to the experimental observations.

In order to assess the influence of the TBC stiffness (effective E-modulus) on the stress response simulations with stiffness modifications were carried out. Thus, the TBC stiffness appears as a key parameter to affect laterally oriented zones of critical tensile stress concentrations which may predetermine delamination crack paths. By contrast, the tensile stresses at the bond coat/TGO interface are affected by only a few percent when the TBC stiffness was modified. By identifying the TBC stiffness as a key parameter for critical tensile stress concentrations in the TBC, the APS process simulation can now be used to tailor the micro or defect structure of the TBC to obtain lower tensile stresses for critical TBC regions. Thus, the developed simulation approaches which cover the manufacturing process as well as the material degradation processes under near-service load profiles may be used for guided optimization of TBCs.

5.7 Conclusions

An integrative modeling and simulation approach has been presented to be applied in the fields of production, heat treatment, degradation, and damage of a ceramic-coated turbine blade. Different modeling and simulation techniques were developed, and data and model interfaces were created to realize a modeling approach of the whole life cycle from the production to the turbine blade failure, particularly the spallation failure of ceramic TBCs. Simulations can be applied in parameter studies to investigate the influence of casting and coating process parameters on the properties of the turbine blade and to design the production processes as an alternative to time- and material-consuming experimental studies. Many physical phenomena during the production and the operational use of the coated turbine blade could be understood and implemented in models and simulation tools. The work is currently continued in the Transferprogram 63 "Industrially relevant modeling tools" supported by the Deutsche Forschungsgemeinschaft (DFG) in close collaboration with gas turbine and coating manufacturers in order to apply the process and material models as well as simulation tools to current applications.

Acknowledgment

The authors gratefully acknowledge the financial support of the Deutsche Forschungsgemeinschaft (DFG) within the Collaborative Research Center (SFB) 370 "Integral Materials Modeling".

References

1 C.T. Sims, N.S. Stoloff, W.C. Hagel, in *Superalloys II*, John Wiley, 1987.
2 M. Durand-Charre, *The Microstructure of Superalloys*, Gordon and Breach, 1997.
3 P.R. Sahm, W. Richter, F. Hediger, *Gießereiforschung*, 35, 1983, 35.
4 D. Ma, U. Grafe, *Int. J. Cast Met. Res.* 13, 2000, 85.
5 I. Steinbach, F. Pezzolla, B. Nestler, M. Seeßelberg, R. Prieler, G.J. Schmitz, J.L.L. Rezende, *Physica D* 94, 1996, 135.
6 J. Tiaden, B. Nestler, H.J. Diepers, I. Steinbach, *Physica D* 115, 1997, 73.
7 U. Grafe, B. Böttger, J. Tiaden, S.G. Fries, *Scripta Mater.* 42, 2000, 1179.
8 N. Warnken, B. Böttger, S.G. Fries, I. Steinbach, in *Modeling of Casting, Welding and Advanced Solidification Processes: X*, ed. D.M. Stefanescu, J.A. Warren, M.R. Jolly, M. Krane, TMS, 2003, p. 21.
9 D. Goldschmidt, *Mat.-wiss. u. Werkstofftech* 25, 1994, 373.
10 R. Herzog, N. Warnken, I. Steinbach, B. Hallstedt, C. Walter, J. Müller, D. Hajas, E. Münstermann, J.M. Schneider, R. Nickel, D. Parkot, K. Bobzin, E. Lugscheider, P. Bednarz, O. Trunova, L. Singheiser, *Adv. Eng. Mater.* 8, 2006, 535.
11 G. Fleury, F. Schubert, PhD thesis, Aachen University, 1997, Berichte des Forschungszentrums Jülich, Jül-3436.

12 J. Müller, M. Schierling, E. Zimmermann, D. Neuschütz, *Surf. Coat. Technol.* 120–121, 1999, 16.
13 U. Eritt, G. von Hayn, E. Lugscheider, J. Müller, D. Neuschütz, *Mat.-wiss. u. Werkstofftech* 33, 2002, 45.
14 J. Müller, Ph.D. thesis, RWTH Aachen University, Shaker Verlag, Aachen, 2004.
15 J. Müller, D. Neuschütz, *Vacuum* 71, 2003, 247.
16 N. Bahlawane, S. Blittersdorf, K. Kohse-Höinghaus, B. Atakan, J. Müller, *J. Electrochem. Soc.* 151, 2004, C182.
17 C. Colombier, J. Peng, H. Altena, B. Lux, *Int. J. Refract. Hard Met.* 5, 1986, 82.
18 PHOENICS-CVD Ver. 3.4, CHAM, Concentration, Heat and Momentum Ltd., Wimbledon Village, London, UK, 2001.
19 J. Müller, D. Neuschütz, in *Chemical Vapor Deposition XVI* and *EUROCVD 14*, ed. M.D. Allendorf, F. Maury, F. Teyssandier, Electrochemical Society, Pennington, NJ, 2003, p. 186.
20 C.R. Kleijn, K.J. Kuijlaars, *PHOENICS J. Computat. Fluid Dynamics Appl.* 8, 1995, 404.
21 C.R. Wilke, *Chem. Eng. Prog.* 46, 1950, 95.
22 P.K. Haff, Z.E. Switkowski, *Appl. Phys. Lett.* 29, 1976, 549.
23 G.W. Reynolds, F.R. Vozzo, R.G. Allas, P.A. Treado, J.M. Lambert, *Nucl. Instrum. Meth. Phys. Res. B* 2, 1984, 804.
24 SPUTY database, https://dbshino.nifs.ac.jp/info.html, National Institute of Fusion Science (NIFS), Japan.
25 J.F. Ziegler, W. Eckstein, SRIM 2003, software package, www.srim.org.
26 U. Eritt, Ph.D. thesis, RWTH Aachen University, Verlag Shaker, Werkstoffwissenschaftliche Schriftenreihe, Bd. 40, Dissertation, 2000.
27 E. Lugscheider, U. Eritt, I.V. Krivtsun, A.F. Muzhichenko, S. Yu, *Paton Welding J.* 12, 2000, 40.
28 N. Papenfuß-Janzen, E. Lugscheider, Proceedings of the Advanced Simulation Technologies Conference 2002, Vol. 34, p. 137.
29 E. Lugscheider, N. Papenfuß-Janzen, R. Nickel, *J. Phys. IV* 120, 2004, 373.
30 E. Lugscheider, K. Bobzin, R. Nickel, Proceedings of the 2005 Business and Industry Symposium of the Spring Simulation Multiconference, Simulation Councils Inc, 2005, p. 13.
31 S. Kundas, T. Kashko, V. Hurevich, E. Lugscheider, N. Papenfuß-Janzen, Proceedings of the International Thermal Spray Conference, DVS Verlag, 2002, p. 765.
32 E. Lugscheider, U. Eritt, V. Hurevich, S. Kundas, A. Kuzmenkov, Proceedings of the National Academy of Sciences of Belarus, Series of Physico-technical Science, 2000, No 1, p. 134.
33 E. Lugscheider, R. Nickel, *Surf. Coat. Technol.* 174–175, 2003, 475.
34 S. Kundas, T. Kashko, E. Lugscheider, G. von Hayn, A. Ilyuschenko, Proceedings of the 15th International Plansee Seminar, 2001, Vol. 3, p. 360.
35 E. Lugscheider, R. Nickel, T. Kashko, S. Kundas, Proceedings of the Business and Industry Symposium of the 2003 Advanced Simulation Technologies Conference, Simulation Councils Inc., 2003, p. 117.
36 E. Lugscheider, R. Nickel, S. Kundas, T. Kashko, Proceedings of the International Thermal Spray Conference, DVS Verlag, 2004, p. 311.
37 E. Trunova, Doctoral thesis, submitted to RWTH Aachen University, Berichte des Forschungszentrums Jülich, to be published 2006.
38 P. Bednarz, Doctoral thesis, submitted to RWTH Aachen University, Berichte des Forschungszentrums Jülich, to be published 2006.
39 P. Majerus, R.W. Steinbrech, R. Herzog, F. Schubert, Proceedings of the 7th Liege Conference, 30 Sept.–2 Oct. 2002, Liege, Belgium, Materials for Advanced Power Engineering, 2002.
40 P. Bednarz, R. Herzog, E. Trunova, R.W. Steinbrech, L. Singheiser, *Ceram. Eng. and Sci. Proc.* 26, 2005, 55.
41 R. Herzog, P. Bednarz, E. Trunova, V. Shemet, R.W. Steinbrech, F. Schubert, L. Singheiser, Proceedings of the 30th International Conference and Exposition on Advanced Ceramics and Composites, Cocoa Beach, 2006, to be published.

6
Deformation Behavior of a Plastics Pipe Fitting (TP C9)

W. Michaeli, E. Schmachtenberg, M. Brinkmann, M. Bussmann, and B. Renner

Abstract

Inner properties of injection-molded parts resulting from the manufacturing conditions have a strong influence on the mechanical behavior of thermoplastic parts. A prediction of the inner properties and their integration into a plastics-oriented material model leads to a suitable and therefore material-saving design. The IKV has developed software tools that predict inner properties locally resolved depending on the process parameters and transfer them to structural analysis software. The mechanical material description in finite element analysis includes the dependency of the material on the local inner properties, temperature, strain, and strain rate. The transfer of the inner properties was tested and validated using microscopic examinations as well as concurrently performed component tests.

6.1
Introduction

The exploitation of the full potential of a material is mainly dependent on the knowledge of the mechanical properties of the material. On the one hand, designing a part that is both safe and material saving is challenging if the mechanical characteristic values are influenced by the manufacturing process and are inhomogeneous within the part. On the other hand, a specific use of these inhomogeneities in part design is possible. For that purpose, a prediction of the locally existing mechanical properties is essential.

Thermoplastic parts manufactured in the injection-molding process show inhomogeneous inner properties both across the wall thickness and along the flow path. These inner properties result from the manufacturing conditions and have a direct influence on the local mechanical properties and, therefore, also on the global part stiffness [1]. This dependency is highly distinctive for short-fiber-

Integral Materials Modeling: Towards Physics-Based Through-Process Models
Edited by Günter Gottstein
Copyright © 2007 WILEY-VCH Verlag GmbH & Co. KGaA, Weinheim
ISBN: 978-3-527-31711-0

reinforced thermoplastics [2]. But also for nonreinforced semicrystalline thermoplastics locally inhomogeneous inner properties can be observed that influence the mechanical properties [3–5]. These are dependent on the polymer, and they are determined by the local rheological and thermal conditions during the filling process and solidification. The layered arrangement of the individual structures and their broadness has an enormous influence on the local mechanical behavior.

6.2
Aims and Procedure

The Collaborative Research Center 370 at RWTH Aachen University in general has the aim of investigating the development of inner properties and their influence on the mechanical properties of different materials (steel, aluminum, plastics) and different manufacturing and processing methods (casting, welding, injection molding). The aim of these investigations is to predict the distribution of inner properties by process simulation and, thus, provide input for the material description of a structural analysis. The result is a more precise reproduction of the mechanical part behavior and thus, a part design suitable for the respective material.

In order to study the influence of manufacturing conditions on the inner and mechanical properties, a material and a part were chosen. The chosen material is a polypropylene made by Sabic, type Sabic PP 505 P. The part chosen is a pipe fitting which can be injection molded using different gate systems. The pipe fitting is shown in Fig. 6.1 with varying gate systems, such as diaphragm, radial, and axial gates. By using mold inserts it is further possible to manufacture the pipe fittings with three different types of wall thickness. Eventually the resulting structure can be influenced by the processing parameters and, therefore, by various rheological and thermal boundary conditions. In this case the most important manufacturing parameters are the injection speed, the melt temperature, and the mold temperature. The multitude of parameters permits one to vary the developing structure widely. The influences of these variations on the real part behavior are examined by testing the parts using a standard tensile testing machine.

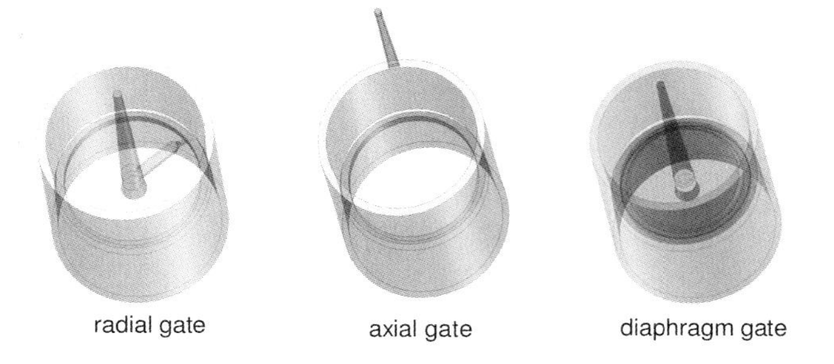

Fig. 6.1 Various gate systems of the exemplary part.

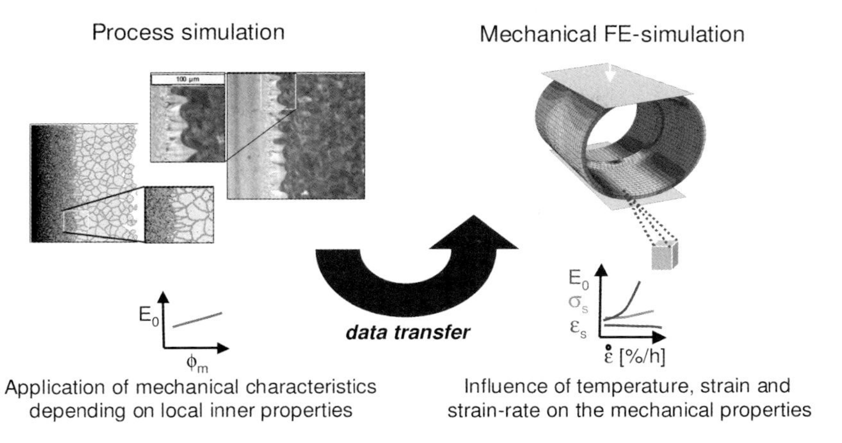

Fig. 6.2 Procedure for the integrative simulation.

To include these effects in the simulation and to make them applicable for part design a fully integrative view of the manufacturing process is necessary. To accomplish this, the procedure shown in Fig. 6.2 is chosen. First a process simulation that calculates local crystallization and molecular orientation data is performed for the part. In the following step this data is transferred to the finite element mesh of the structural analysis using an interface. The interface permits the variation of the meshes of process simulation and structural analysis independently and, thus, to optimize them with regard to their special requirements. The local mechanical properties are then concluded from these inner properties.

6.3 Calculation of Local Inner Part Properties Using Extended Process Simulation

Only a few injection-molding simulation codes are currently able to simulate the real processing of inner part properties development during injection molding of semicrystalline thermoplastics [6]. These in reality anisotropic properties are normally considered as integral values given for a part cross-section or for the whole molding.

For this reason a process simulation software was developed and extended to simulate the development of inner part properties under injection-molding conditions [7, 8].

6.3.1 Developed Software

For several years a variety of projects have been carried out at the Institute of Plastics Processing (IKV) to enable the simulation of morphology development and

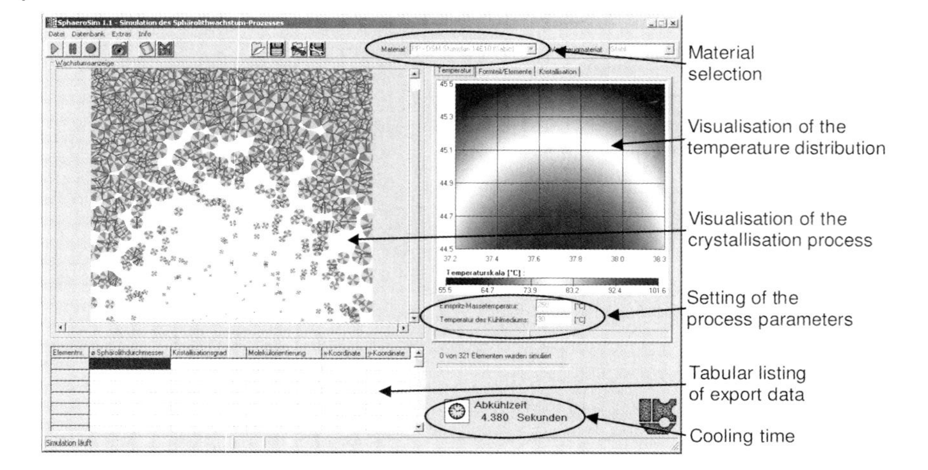

Fig. 6.3 User interface of the SphäroSim software.

the prediction of part properties during polymer solidification via an in-house-developed software [9, 10]. Extensive experience is at hand at the IKV in the field of injection molding simulation and material data determination. In addition to the direct prediction of the resulting part dimensions, current works deal with the calculation of inner part properties. These are molecular and fiber orientation as well as the crystallization behavior of semicrystalline thermoplastic resins.

To reach the aim of an integrative simulation Hoffmann et al. introduced the software SphäroSim in 2002 within the scope of project A2 of the SFB 370. This software tool enables an online simulation of the microstructure development [11, 12]. It uses the Cellular Automata Method to calculate the growth in quiescent crystallizing polymer melts during cooling down in two-dimensional part cross-sections [8]. A further development in the last project phase enables the software to include the dynamics of the injection-molding process and to describe the resulting material behavior close to reality [13]. In the following the subroutines of the software will be explained and selected results will be presented.

Figure 6.3 shows the final version of the graphical user interface of the software. The interface enables the user to enter the process parameters, to observe the online visualizations of the spherulite growth and the temperature field calculation, and to examine the calculated value triples of spherulite diameter, degree of crystallization, and molecular orientation.

6.3.2 Temperature Field Calculation

Because of limited computer performance and the required resolution the software calculates only cross-sections of parts. The calculation of such a cross-section works two-dimensionally with a direct online visualization of the simulation results. A real injection-molding part does not only consist of coplanar walls. Possible geometries may be corners, cavity walls with varying flow channel heights along the flow path, roundings, and combinations of these.

To simulate the cooling and crystallization process of real parts like the chosen pipe fitting the underlying temperature field calculation was adapted to these requirements. By the local temperature distribution and the cooling conditions local nucleation and spherulite growth is mainly affected. To obtain an exact calculation based on temperature- and pressure-dependent material data with high resolution, an in-house-developed temperature calculation was implemented into the software. This calculation is able to consider temperature- and pressure-dependent material data as well as nearly all geometrical boundary conditions of a molded cross-section [13]. Thus, real-time simulation of local temperatures and cooling processes for each time step and each element is possible.

The necessary material data were determined for the chosen polypropylene Sabic PP 505 P and provided to the software via a database [14, 15].

6.3.3
Calculation of Inner Properties

The process of spherulite growth is mainly affected by local cooling rates. By this not only thermal nucleation is locally affected but also the spherulite growth rate [16, 17]. The process of growth is simulated with the aid of a temperature calculation which includes the dynamics of the injection-molding process.

To deliver a realistic data input to the software two tests were carried out. One test ascertains the crystallization interval of the material. This test was performed by DSM Research, Geleen NL, by high-speed differential scanning calorimetry (DSC), as shown in Fig. 6.4.

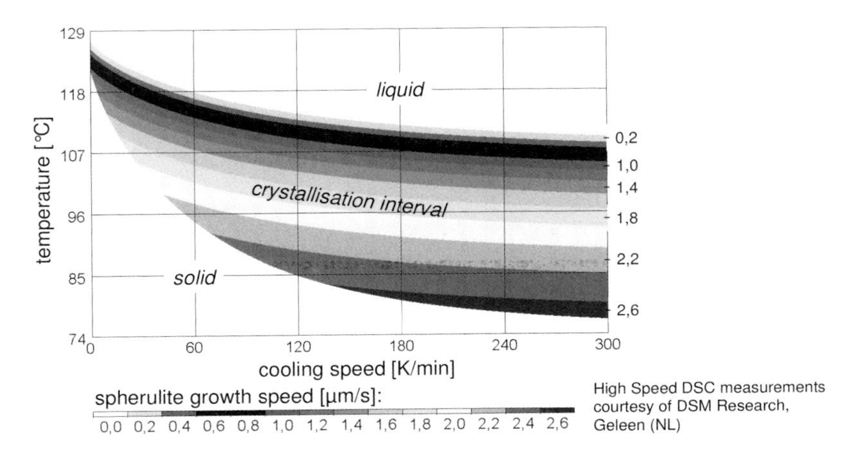

Fig. 6.4 Crystallization interval depending on cooling speed.

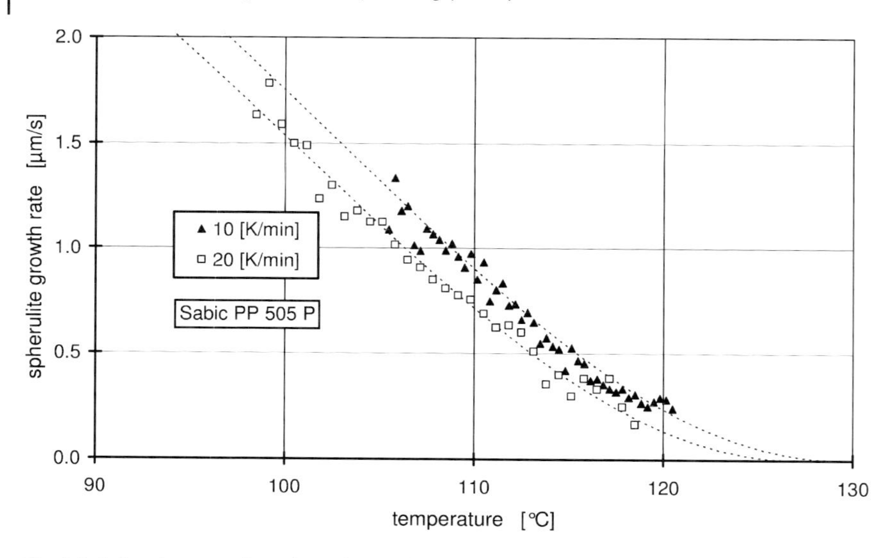

Fig. 6.5 Spherulite growth in dependence of different cooling speeds.

In Fig. 6.4 it is clear that with increasing cooling speed the starting and the ending point of the crystallization interval is shifted to lower values. Furthermore, the crystallization interval in general broadens with increasing cooling speed. In contrast to these exact measurements regular DSC in this case only allows qualitative conclusions [14].

Concurrent spherulite growth rates at different cooling speeds were experimentally determined with a Mettler Toledo heating desk, type FP800HT, and the use of a polarization light microscope [13, 14]. Selected growth speed results are presented in Fig. 6.5.

By an extrapolation of these measured growth rates and the linkage with the measured crystallization interval temperature shift a cooling speed-dependent crystallization interval modeling was obtained. Figure 6.6 shows a simulation result. Visible is a part cross-section with a small amount of thermal nucleation in the transient layer. Resulting from a high-temperature gradient distorted spherulites grow in this area, as shown in the comparable real part sections of Fig. 6.6.

In the transition layer between the fine-grained edge zone and the coarse-grained core layer of a part cross-section flow-induced nuclei arise depending on height and duration of the rheological load on the polymer melt. These nuclei then grow spherically. The resulting flow-induced structures are the so-called Shish Kebabs [18, 19]. The probability of flow-induced structure development depends solely on the molecular architecture of the polymer used.

Besides the spherulite diameter the degree of crystallinity plays a significant role in the characteristics of a semicrystalline structure. Since in the literature the degree of crystallinity is rarely modeled, a semiempirical approach was chosen here. This approach is based on the correlation between the experimentally

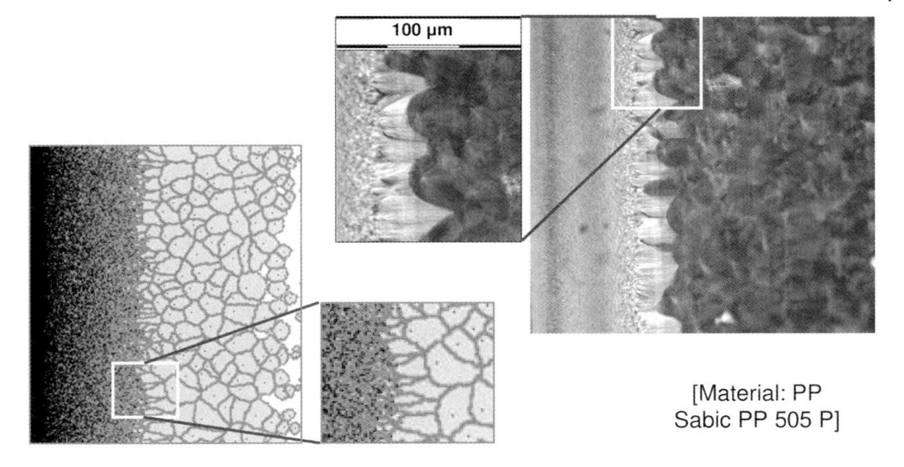

Fig. 6.6 Distorted growing spherulites in the transition region of a part cross-section.

determined spherulite diameter and the locally present degree of crystallinity. By implementing this direct connection which takes into account the cooling dynamics of the process it is possible to calculate both local crystallization values with a low computational effort.

In addition to the already mentioned inner part properties, degree of crystallinity and spherulite diameter, the molecular orientation enormously affects the final global part behavior [20]. In comparison to molecular orientation simulations of amorphous thermoplastics the prediction of molecular orientation of semicrystalline polymers is more difficult. This is primarily based on the lack of transparency that prevents the observation of the morphology and secondly the superposition of crystallization during solidification.

A prediction of molecular orientation was finally realized by the introduction of a correlation between shear rate and molecular orientation of each morphology layer in a part cross-section [21]. A validation was carried out by Fourier transformed infrared (FTIR) examinations of injection-molded samples [22].

6.3.4 Procedure of Simulating Inner Properties

To transfer the described values of a real part to a subsequent structure analysis the part geometry is segmented. Depending on the position of the gate system one or more cross-sections of the pipe fitting are simulated at different rotational locations and flow length sections. Each segment is subdivided into cells to calculate the cell growth with the Cellular Automata Method.

During simulation every segment is calculated numerically, and the results of the local temperature distribution, the morphology, and the degree of crystallinity are displayed. In the final stage the values so determined are stored in a transfer file as averaged inner property values of the simulated segment. Thus, an input file is transferred to structure analysis that includes information about the barycenter of each element and the matching value triplet of the inner properties.

6.4 Integration of Inner Properties into Structural Analysis

Based on the value triplets transferred from the data file containing the local inner properties, conclusions concerning the local mechanical properties are drawn. For this purpose extensive investigations were performed on specimens with different inner properties. As a result correlations were determined between the degree of crystallization, spherulite diameter, and orientations on the one hand and initial Young's modulus E_0, yield stress σ_S, and yield strain ε_S, on the other hand [23]. These three parameters lead to a modified version of the three-parameter approach, which forms the basis of the material description in the subroutine [24]:

$$\sigma = \varepsilon \frac{\sigma_S}{\varepsilon_S} \frac{E_0 \varepsilon_S^2 - \sigma_S \varepsilon}{\sigma_S \varepsilon_S + (E_0 \varepsilon_S - 2\sigma_S)\varepsilon} \tag{6.1}$$

The values for E_0, σ_S, and ε_S are constant, but different for each element, according to the existing local inner properties. The result is a different dependency of the element stiffness on the strain for each element.

In addition to the strain dependency of the local stiffness gained from the three-parameter approach a strain-rate dependency is applied. Therefore, the current strain rate in an element, consisting of strain- and time-increment, is determined and subsequently the respective values E_0, σ_S, and ε_S obtained from calibration measurements are assigned to this element. The temperature dependency of the material behavior is connected with the strain rate by using the principle of time/temperature shift with the Arrhenius constant k:

$$\log\left(\frac{\dot{\varepsilon}_{\text{ref}}}{\dot{\varepsilon}}\right) = k\left(\frac{1}{T} - \frac{1}{T_{\text{ref}}}\right) \tag{6.2}$$

The final result is a tangential modulus E_T that depends on the temperature, on the inner properties of each finite element (spherulite diameter Sph, degree of crystallinity K, and degree of molecular orientation Or), and on the strains and strain rates appearing in the element [25]:

$$E_T = E(T, \varepsilon, \dot{\varepsilon}, \text{Or}, K, \text{Sph}) \tag{6.3}$$

In order to carry out the transfer of the inner properties and the material data derived from these in a traceable way, they are saved and can be visualized [26]. The assignment of the inner properties to the pipe fitting manufactured with a diaphragm gate is shown in Fig. 6.7, and the resulting stiffness and stresses are

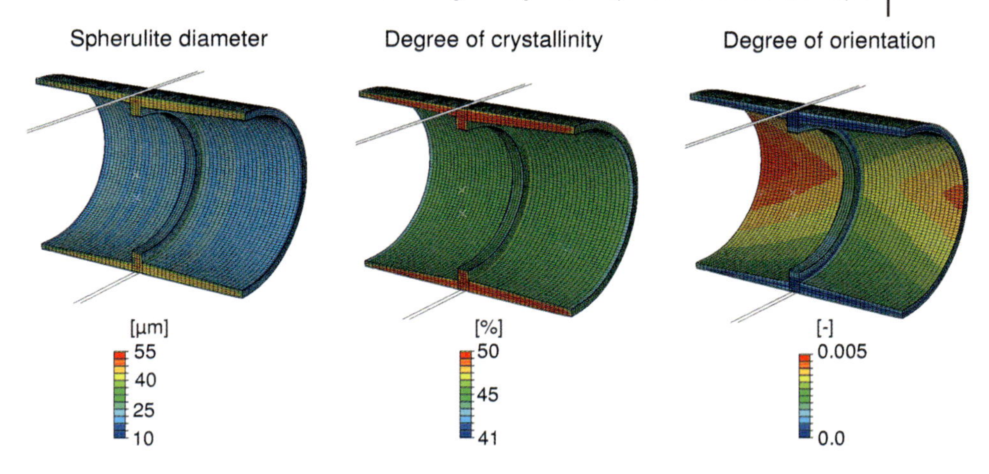

Fig. 6.7 Visualization of inner properties in structural analysis.

shown in Fig. 6.8. Figure 6.7 presents the locally resolved spherulite diameter, the degree of crystallization, and the amount of molecular orientation in the form of the Herrmann orientation value. The increasing spherulite diameter towards the interior of the part can easily be seen. Also a local degree of crystallization between 45 and 50% can be observed that is typical for this material processed by injection molding. There are no orientations in the interior of the part, while the orientation value is very high in the intensely sheared and quickly cooled edge areas. The left side of Fig. 6.8 shows the initial modulus resulting from the dependency on the inner properties and local strain rates. The initial modulus of elasticity is converted into a tangential modulus according to the current local

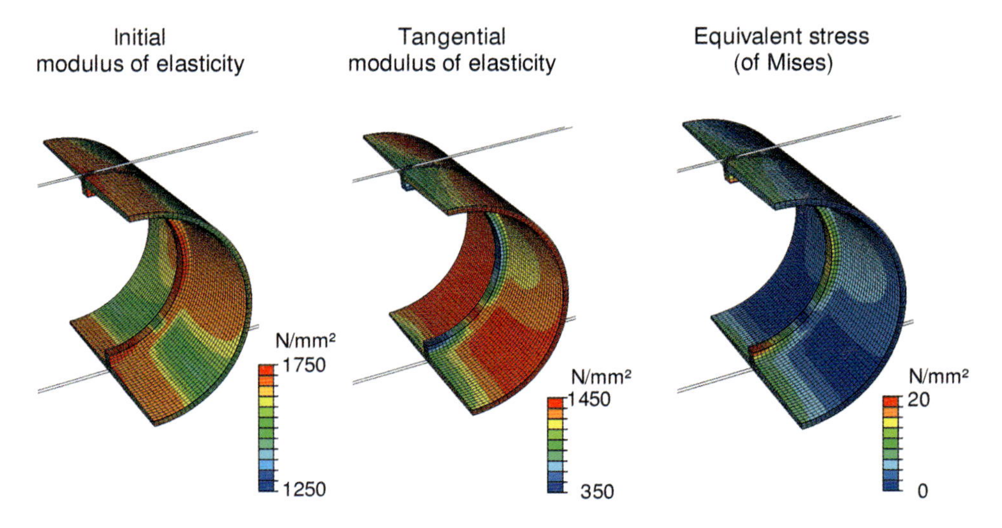

Fig. 6.8 Visualization of local stiffness and stresses.

strain. The tangential modulus is shown in the middle of Fig. 6.8. Finally, the right side of Fig. 6.8 shows the stress distribution caused by the load as the equivalent stress according to von Mises.

6.5 Conclusions and Perspectives

It is possible to predict the inner properties of semicrystalline thermoplastics by an extended process simulation, to generate mechanical characteristics from these by means of appropriate calibration measurements, and to use these characteristic values as a basis for a material description.

The obstacle that has to be overcome in order to utilize integrative simulation is an extraordinary operating expense. A process simulation enhanced by a crystalline growth simulation requires a much higher computing time than a simple injection-molding process simulation. The simulation of crystalline growth represents a simulation step that has to be made in addition to the injection-molding process simulation and the mechanical FEA. Finally, the determination of the correlation between inner properties and mechanical properties requires a huge effort regarding measuring and testing.

However, a growing interest in a more specific design for plastics parts could enlarge the amount of material data available. In the future, the greater expense in data processing will be compensated by improved algorithms on the one hand and more powerful processing equipment on the other hand. Finally, the additional third step of crystalline growth simulation will be avoidable by integrating it directly into the injection-molding process simulation.

Acknowledgments

The authors gratefully acknowledge the financial support of the Deutsche Forschungsgemeinschaft (DFG) within the Collaborative Research Center (SFB) 370 "Integral Materials Modeling".

We also would like to extend our thanks to our partners who supported us with polymer resin, information, and equipment. These are DSM Research BV, SABIC EuroPetrochemicals BV, Abaqus Inc., simcon kunststofftechnische Software GmbH, SIGMA Engineering GmbH, and HASCO Hasenclever GmbH & Co. KG.

References

1 E. Schmachtenberg, M. Brandt, M. Brinkmann, Conference Proceedings, Erlanger-Bayreuther Kunststofftage, 2005.
2 M. Brinkmann, diploma thesis, RWTH Aachen University, 2004.
3 G.H. Michler, *Kunststoffmikromechanik – Morphologie, Deformations- und Bruchmechanismen*, Carl Hanser Verlag, Munich, 1992.
4 G.H. Michler, *Polymer Science* 35, 1993, 1559.
5 G. Menges, E. Haberstroh, W. Michaeli, E. Schmachtenberg, *Werkstoffkunde Kunststoffe*, Hanser Verlag, Munich, 2002.
6 URL: www.sigmasoft.de, Sigma Engineering GmbH, 2005.
7 W. Michaeli, M. Bussmann, Conference Proceedings, 22nd International Plastics Technology Colloquium, IKV Aachen, 2004, p. 8.
8 W. Michaeli, M. Bussmann, *Journal of Polymer Engineering*, 2005, in press.
9 S. Hoffmann, Doctoral thesis, RWTH Aachen University, 2002.
10 J. Zachert, Ph.D. thesis, RWTH Aachen University, 1998.
11 W. Michaeli, S. Hoffmann, Conference Proceedings, Annual Technical Conference (ANTEC), Dallas, TX, 2001.
12 W. Michaeli, S. Hoffmann, Conference Proceedings, PPS-17, Montreal, 2001.
13 W. Michaeli, M. Bussmann, B. Renner, Conference Proceedings, Annual Technical Conference (ANTEC), Boston, MA, 2005.
14 W. Michaeli, M. Bussmann, B. Renner, Conference Proceedings, PPS-21, Leipzig, 2005.
15 X. Wang, diploma thesis, RWTH Aachen University, 2005.
16 W. Michaeli, M. Bussmann, B. Renner, First International Conference on Multi-Material Micro Manufacture, Book of Abstracts, Poster Presentation, Elsevier Verlag, 2005, p. 17.
17 A. Cramer, study thesis, RWTH Aachen University, 2002.
18 H. Janeschitz-Kriegl, *Process in Colloid & Polymer Science* 87, 1993, 117.
19 S. Vleeshouwers, H. Meijer, *Rheological Acta* 35, 1996, 391.
20 Final Report, Collaborate Research Center 106, RWTH Aachen University, 1988.
21 T. Nguyen-Chung, G. Mennig, *Rheological Acta* 40, 2001, 67.
22 Y. Zhang, diploma thesis, RWTH Aachen University, 2005.
23 D. Grunewald, diploma thesis, RWTH Aachen University, 2001.
24 D. Gutberlet, Doctoral thesis, RWTH Aachen University, 2000.
25 E. Schmachtenberg, M. Brinkmann, Conference Proceedings, Fachbeiratsgruppe FAWT, Institute of Plastics Processing, RWTH Aachen University, 2005.
26 G. Weinhold, study thesis, RWTH Aachen University, 2005.

7
Modeling of Flow Processes During Solidification (TP A1)

M. Bussmann, B. Renner, W. Michaeli, B. Pustal,
A. Bührig-Polaczek, V. Pavlyk, O. Mokrov, and U. Dilthey

Abstract

The aim of the study (project A1) is the calculation of flow and solidification processes during and after the mold process of casting, injection molding, and arc welding. Thus, the project constitutes an important link within the chain of an integrative process simulation.

By calculating melt distribution, melt flow, and actual cooling conditions, varying part properties can be explained. Intensive work was performed by the participating institutes of this project to model the complex interaction of flow and solidification. To realize the simulation of flow and solidification processes commercial software was used and expanded; moreover, new software was developed to meet special needs.

To validate the implemented models several experiments were performed and parts were produced. The main basis for all participants of this subproject was the solution of mass, momentum, and energy conservation equations for each flow and solidification process.

In addition to the topics mentioned, the project was also partly responsible for the solidification processing of parts, which were provided for other projects of SFB 370 for validation tests and experiments.

7.1
Introduction

The work presented in this chapter was conducted by a collaboration of the Foundry Institute (GI), the Institute of Plastics Processing (IKV), and the Welding and Joining Institute (ISF). It is subdivided into the topics of casting (GI), injection molding (IKV), and arc welding (ISF). The teamwork and the information exchange are displayed in Fig. 7.1.

Integral Materials Modeling: Towards Physics-Based Through-Process Models
Edited by Günter Gottstein
Copyright © 2007 WILEY-VCH Verlag GmbH & Co. KGaA, Weinheim
ISBN: 978-3-527-31711-0

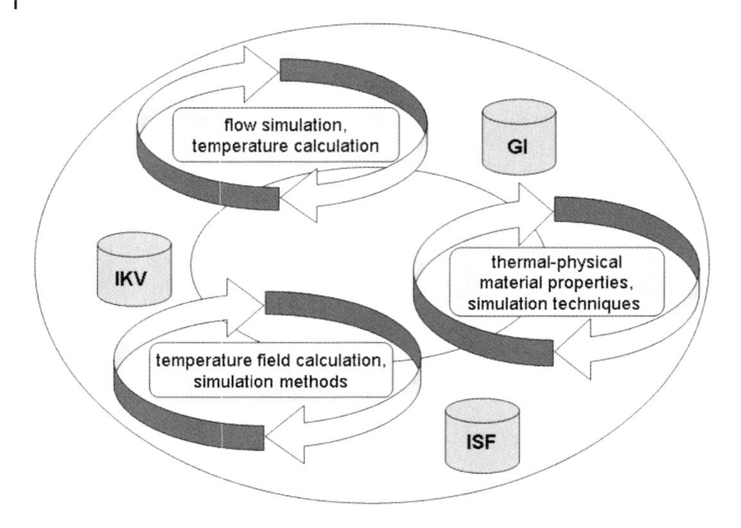

Fig. 7.1 Teamwork and exchange topics of subproject A1.

7.1.1 Aluminum Cup

Melt convection and grain movement play an important role in casting solidification processes, which greatly influence the formation of grain structures and solute segregation. In general, the melt convection and grain movement are a result of gravitational forces. The density within the melt varies due to temperature and concentration differences, which cause under gravity melt convection and is called thermosolutal convection. Similarly, the density differences between the solid grains and the bulk melt cause grain movement, known as solid sedimentation or grain floating. An additional driving force for melt convection and grain movement is solidification shrinkage. It is well known that solidification shrinkage is the main reason for the formation of a shrinkage cavity and one of the dominant reasons for porosity, hot/cold tearing, stress, and deformation. The shrinkage pressure affects the flow pattern and the macrosegregation formation, as already described by Campbell [1]. Yet for simplicity, the vast majority of casting solidification models neglects the shrinkage-induced flow (feeding flow). Thermosolutal convection modeling without consideration of the density difference in the mass conservation equation is generally carried out using the well-known Boussinesq approximation [2]. This considers the density difference due to thermosolutal effects only for the momentum equations to model the thermosolutal convection. The shrinkage effect was ignored in previous work [3–6].

To consider shrinkage flow, modeling of the free surface is necessary because the shape of free surfaces and the physical properties close to free surfaces change dynamically during solidification. Only a few publications have dealt with the prediction of moving free surface formation so far. A review is given by Ehlen [7]. An excellent algorithm was used by Mostaghimi et al. [8]. Impact and solidification of a tin droplet on a steel plate was simulated, and good agreement was found when the results were compared to experiments. Other publications report on the modeling of sequential impact and solidification of two molten droplets on a solid surface or the surface cooling by an impinging water drop.

In this study, the moving free surface caused by solidification shrinkage was tracked using a three-phase volume averaging approach. In addition, the melt density was assumed to be temperature and concentration dependent, so that both the shrinkage flow and thermosolutal convection could be taken into account.

7.1.2 Plastics Pipe Fitting

To predict the flow behavior of polymer melts the mass, momentum, and energy conservation equations have to be solved. An important material property for the flow behavior of polymer melts is the polymer viscosity. The viscosity of polymers depends on temperature, pressure, and shear rate. It is in general much higher than that of liquid metals. Depending on the part geometry a consideration of inertia effects during the filling phase is necessary to account for the resulting effects in the simulation.

A special metal casting process, the so-called thixo-molding, is a casting process comparable to injection molding. Analogies like shear thinning material behavior are utilized during this subproject for software improvements. A further analogy to casting and welding is, for example, the free surface at flow fronts. Thus, these analogies were used to optimize software tools for a better prediction of flow and solidification processes during injection molding.

The common simulation method based on shell elements is insufficient because some physical effects, such as melt movement in the thickness direction, fountain flow, and inertia effects, are neglected [9, 10]. For thin-walled parts good results are obtainable with a $2\frac{1}{2}$D software, because the neglected effects have less impact. But in plastics parts with variations in wall thickness or sections of branching (e.g. ribs) there is melt flow in the thickness direction which has to be considered in the simulation. For this purpose a complete three-dimensional calculation software is required.

7.1.3 Steel Profile

The main aim of macroscopic modeling of heat and mass transfer during welding is the prediction of temperature field and weld seam dimensions. For gas–metal arc welding (GMAW) with consumable melting electrode, the heat input of the process, the heat flow in the workpiece, and the influence of the curved shape of the weld pool free surface must be considered [11]. In our model, the energy conservation equation in the melting wire is solved in a two-dimensional axisymmetric formulation, with accounts for ohmic heating, contact resistance, and heat input at the anode spot. The metal transfer is modeled, assuming the drop to move as a rigid body under the action of an electromagnetic force. Thus, the momentum equation is solved analytically [12]. The nonlinear heat transfer equation in the workpiece is solved in a steady-state formulation, taking into account the deformation of the free weld pool surface and neglecting the heat diffusion in the welding direction, which is a good approximation for welding speeds usually used when welding thin steel plates. This allows one to obtain a fast three-dimensional solution using the finite difference method. The deformation of the weld pool is calculated by taking into account the arc pressure distribution, the mass balance, and the melting boundary [12]. The simulation of multipass welding was performed on the basis of the software package SimWeld, which was extended by taking into account the geometry of edge preparation and preheating [11].

7.2 Software Development

7.2.1 Aluminum Cup

Within this project a model was developed to describe grain nucleation and kinetic growth under the influence of multiphase flow for globular equiaxed solidification structures. The standard conservation equations for mass, momentum, species, and energy are solved using a fully implicit and control-volume-based finite difference method. The computational fluid dynamics (CFD) software FLUENT Rev. 6.1 is used here. Source terms appearing in the conservation equations are modeled by means of a user defined scalar (UDS) and a user defined memory (UDM) to deal with solidification. There are three phases involved: the liquid melt, the solidifying grains, and the air. A volume averaging approach to formulate the conservation equations for each of the three phases is employed.

During recent years the two-dimensional FORTRAN code accounting for only two phases has been extended to a three-dimensional C code to account for liquid, solid, and air. In this model, the moving free surface caused by solidification shrinkage is tracked using a Eulerian three-phase volume averaging approach. In addition, the melt density is assumed to be temperature and concentration dependent so that both the shrinkage flow and thermosolutal convection can be taken into account. The moving free surface is defined as the interface between melt and air. Particular attention is paid to the change of heat, momentum, and pressure near the free surface. Meanwhile, the model considers nucleation and growth of globular equiaxed grains, solute transport by diffusion, and convection. For liquid–air and solid–air mixtures, the source terms for the momentum exchange are calculated by

$$\mathbf{U}_{la} = -\mathbf{U}_{al} = \mathbf{U}^{d}_{la} = K_{la}(\mathbf{u}_{l} - \mathbf{u}_{a}) \tag{7.1}$$

$$\mathbf{U}_{sa} = -\mathbf{U}_{as} = \mathbf{U}^{d}_{sa} = K_{sa}(\mathbf{u}_{s} - \mathbf{u}_{a}) \tag{7.2}$$

where $\mathbf{u}$ is the velocity of the liquid, solid, or air. The Schiller–Naumann model is used to determine the momentum exchange coefficient K_{la} [13]:

$$K_{la} = 3\mu_{l} f_{l} f_{a} C_{D} R_{e}/(4d_{a}^{2}) \tag{7.3}$$

where μ_l is the viscosity, f_l is the liquid viscosity, and f_s is the fraction of the solid phases. d_a is a virtual characteristic diameter of air which is used for calculating the drag force only. A smaller air diameter represents a larger drag force. C_D is the drag coefficient that is based on the relative Reynolds number R_e.

$$C_D = \begin{cases} 24(1 + 0.15R_e^{0.687})/R_e & R_e \leq 1000 \\ 0.44 & R_e > 1000 \end{cases} \tag{7.4}$$

with

$$R_e = \frac{\rho_l |\mathbf{u}_a - \mathbf{u}_l| d_a}{\mu_l} \tag{7.5}$$

A symmetric model is used to calculate K_{sa}. This is similar to the Schiller–Naumann model [13]. The difference is that the average grain diameter of air and solid $(d_a + d_s)/2$, the mixed viscosity $\mu_a f_a + \mu_s f_s$, and the mixed density $\rho_a f_a + \rho_s f_s$ are used in the calculation instead of using the properties of air only.

The reader is referred to our previous work [3–6, 14–16] for more information on the model and the source terms used to deal with solidification phenomena like nucleation, growth kinetics and mass transfer rate definition, solute partitioning at the liquid–solid interface, momentum, and enthalpy transfer among the phases.

7.2.2
Plastics Pipe Fitting

Within this project different simulation tools, like FIDAP (FEM based), MAGMASOFT (FVM based), Cadmould ($2\frac{1}{2}$D, FEM based), and the in-house developed FEM-based IKV software, have been compared and tested for injection-molding simulation. Additionally, different approaches for mold filling were tested.

The calculation of mold filling and melt solidification were initially simulated using the $2\frac{1}{2}$D FEM software Cadmould with shell elements. These results were used as a database for the simulation of inner part properties in project A2.

The three-dimensional FVM thixo-molding software tool MAGMASOFT was applied for the simulation of three-dimensional melt flow and temperature distribution calculation. For this purpose, new interfaces, stable solvers, and alternative material functions were developed and implemented.

An isolated analysis of melt front flow behavior was conducted with the software FIDAP. Additional three-dimensional unsteady temperature fields of injection molds were simulated with boundary element method software [17].

Finally, IKV has generated 3D-FEM software within this project. This software uses tetrahedron elements for the calculation of flow and solidification processes during cavity filling [18–20].

The influences of inertia and gravitation on melt flow were analyzed, simulated, and practically evaluated with different part geometries. Thus, mathematical models to calculate melt flow and solidification processes during injection molding were simplified and validated. The results were reasonable at acceptable simulation times and in good accordance to experimental results [21].

To get a closer approximation for the whole injection-molding process the FEM algorithm was expanded to simulate the compression phase as well. This was possible by implementing compressible material behavior during flow and solidification.

7.2.3 Steel Profile

The following interrelated problems were consecutively solved for calculation of the temperature field and form of the weld. The main components of these problems were described previously [11, 12, 22]: finding the integral characteristics of the heat source (arc), finding the temperature distribution in the welded joint, and finding of the configuration of the weld pool free surface. The simulation of multipass welding differs in two basic aspects from single-pass welding. Firstly, it considers preheating of the work piece and heating by the previous passes and, secondly, geometry of edge preparation and previous beads as well as their effect on the distribution of heat introduced from the arc source. These aspects were taken into account when extending the simulation package SimWeld to multipass welding [22].

7.3 Experiments and Results

7.3.1 Aluminum Cup

The experiments necessary to validate the model were carried out in a different project entitled "Microstructure Modeling during Solidification of Castings", which can be found in Chapter 8. An example of a three-dimensional result showing the grain diameter is presented in Chapter 3. The grain diameter is a characteristic microstructural parameter passed on to the subsequent processing step, hot rolling.

Here a study is presented of a two-dimensional benchmark test casting comparing the Boussinesq approach with the developed three-phase model. The Boussi-

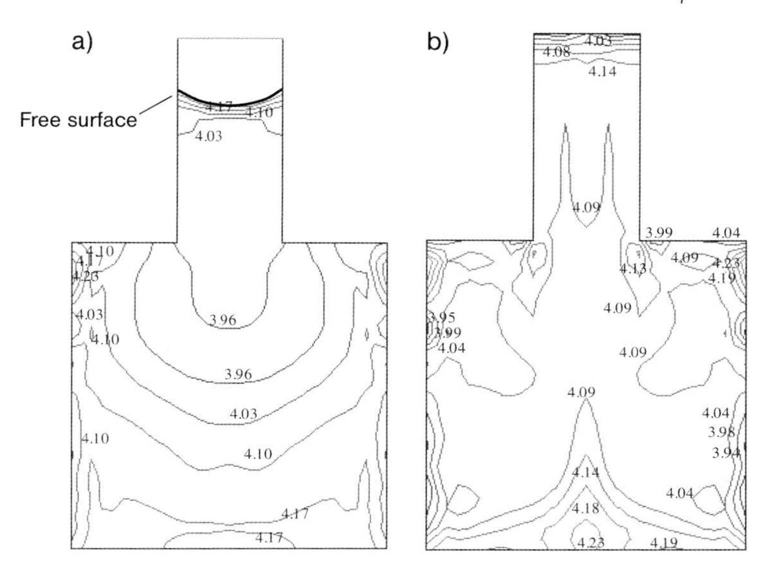

Fig. 7.2 Macrosegregation maps near the solidification end considering thermosolutal convection, and grain movement (a) with and (b) without shrinkage flow (c_{mix}: 4.1–4.3 wt% Cu).

nesq approach accounts for two phases, solid and liquid, while the density difference between solid and liquid is accounted for in the momentum conservation equation but neglected in the mass conservation equation. Figure 7.2 shows the influence on the macrosegregation pattern of copper in an AlCu4 alloy formed with (three-phase model) and without (Boussinesq approach) shrinkage flow.

7.3.2 Plastics Pipe Fitting

To validate the simulation models different plastics pipe fittings were produced, as shown in Fig. 7.3.

The developed injection mold permits a pipe fitting production with different thermal and rheological boundary conditions [23]. The produced parts were transferred to project A2 for validation and development of morphology and orientation modeling as well as to project C9 for analysis of microstructures. In addition, three-dimensional melt movements were visually analyzed during the filling phase by the aid of an observation mold.

For all gate positions melt flow and solidification simulations were performed three-dimensionally. Based on the numerical results a finer mesh of the part geometry was possible. Thus, the resulting time steps had positive effects on the prediction of melt distribution and the calculated temperature field spreading. Results like this were generated and transferred to project A2 for the simulation of inner properties.

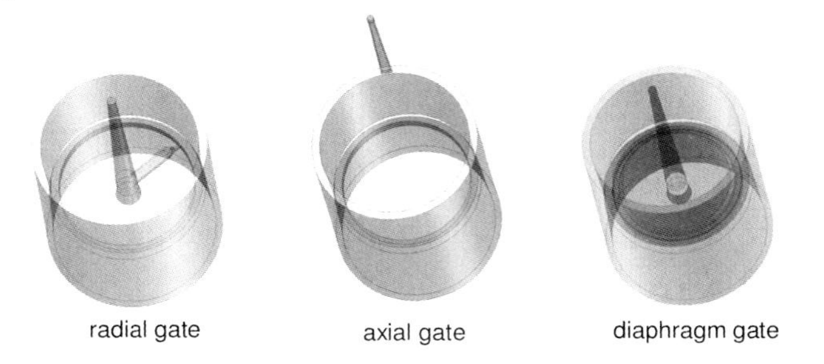

Fig. 7.3 Pipe fittings with different gate positions.

To predict final part properties, like shrinkage and warpage, the accurate temperature field distribution has to be known. Calculation errors can be mostly attributed to inaccurate thermal material data and an insufficient description of heat transport phenomena between part and mold. Most calculation models and simulation software tools often use constant thermal material data. By contrast, tests show a distinctive temperature and pressure dependence of thermal material properties. Especially the temperature dependence of the heat transfer coefficient is essential to account properly for heat exchange. In extensive tests the heat transfer coefficient was determined and its temperature and pressure dependence analyzed [18, 24]. The modeled value was then provided for the flow simulation and the morphology simulation in project A2.

7.3.3
Steel Profile

A diagram of the weld and the welding parameters are shown in Fig. 7.4. To provide full penetration and to prevent formation of defects in the weld root the

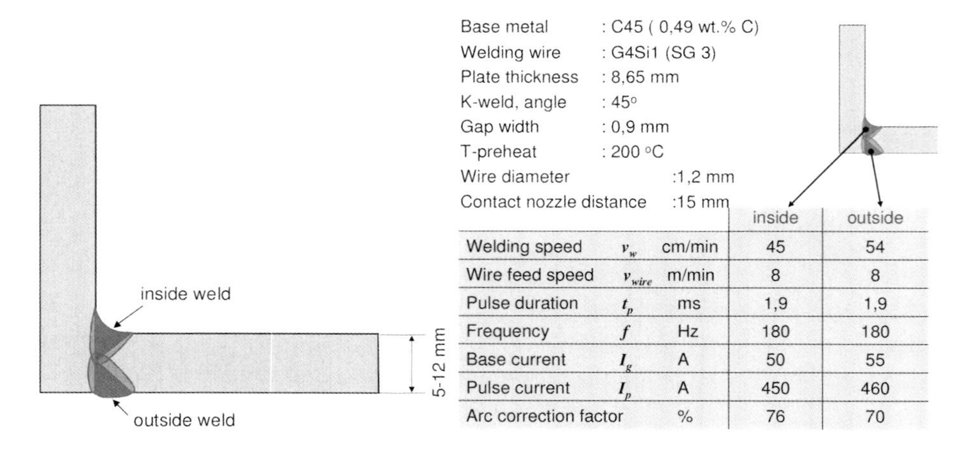

			inside	outside
Welding speed	v_w	cm/min	45	54
Wire feed speed	v_{wire}	m/min	8	8
Pulse duration	t_p	ms	1,9	1,9
Frequency	f	Hz	180	180
Base current	I_g	A	50	55
Pulse current	I_p	A	450	460
Arc correction factor		%	76	70

Fig. 7.4 Scheme of fillet joint with K-groove welded in two passes and welding parameters of GMA pulse welding with Ip/Ig modulation.

K-groove was used. The first pass was carried out from the inside. Preheating to 200 °C and constant inter-pass temperature were used to prevent formation of hardening cracks. The modern digital inverter power supply CLOSS Quinto II of the CLOSS Schweisstechnik Company in the pulse regime with I/I modulation was applied for welding, which warrants a high stability of the process.

7.4 Discussion

7.4.1 Aluminum Cup

The developed three-phase volume averaging model is applied to simulate a two-dimensional benchmark ingot casting using the alloy AlCu4 and to simulate the AA3104 ingot in three-dimensions. The results show that the presented model is capable of modeling the thermosolutal convection together with grain movement and shrinkage flow for globular equated solidification. The following conclusions can be drawn from the calculations performed:

1. For the benchmark casting, comprehensive flow patterns are first dominated by thermal convection and shrinkage flow, then by thermosolutal convection, and finally by shrinkage flow during solidification.
2. The absolute values for macrosegregation are not very pronounced in the calculated example, but a significant difference in segregation patterns is found when comparing models with and without shrinkage flow. The isoconcentration lines are convex in the absence of shrinkage flow and concave in the presence of shrinkage flow.

For the AA3104 ingot casting a comparison of experiments and simulations for the relevant microstructure parameter for subsequent processing, the grain size, showed a good qualitative agreement. The absolute values predicted by the model are far too small, which is a consequence of unknown nucleation parameters. To predict quantitative values these parameters should be determined by measuring the undercooling for different cooling conditions.

7.4.2 Plastics Pipe Fitting

The melt flow and solidification simulations with the aid of the in-house-developed software and modified commercial software are promising. Validation with filling phase experiments showed good agreement. The consideration of compressible material behavior offers also a high potential for optimization of the simulation software.

Three-dimensional flow and solidification calculations based on volume elements require time-consuming computation capacity. By numerical optimization of the solution algorithm the calculation time was reduced. Due to the rapidly improving processor technology the aspects of three-dimensional melt flow and solidification simulation will advance in the future and improve the simulation quality.

In the frame of the cooperation of the Institute of Plastics Processing (IKV) and the Foundry Institute (GI) mathematical models for the description of plastic melt flow and solidification were developed and implemented in the three-dimensional thixo-molding part of MAGMASOFT. This modified software for plastics melt flow and solidification simulation has been commercially available since 2001 and is distributed as SIGMASOFT. SIGMASOFT-generated simulation results show a good correlation to experiments. Improvements to existing $2\frac{1}{2}$D simulations are clearly visible.

7.4.3 Steel Profile

The simulated temperature field, weld shape, and free surface deformation for the inner pass are demonstrated in Fig. 7.5. Comparison of simulated and experimental temperature cycles and weld seam shapes, shown in Fig. 7.6, demonstrates good agreement between modeling and experiments. The peak temperatures and the cooling time from 800 to 500 °C are used as input parameters for calculation of the microstructure and the mechanical properties in project C7, based on the experimentally determined TTT diagram.

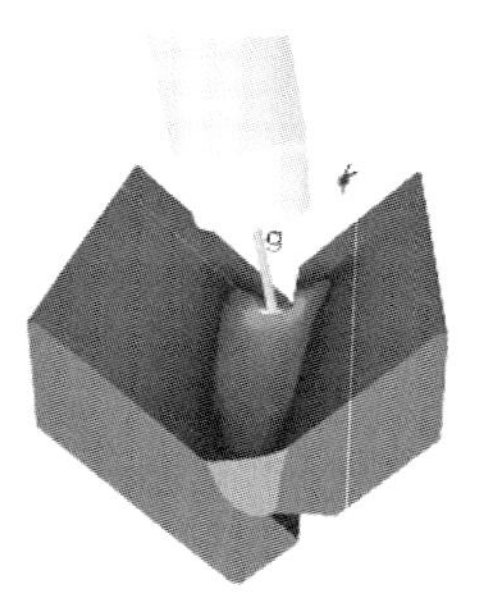
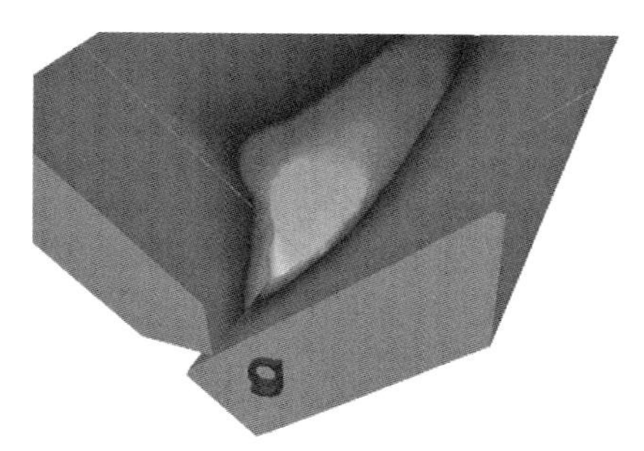

Fig. 7.5 Simulated temperature field, weld shape, and free surface deformation for inner pass.

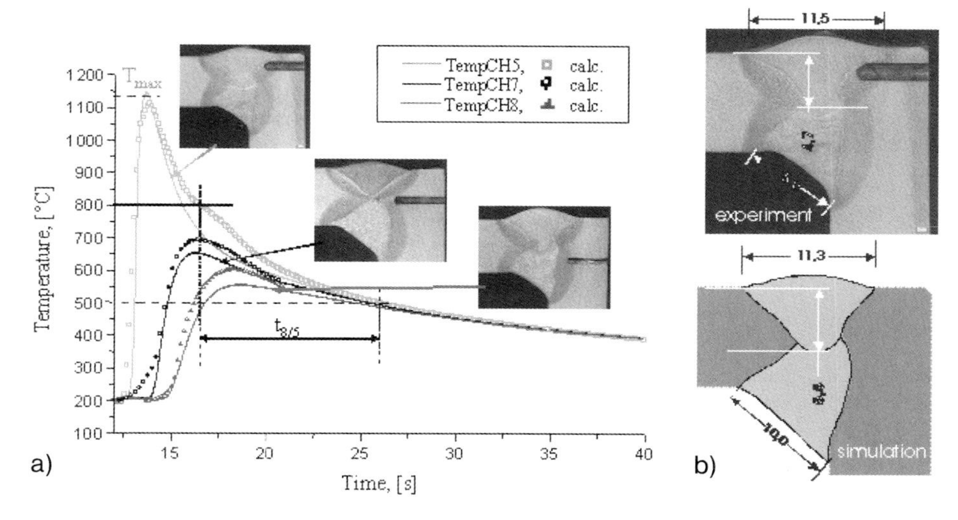

Fig. 7.6 Comparison of simulated and experimental temperature cycles (a) and weld seam shapes (b).

Acknowledgments

The authors gratefully acknowledge the financial support of the Deutsche Forschungsgemeinschaft (DFG) within the Collaborative Research Center (SFB) 370 "Integral Materials Modeling".

The authors thank Hydro Aluminum for providing the material for the process line aluminum cup. We are grateful to the company CLOOS Schweißtechnik for providing the welding power source. Special thanks go to the companies Sabic, Geleen, (NL) for providing the polypropylene resin, as well as DSM Research, Geleen (NL) for the high-speed DSM measurements.

References

1 J. Campbell, *Castings*, Butterworth-Heinemann, Oxford, 1991.
2 T. U. Kaempfer, M. Rappaz, Proc. 9th Conf. on Modelling of Casting, Welding and Advanced Solidification Processes, Aachen, Germany, TMS, 2000, p. 640.
3 M. Wu, A. Ludwig, *Adv. Eng. Mater.* 5, 2003, 62.
4 M. Wu, A. Ludwig, A. Bührig-Polaczek, M. Fehlbier, P. R. Sahm, *Int. J. Heat Mass Transfer* 46, 2003, 2819.
5 A. Ludwig, M. Wu, T. Wang, A. Bührig-Polaczek, Proc. 3rd Conference on Computational Modelling and Simulation of Materials, Sicily, Italy, TMS, in press.
6 A. Ludwig, M. Wu, *Metall. Mater. Trans.* 33A, 2002, 3673.
7 G. Ehlen, Ph.D. thesis, Aachen, Shaker-Verlag, 2004.
8 J. Mostaghimi, S. Chandra, R. Ghafouri-Azar, A. Dolatabadi, *Surf. Coat. Technol.* 163, 2003, 1.
9 J. F. Hétu, D. M. Gao, A. Garçia-Rejon, Proc. Annual Technical Conference (ANTEC), Indianapolis, IN, 1996.

10 F. Dupret, R. Keunings, Polymer Processing: Process Simulation, EUPOCO lecture, Leuven, 1992–1993.

11 U. Dilthey, I. Dikshev, O. Mokrov, V. Pavlyk, *Proc. Int. Conf. on Mathematical Modelling and Information Technologies in Welding and Related Processes*, ed. V. Machnenko, Paton Welding Institute, Kiev, 2002, p. 201.

12 U. Dilthey, V. Pavlyk, O. Mokrov, I. Dikshev, in *Mathematical Modelling of Weld Phenomena 7*, Verlag der TU Graz, 2005, p. 1057.

13 'FLUENT 4.4 User's Guide, Fluent Inc., Vol. 2, Chapter 9, 1998, p. 16.

14 T. Wang, M. Wu, A. Ludwig, M. Abondano, B. Pustal, A. Bührig-Polaczek, *Int. J. Cast Metals Res.* 18, 2005, 221.

15 T. Wang, B. Pustal, M. Abondano, T. Grimmig, A. Bührig-Polaczek, M. Wu, A. Ludwig, *Trans. Nonferrous Metals Soc. China* 15, 2005, 389.

16 T. Wang, M. Wu, A. Ludwig, M. Abondano, B. Pustal, A. Bührig-Polaczek, 6th Pacific Rim Int. Conf. on Modeling of Casting and Solidification Processes, Kaohsiung, Taiwan, 2004, p. 181.

17 D. Trinczek, Diploma thesis, Institute of Plastics Processing, 1996.

18 W. Michaeli, S. Hoffmann, M. Kratz, K. Webelhaus, *Int. Polym. Process.* 16, 2001, 398.

19 K. Webelhaus, Presentation at the seminar Optimierte Formteilentwicklung, IKV, Aachen, 1998.

20 K. Webelhaus, Ph.D. thesis, Aachen, Verlag Mainz, 2000.

21 J. Zachert, Ph.D. thesis, Aachen, Verlag Mainz, 1998.

22 U. Dilthey, O. Mokrov, V. Pavlyk, *Paton Welding J.* 4, 2005, 2.

23 U. Tischer, Student thesis, Institute of Plastics Processing, 1995.

24 X. Wang, Diploma thesis, Institute of Plastics Processing, 2005.

8
Microstructure Modeling During Solidification of Castings (TP A2)

B. Pustal, A. Bührig-Polaczek, N. Warnken, I. Steinbach, M. Bussmann, B. Renner, W. Michaeli, A.-P. Hollands, D. Senk, C. Walter, B. Hallstedt, and J. M. Schneider

Abstract

In this chapter a review is given of work carried out to couple efficiently the macroscopic process scale and the microscopic material scale for solidification processes. Different models were developed or extended to predict the microstructure evolution or significant microstructure parameters. These models were coupled more or less intensely to macroscopic software tools depending on the calculation efforts of each microscopic model. The software packages were validated by designed casting experiments on a laboratory scale. Casting was the first process step within different process lines using different material groups. Numerical techniques and data were exchanged among the partners to carry out the calculations and to develop the models, which comprise the phase field method, the cellular automata method, a microsegregation model, and a semiempirical approach.

8.1
Introduction

During the past 12 years new methods have been developed to describe the evolution of microstructure or characteristic parameters of the microstructure during solidification. The focus of the current work was to couple process scale and material scale where remarkable dependencies and interactions are present. To evaluate these dependencies and to gain an efficient coupling, experiences and numerical techniques have been exchanged among the partners working with different material groups and processes. A successful example for the interdisciplinary transfer was the adoption of the cellular automata method from welding simulations in steels to the morphology evolution in semicrystalline plastics.

Integral Materials Modeling: Towards Physics-Based Through-Process Models
Edited by Günter Gottstein
Copyright © 2007 WILEY-VCH Verlag GmbH & Co. KGaA, Weinheim
ISBN: 978-3-527-31711-0

The methods used to calculate the evolution of the microstructure were the phase field method and the cellular automata method. With these techniques it is possible to describe microsegregation, phase fractions, and the morphology evolution qualitatively. First quantitative phase field models have been presented for two-component systems. The disadvantage of these methods is the enormous computational effort.

To obtain quantitative parameters of the microstructure and to realize an effective macro–micro coupling more simple approaches were developed and enhanced. Using these methods the solidification morphology was assumed and the microsegregation field was calculated in a one-dimensional domain. These sharp interface models were coupled directly to thermodynamic calculations and to macroscopic temperature solvers to improve the temperature solution on the process level.

Besides physical models, semiempirical approaches exist to obtain knowledge about the microstructure. The parameters of these models were deduced from experiments. Once a valid set of parameters for a certain material was found, these correlations yielded reliable predictions. For well-known material classes such as steels, this is a promising and practical approach to couple process and material scale.

Like metals, plastics also form inner microstructures during solidification. The degree of crystallinity depends on the local cooling conditions and the local inner properties. Both the local microstructure and the molecular orientation are anisotropic in a molded part. To describe the complex interaction between these local structures and the global part properties, software tools were produced to predict the local distribution of these values by a chain of complex process simulations and structural analysis.

While the developed models were successfully tested for each process step, test parts were designed to simulate complete process lines. To validate the results for each process step, the simulated parts were also processed experimentally. Four test parts in four different process lines were simulated: a turbine blade, an aluminum cup, a plastics pipe fitting, and a steel profile.

Each of the following sections of this paper is split into these four process lines. The authors intend to give an overview on the work carried out during recent years. An overview on the process line is given in separate chapters within this book. For more detailed information about the models and results the reader will find references in each section to further papers published by the authors.

8.2 Experiments

8.2.1 Turbine Blade

Superalloys are commonly used in high-temperature applications, such as turbine blades. The creep strength at high temperatures is improved due to the ab-

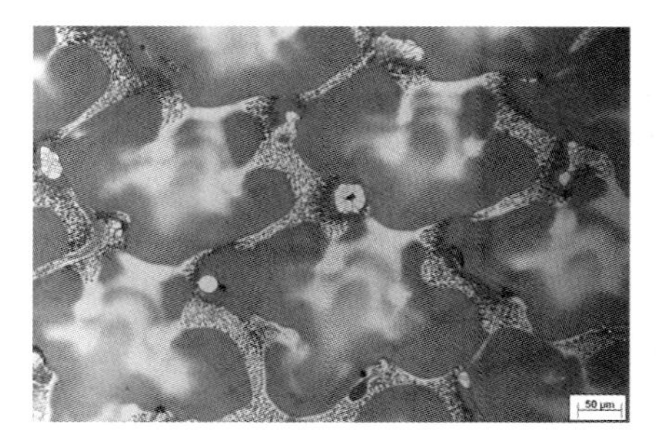

Fig. 8.1 Typical as-cast microstructures of CMSX-4: overview and close-up of an interdendritic region. The primary γ dendrites and interdendritic γ' can be seen.

sence of grain boundaries (single crystals). These single-crystal cast turbine blades are widely made from Ni-based alloys, alloyed mainly with Al and additions of Ta, Ti, W Cr, Re, and others [1]. The desired microstructure consists of the γ matrix with finely dispersed γ' precipitates. However, the as-cast microstructure consists of a large number of parallel aligned γ dendrites, with interdendritic γ' precipitates. Figure 8.1 shows typical transverse sections of directionally solidified dendrites in superalloys. Due to heavy segregation the distribution of the alloying elements is very inhomogeneous. Subsequent process steps aim at dissolving the as-cast microstructure.

In order to obtain the desired single-crystal structure, the turbine blades are manufactured according to the investment casting route involving directional solidification through the Bridgman process. A directed heat flux forces the growing crystals to grow in a predefined direction. This is achieved by withdrawing the investment cast shell, filled with the melt, from a heated zone to a cooled zone within the furnace. Both zones are separated by an insulation layer, the so-called baffle. Grain selection techniques ensure that only one grain is present in the final turbine blade. For the present work the commercially available alloy CMSX-4 was investigated [2]. Directional solidification experiments were carried out using a simplified turbine blade geometry.

8.2.2 Aluminum Cup

Two alloys were used to carry out the experiments. The first alloy was the commercial wrought alloy AA3104. It is typically used for sheet forming applications. The second one was a model alloy of AA3104 containing only the five major elements (given in wt%, AlMg1Mn1Fe0.45Si0.2) to validate the models. The liquidus temperature is 651 °C where the fcc_a1 phase starts to form followed by $Al_6(Mn,Fe)$ at about 650 °C. Other intermetallic phases might appear depending on the local cooling conditions, namely $Al_{13}Fe_4$ and Mg_2Si. A Gulliver–Scheil calculation yields an end temperature of solidification of about 592 °C.

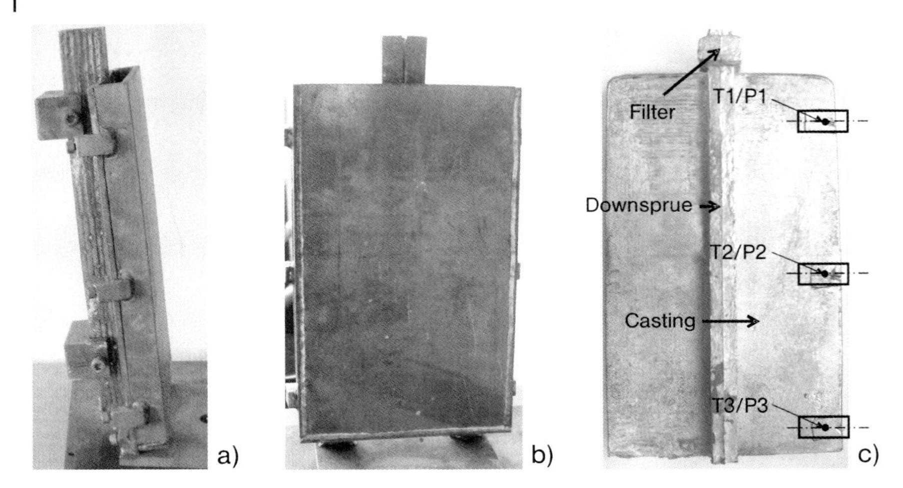

Fig. 8.2 The steel chill (a, b) used for (c) the ingot casting.

To validate the various models used within the process line a die casting chill was designed to produce a sound and homogenous ingot for the subsequent homogenization and hot rolling process. The ingot casting was of rectangular shape as depicted in Fig. 8.2, where the chill itself and the actual casting are shown. The quality criteria for the ingot were: avoiding porosity, oxides, or other inclusions, as well as a microstructure gradient from the bottom to the top (direction of hot rolling) of the casting, which was characterized by measuring the dendrite arm spacing (DAS) and the grain size at the positions P1 through P3.

The gating system consisted of a filter, a downsprue, and a thin horizontal ingate with a 2 mm cross-section ranging from the bottom to the top of the casting to connect the downsprue with the actual casting. Precise information about the casting conditions is given in Section 8.4. The casting had 15° deviation from the gravity direction. A filter, argon gas, a low casting temperature, and a low filling rate ensured that the ingot was free of porosity and oxide films, which was confirmed by an x-ray analysis. Using this technique, no significant macrosegregations were found as pointed out in Chapter 7.

8.2.3
Plastics Pipe Fitting

Tests performed in previous projects and in this project show clearly that global mechanical part behavior depends on the local composition of inner properties [3]. Besides local crystalline conditions, like the degree of crystallinity and spherulite diameter, the local molecular orientation is very important. All these structures are mainly formed during melt flow and solidification according to local thermal and rheological conditions.

The plastic parts were injection molded for validation and material data measurements (Chapter 7). All parts were molded of the chosen semicrystalline isotactic polypropylene, type PP 505 P, provided by Sabic, Geleen (NL). During the production the process parameters were modified to obtain different compositions of thermal and rheological boundary conditions during solidification of the moldings.

The resulting composition of the inner properties was analyzed and, eventually, used to model and validate the microstructure simulation.

8.2.4 Steel Profile

For the experiments an unalloyed carbon steel C45 ([C] = 0.45 wt%, [Si] = 0.3 wt%, [Mn] = 0.8 wt%, [S] = 0.02 wt%, [P] = 0.02 wt%, [Al] = 0.04 wt%) was used. Owing to a minimum carbon content of 0.4 wt% and a Mn content of about 0.8 wt% during solidification only austenite was formed and no peritectic reaction occurred as for lower C contents.

Two water-cooled Cu molds as depicted in Fig. 8.3 were constructed to simulate the solidification during the continuous casting process in laboratory experiments. The experiments were carried out to validate calculation of the resulting microstructure. To analyze the influence of heat conductivity on the solidification structure a two-side Cu mold was coated on one side with Ni and on the other side the Cu was placed in a chamotte fitted steel shell to obtain an intense unidirectional solidification. The solidification velocity could be changed by using different spacers between the mold surfaces. In order to measure the local cooling rate and the temperature field some thermocouples were located within the Cu mold. Subsequent to the casting experiments the ingots were analyzed metallographically for the dendrite arm spacings and microsegregations as well as the average grain size of primary austenite.

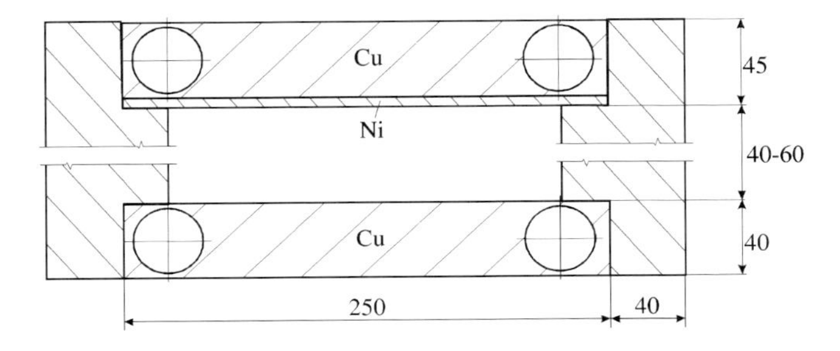

Fig. 8.3 Constructed Cu mold with different surface coatings for solidification experiments with different heat extraction rates.

8.3 Models

8.3.1 Turbine Blade

The microstructure formation in the directionally solidified turbine blade was simulated using a three-stage model. First the formation of phases and the segregation of the alloying elements were investigated using thermodynamic and kinetic calculations according to the CALPHAD approach, involving the software packages ThermoCalc and DICTRA [4]. From the calculations a thermophysical dataset was derived which was used to represent the alloy properties within the second stage, the macroscopic simulation of the Bridgman process. Thus, calculations were performed using the CASTS [8] package. From the calculated temperature fields the primary dendrite spacing was calculated according to [5]. The array of growing dendrites exhibits a fairly regular spatial arrangement. Thus, in the third stage a unit cell based on the primary dendrite spacing was defined [6]. The actual microstructure evolution was simulated within this unit cell, using the phase-field code MICRESS [7], coupled to ThermoCalc via a TQ-interface [4].

8.3.2 Aluminum Cup

To model the temperature field on the process scale and the microsegregation field three software components were coupled: the finite element method (FEM) temperature solver CASTS Rev. 14.1 [8], the finite volume method (FVM) microsegregation model MicroPhase Rev. 2.3 [9], and the TQ-interface of the thermodynamic software Thermo-Calc Rev. tcq [4]. At each FEM node CASTS updates MicroPhase with the current temperature and the time discretization. Together with the current composition at the interface the equilibrium is calculated using Thermo-Calc. In an iteration the interface position and the concentrations are determined. The latent heat released during each time-step is passed back to CASTS. The equilibrium calculation at the interface was substituted by a short-range diffusion model developed by Pustal within this project based on previous work by Ludwig [10] and Ludwig et al. [11]. Earlier work on this topic has been published elsewhere [12].

8.3.3 Plastics Pipe Fitting

Models were applied to predict inner properties like the degree of crystallinity according to Avrami [18, 19] and the crystallization process according to Nakamura [20, 22] and modified to account for the influence of the cooling velocity [23, 24]. All models were implemented in the in-house developed simulation software SphaeroSim. After first results of validating tests the spherulite growth was optimized in order to include realistic cooling velocities [25]. Furthermore, a modified temperature calculation described in Chapter 7 was integrated into the software to calculate local temperatures faster, and with better spatial resolution by considering temperature and pressure dependent material data [25].

8.3.4 Steel Profile

For the simulation of the continuous casting process a VisualBasic software tool [26] was adapted to calculate a two-dimensional temperature field from process parameters of the material and of the continuous casting mold (enthalpy, specific heat capacity, density, geometry, heat conductivity, and transmission coefficient) using the FDM method (Fig. 8.4). The cooling rate and the local solidification time were calculated. The necessary enthalpies for the C45-steel were calculated using ThermoCalc and Factsage.

In this macro model it is possible to vary, for instance, the heat transfer between the mold and strand (heat conductivity of Cu and flux powder) as well as the casting velocity and liquidus/solidus temperatures. Furthermore, the user is able to change the properties of the Cu mold (dimensions, conductivity, surface coating, velocity of cooling water, water conduits) for the simulation. The results are exported to data packets and can be plotted in various dimensions.

Subsequent to the calculation of the local cooling rate and temperature gradient (macro model) several microscopic operational detail processes, e.g. secondary

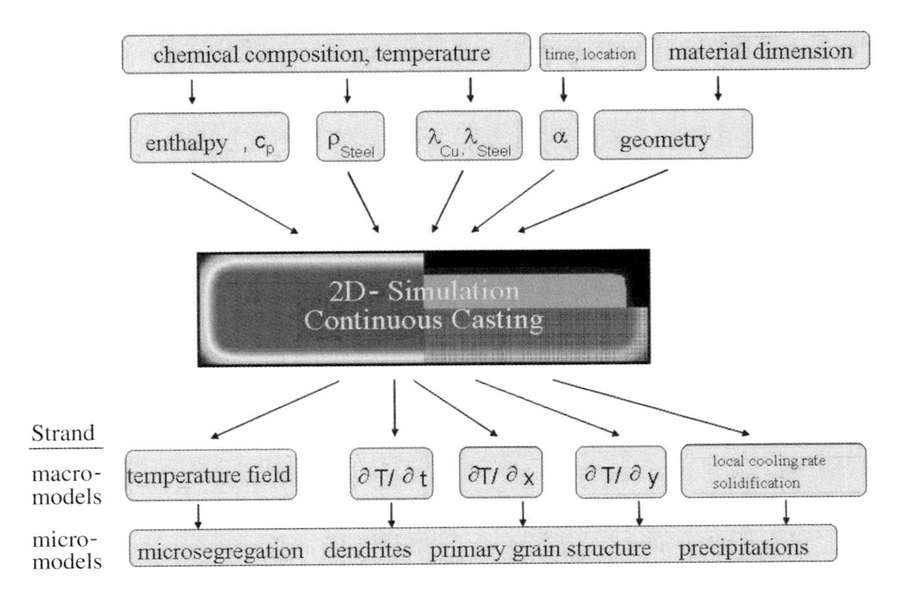

Fig. 8.4 Input and output parameters of the simulation.

dendrite arm spacing (SDAS, λ_2), were simulated. The two-dimensional equivalent grain sizes were modeled using a Vertex simulation.

8.4
Simulations and Results

8.4.1
Turbine Blade

Calculations with DICTRA, using the NIST thermodynamic and kinetic databases [27, 28], showed a significant deviation from results obtained by Scheil and lever-rule calculations, concerning the freezing range and the temperature at which the formation of interdendritic γ' starts (Fig. 8.5) [29]. With the thermophysical data

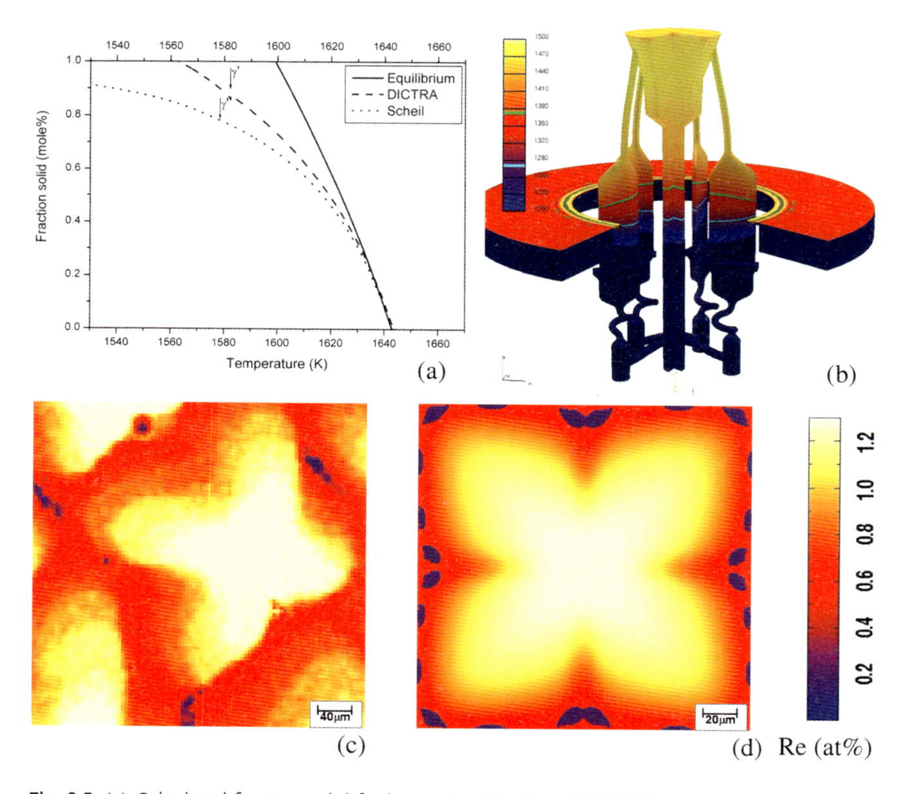

Fig. 8.5 (a) Calculated fraction solid for lever-rule, Scheil, and DICTRA models; (b) calculated temperature field of a half solidified turbine blade cluster; (c) measured element distribution of Re in an as-cast microstructure on a cross-section of a dendrite; (d) corresponding calculated element distribution.

set obtained from DICTRA the macroscopic solidification of the turbine blades was simulated. Figure 8.5 shows the calculated temperature field for a time when the turbine blade is approximately 50% solidified. The green and light blue lines, respectively, represent the solidus and liquidus temperatures as obtained from the DICTRA calculations.

The formation of the as-cast microstructure was simulated in a unit cell, which was defined according to the primary dendrite spacing and cooled according to the corresponding cooling rate. Additionally, WDX measurements were performed on directionally solidified CMSX-4 samples in order to obtain the element distribution for the alloying elements on a cross-section of the primary dendrite trunk. Figure 8.5c and d show the measured and the calculated element distributions for rhenium. The color bar is scaled to an equal range of values for an easy comparison of the results. The dendritic morphology of the primary phase is easily visible. It can be clearly seen that rhenium enriches in the dendrite cores. In the interdendritic regions γ' is found, which shows very low Re concentrations.

As for the simulation of the as-cast microstructure, knowledge of the nucleation of the interdendritic γ' is essential. Further investigations showed that the so-called eutectic islands, although the reaction is a peritectic one [30], nucleate from the existing γ-dendrites. A detailed description is given in [31].

8.4.2 Aluminum Cup

Figure 8.6a shows a simulation result of the temperature field after 84.5 s using CASTS. The simulated part is a test casting with a shape similar to the actual ingot used for further processing. To obtain a strong microstructure gradient the bottom part is made of commercially pure copper, and the top part is an insulation mat.

The material surrounding the test casting is a typical die steel. The initial conditions were 690 °C for the alloy (AlMn1Mg1Fe0.45Si0.2 wt%), 270 °C for the die, and 105 °C for the copper plate and the insulation. Between the materials Cauchy boundary conditions were applied, i.e. linear relationships between heat flux and temperature. Constant heat transfer coefficients (HTCs) of 1000 W m^{-2} K^{-1} were assumed between steel–copper and steel–insulation. The HTC for steel-alloys is given by 800 W m^{-2} K^{-1} above 571 °C and 400 W m^{-2} K^{-1} below 530 °C. The HTC for copper alloys is 2800 W m^{-2} K^{-1} above 617 °C and 600 W m^{-2} K^{-1} below 535 °C. The HTCs were deduced by an inverse analysis using temperature measurements at various positions in the casting. For ambient (20 °C) a Robin-type boundary condition was used, i.e. a relationship between heat flux and temperature caused by air flow at the free surface, assuming a heat transfer of 10 W m^{-2}.

Figure 8.6b shows the representative volume element (RVEs) for the microsegregation calculation at a given time. The RVE is for clarity split into two parts. The length of the RVE is equal to half of the dendrite arm spacing. On the left ordinate the local relative phase fraction is plotted, and on the right the concen-

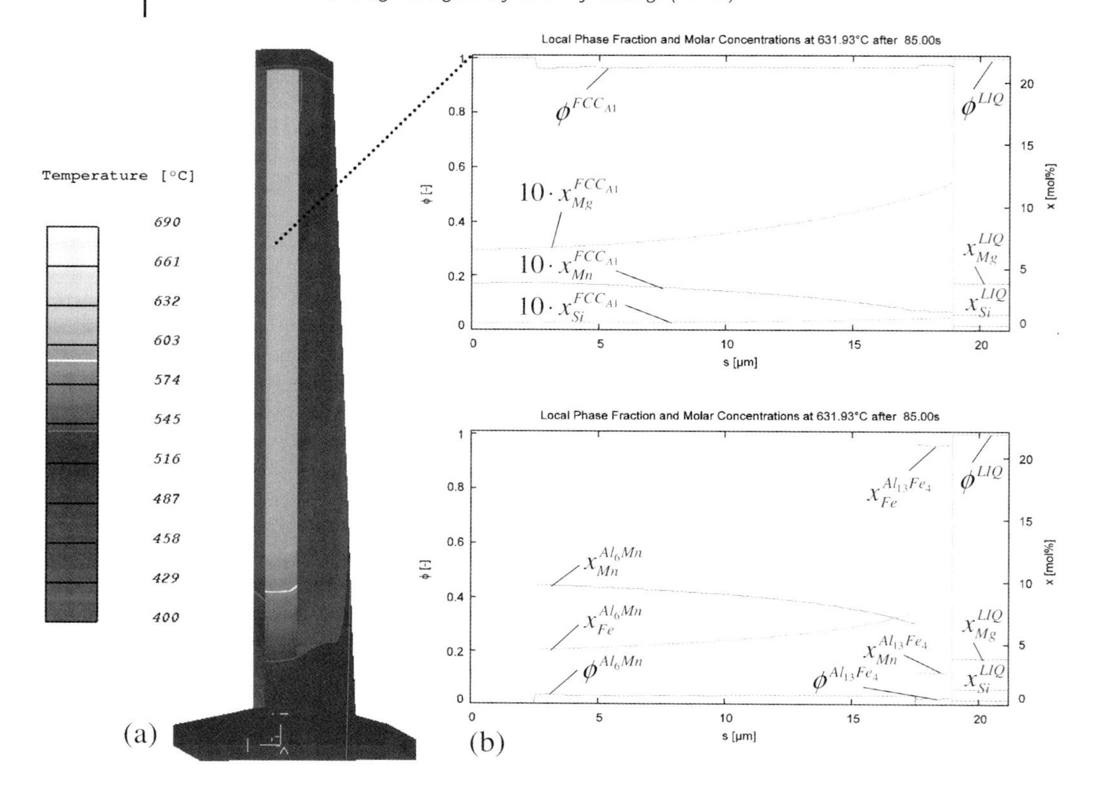

Fig. 8.6 (a) Temperature field calculated using CASTS; (b) representative volume element of MicroPhase, the local coordinate s varies within $0 \leq s \leq \lambda/2$, where λ is the dendrite arm spacing.

trations are given in mole fractions. At the western boundary of the solid phases (very left side of the RVE) and at the eastern boundary of the liquid phases (very right side of the RVE) a homogeneous Neumann boundary condition, i.e. zero flux, was applied. At the solid–liquid interface a Dirichlet boundary condition was assumed, i.e. fixed concentrations as given by the thermodynamic software in each iteration. The thermodynamic database used was an extension based on the COST 507 project [32].

8.4.3 Plastics Pipe Fitting

To predict local part properties software was developed to simulate spherulite nucleation and growth. This was realized by the creation of SphäroSim, an in-house developed software based on the Cellular Automata Method. A Cellular Automata Model used to simulate grain growth and recrystallization processes during metal welding [33] was adopted for this purpose. Since there are similarities between the material crystallization processes the description and the simulation based on Cellular Automata were adapted to polymer solidification [34]. This software permits a two-dimensional simulation of nucleation and growth processes of a solidifying polymer part cross-section depending on the actual thermal boundary conditions. Consequently, spherulite diameter distribution and spherulite growth processes can be simulated for the chosen material polypropylene.

To calculate a whole injection-molded part the temperature field calculation, nucleation, and spherulite growth descriptions were extended to accept all geometrical boundary conditions of an injection-molded part. By integrating flow path dependencies and the interpretation of rheological input a simulation of microstructure growth of the plastic pipe fitting is possible.

Additionally, thermal and flow-induced crystallization has to be taken into account to describe the whole crystallization process. This is especially required in the injection-molding process. The edge sections of an injection-molded part's cross-section solidify under shear influence. Thus, the software was extended to include flow influences during solidification. For this flow information Cadmold simulations were used according to Chapter 7 to solve the implemented calculation models [24, 25]. Intensive work in this field has generated different models [34, 37]. A summary of the chosen and implemented approach is presented in [24, 25]. Simulation results show good agreement with cross-sections experimentally analyzed by microscopy.

An extension to a three-dimensional calculation was not realized since it leads to no additional information but to drastically increased computation times and problems in displaying the results. The microstructure simulation, such as the prediction of orientation is realized two-dimensionally in different cross-sections of the parts. By this a high-quality transfer of simulation results was obtained to feed the subsequent process step with sufficient input at adequate computation time.

The prediction of molecular orientation in semicrystalline polymers is more difficult than that of amorphous polymers. On the one hand semicrystalline parts are nontransparent in the solid state; thus, a light-optical observation is difficult. On the other hand crystallization affects the molecular orientation during solidification. Therefore, birefringence and Fourier transform infrared (FTIR) tests were performed on small samples of the casting. Initially existing models of amorphous molecular orientation prediction were tested. The validation showed that molecular orientation is not easy to model. Therefore, short fiber orientations of fiber-filled polymers were observed, three-dimensionally modeled, and transferred to the subsequent process step [38, 39]. However, the modeling of molecular orientation is comparatively difficult because molecular architecture, molecular length, and weight distribution affect this property enormously. Finally, the molecular orientation was described by layer models [40].

8.4.4 Steel Profile

The primary austenite grain structure was calculated by a model of Andersen and Grong [41], but additionally accounting for mobility terms of grain boundaries

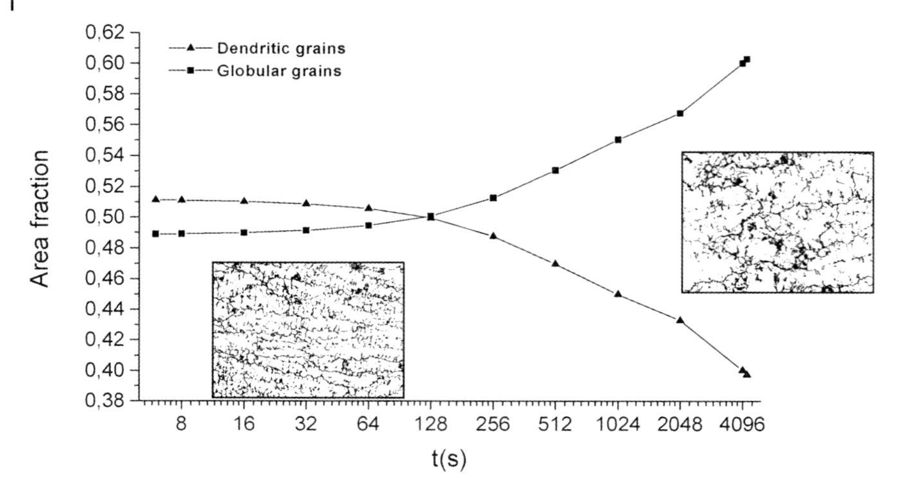

Fig. 8.7 Coarsening of primary austenite grains at lower cooling rates calculated by a Vertex simulation.

and microsegregations. It was found by metallographical analysis that the primary grain structure adapted to the primary dendrites by a columnar solidification microstructure. Therefore, a shape factor was defined to describe the grain shape besides the grain diameter. The grains had an oblong morphology because of the columnar primary dendrites. In an equiaxed dendritic structure the grains appear spheroidally and larger in equivalent diameter. The equivalent grain diameter and the shape factor were proportional to each other. The most important influence on grain size and shape factor is the cooling rate during solidification. In the case of cooling rates below 6 K s^{-1} a coarsening effect was found which deformed former columnar grains to spheroidal grains (Fig. 8.7).

8.5 Summary

Different numerical approaches have been presented for different material classes used to simulate the evolution of microstructure or relevant microstructure parameters during solidification. These models have been developed in close cooperation and used within the context of different process lines.

8.5.1 Turbine Blade

A concept of calculating microstructures of complex superalloys was developed and tested. The model spans calculations from the constitution of the alloy over the process to the final microstructure. Good agreement between the final calculated microstructures and experimentally observed ones is found. The calculated as cast microstructures were used as an initial condition for the subsequent homogenization treatment. Details of the heat treatment simulations are given in Chapter 12.

8.5.2 Aluminum Cup

A macroscopically and thermodynamically coupled microsegregation model was developed to carry out simulations on the casting process scale and materials scale with a strong coupling. The model was validated [9] and successfully applied to three-component and five-component aluminum-based alloys. Relevant parameters of the microstructure were passed on to the subsequent process step for through-process modeling.

8.5.3 Plastics Pipe Fitting

The integrative simulation of the plastic pipe fitting was realized based on an analysis of inner part properties as well as a choice and an extension of mathematical models to predict the needed local inner properties. Parallel to this a software tool based on the Cellular Automata Method was adopted and applied successfully to injection molding boundary conditions.

8.5.4 Steel Profile

A simulation code based on macroscopic and microscopic models was developed which enables the user to calculate parameters of the solidification structure, e.g. dendrite arm spacing, microsegregation, and primary grain structure in continuous casting processes. The simulation was validated by laboratory experiments and successfully applied to an unalloyed carbon steel C45.

Acknowledgments

The authors gratefully acknowledge the financial support of the Deutsche Forschungsgemeinschaft (DFG) within the Collaborative Research Center (SFB) 370 "Integral Materials Modeling".

The authors thank Hydro Aluminum for providing the AA3104 as well as the company Sabic for providing the polypropylene resin PP 505 P. We are also grateful for measurement of material data at high cooling speeds carried out by DSM Research.

References

1 M. Durand-Charre, Gordon and Breach, 1997.
2 D.J. Frasier, J.R. Whetstone, K. Harris, G.L. Erickson, R.E. Schwer, *Proc. Conf. on High Temperature Materials for Power Engineering*, ed. P. Esslinger, J. Ewald, Kluwer Academic, 1990.
3 W. Michaeli, M. Bussmann, B. Renner, *4M 2005: 1st International Conference on Multi-Material Micro Manufacture*, Elsevier Verlag, 2005, p. 17.
4 J.-O. Andersson, T. Helander, L. Höglund, P. Shi, B. Sundman, *Calphad*, 26, 2002, 273.
5 W. Kurz, D.J. Fischer, *Fundamentals of Solidification*, 4th edn, Trans Tech Publications, 1998.
6 D. Ma, U. Grafe, *Int. J. Cast Met. Res.* 13, 2000, 85.
7 www.micress.de.
8 G. Laschet, J. Neises, I. Steinbach, Lecture Notes of 4iéme Ecole d'été du CNRS, Porquerolles, C8, 1998, p. 1.
9 B. Pustal, B. Böttger, A. Ludwig, P.R. Sahm, A. Bührig-Polaczek, *Metall. Mater. Trans. B* 34B, 2003, 411.
10 A. Ludwig, *Physica D* 124, 1998, 271.
11 A. Ludwig, B. Pustal, D.M. Herlach, *Mater. Sci. Eng.* A318, 2001, 337.
12 M. Fackeldey, A. Ludwig, P.R. Sahm, *Comp. Mater. Sci.* 7, 1996, 194.
13 M. Fackeldey, Ph.D. thesis, RWTH Aachen, Fachbereich Metallurgie und Werkstofftechnik, 1998.
14 K. Greven, M. Fackeldey, A. Ludwig, T. Kraft, M. Rettenmayr, P.R. Sahm, Proc. 8th International Conference on Modelling of Casting, Welding and Advanced Solidification Process, San Diego, CA, 1998, p. 187.
15 K. Greven, A. Ludwig, T. Hofmeister, P.R. Sahm, Proc. Solidification of Metals: Research and Foundry Praxis, Weinheim, Germany, 1999, p. 119.
16 K. Greven, Ph.D. thesis, RWTH Aachen, Fachbereich Metallurgie und Werkstofftechnik, 2000.
17 T. Hofmeister, Ph.D. thesis, RWTH Aachen, Fachbereich Metallurgie und Werkstofftechnik, 2002.
18 M. Avrami, *J. Chem. Phys.* 7, 1939, 1103.
19 M. Avrami, *J. Chem. Phys.* 9, 1941, 177.
20 K. Nakamura, K. Katayama, T. Amano, *J. Appl. Polym. Sci.* 17, 1973, 1031.
21 K. Nakamura, T. Watanabe, K. Katayama, T. Amano, *J. Appl. Polym. Sci.* 16, 1972, 1077.
22 K. Nakamura, T. Watanabe, K. Katayama, T. Amano, *J. Appl. Polym. Sci.* 18, 1974, 615.
23 S. Hoffmann, Ph.D. thesis, RWTH Aachen University, 2002.
24 W. Michaeli, M. Bussmann, B. Renner, Proceedings of PPS-21, Leipzig, 2005.
25 W. Michaeli, M. Bussmann, B. Renner, Society of Plastics Engineers (SPE), Proceedings of Annual Technical Conference (ANTEC), Boston, MA, 2005.
26 D. Senk, M. Safi, H. Hamadou, Proc. 5th European Continuous Casting Conf., Nizza, 2005.
27 U.R. Kattner: Turchi, A. Gonis, R.D. Shull (eds.), *CALPHAD and Alloy Thermodynamics*, TMS, Warrendale, PA, 2002, p. 147.
28 C.E. Campbell, W.J. Boettinger, U.R. Kattner, *Acta Mater.* 50, 2002, 775.
29 C. Walter, B. Hallstedt, N. Warnken, *Mater. Sci. Eng. A* 397, 2005, 385.
30 K. Hilpert, D. Kobertz, V. Venugopal, M. Miller, H. Gerads, F.J. Bremer, H. Nickel, *Z. Naturforsch.* 42A, 1987, 1327.
31 N. Warnken, D. Ma, M. Mathes, I. Steinbach, *Mater. Sci. Eng. A* 413–414, 2005, 267.
32 Report of the European Community, Project COST 507.
33 U. Dilthey, V. Pavlik, T. Reichel, in *Mathematical Modelling of Weld Phenomena 3*, ed. H. Cerjak, Institute of Materials, London, 1997.
34 W. Michaeli, M. Bussmann, IKV-International Colloquium 2004, Aachen, 2004, p. 8.
35 W. Tietz, Ph.D. thesis, RWTH Aachen University, 1994.

36 G. Eder, H. Janeschitz-Kriegl, *Materials Science and Technology*, VSH Verlagsgesellschaft, Weinheim, 1997, Chapter 5.
37 A.I. Isayev, X. Guo, L. Guo, *Polym. Eng. Sci.* 39, 1999, 2096.
38 W. Michaeli, M. Kratz, M. Bussmann, *J. Polym. Eng.* in press.
39 H. Koeppen, student thesis, Institute of Plastics Processing, RWTH Aachen, 2005.
40 Y. Zhang, diploma thesis, Institute of Plastics Processing, RWTH Aachen University, 2005.
41 I. Andersen, Ø. Grong, *Acta Metall. Mater.* 43, 1995, 2673.

9
Coating of Turbine Blades (TP A3)

B. Hallstedt, J. Müller, D. Hajas, E. Münstermann, J.M. Schneider,
R. Nickel, D. Parkot, K. Bobzin, and E. Lugscheider

Abstract

In this chapter the simulation of a number of coating processes used for gas turbine blades is described. These processes comprise chemical vapor deposition for a diffusion barrier coating, physical vapor deposition (magnetron sputtering) for an oxidation protection coating (bond coat), and atmospheric plasma spraying for a thermal barrier coating.

9.1
Introduction

In order to withstand the severe thermal and oxidative load in modern gas turbines, several coatings are applied on the base superalloy that forms the core of the turbine blade. Generally at least an oxidation protection coating (or bond coat) and a thermal barrier coating are applied. The oxidation protection coating consists either of an MCrAlY (M = Co, Ni, Fe) alloy or a nickel aluminide, and the thermal barrier coating typically consists of yttria-stabilized zirconia. In this work an Al_2O_3 diffusion barrier coating between the base alloy and the oxidation protection coating is also considered. We describe the modeling and simulation of three different coating processes: chemical vapor deposition (CVD) of an Al_2O_3 diffusion barrier coating, physical vapor deposition (PVD) by magnetron sputtering of an NiCoCrAlY oxidation protection coating, and atmospheric plasma spraying of an yttria-stabilized zirconia thermal barrier coating. The simulation results are supported by parallel experimental investigations of the coating processes.

9.2
Modeling and Simulation of Al_2O_3 Chemical Vapor Deposition

At high temperature there is a strong tendency to interdiffusion between the MCrAlY oxidation protection coating and the substrate material. This can lead to

Integral Materials Modeling: Towards Physics-Based Through-Process Models
Edited by Günter Gottstein
Copyright © 2007 WILEY-VCH Verlag GmbH & Co. KGaA, Weinheim
ISBN: 978-3-527-31711-0

Al loss from the oxidation protection coating, and thus loss of oxidation resistance. In some cases intermetallic phases or Kirkendall porosity may form in the interdiffusion zone, possibly leading to embrittlement. By applying a thin intermediate Al_2O_3 layer, interdiffusion between the oxidation protection coating and substrate can be completely or almost completely inhibited [1–4]. The Al_2O_3 layer should be as thin as possible in order to minimize thermomechanical load from lattice mismatch, but if it is less than about 3 μm thick it may become unstable during high-temperature exposure and not block interdiffusion completely [1].

CVD of Al_2O_3 on metallic substrates may lead to whisker growth [1–3, 5, 6]. To function as a diffusion barrier the Al_2O_3 layer needs to be dense, and whisker growth cannot be accepted. The mechanism of whisker growth is a matter of discussion [3]. Whisker growth can, however, be effectively stopped by applying a thin intermediate layer of TiN between the substrate and the Al_2O_3 layer [1–3]. TiN also appears to act as a buffer to adapt the different thermal expansions of substrate and α-Al_2O_3 layer [1]. Whisker growth is pronounced in the $AlCl_3$–CO_2–H_2 precursor system and can be reduced by changing $AlCl_3$ to $AlBr_3$ or AlI_3 [6].

The α-Al_2O_3 deposition kinetics from an $AlCl_3$–CO_2–H_2–HCl precursor system was studied experimentally and by simulation using the PHOENICS-CVD [7] code [3, 8]. The overall reaction is described as a hydrolysis of $AlCl_3$ by the CO_2–H_2 mixture. HCl is produced by the reaction, but is also added to the precursor gas in order to achieve more uniform deposition rates throughout the CVD reactor by slowing down the overall reaction rate. The deposition kinetics turns out to be very complex without any identifiable rate-controlling reaction step. Al_2O_3 deposition takes place by the overall reaction

$$2AlCl_3 + 3H_2 + 3CO_2 \rightarrow Al_2O_3 + 3CO + 6HCl \tag{9.1}$$

and it has been commonly assumed that the deposition rate is controlled by the water-gas shift reaction:

$$H_2 + CO_2 \leftrightarrow H_2O + CO \tag{9.2}$$

However, Schierling et al. [9] showed that the water-gas shift reaction alone cannot explain the Al_2O_3 deposition kinetics, but that there must be alternative reaction paths. Unfortunately, these alternative paths could not be identified. Schierling et al. [9] also suggested AlOCl(ad) as a possible intermediate rate-determining adsorbed species. The only work so far attempting a simulation including also surface reaction mechanisms is by Nitodas and Sotirchos [10]. Important adsorbed intermediate species suggested are $AlHO_2$(ad) and AlOCl(ad). Tan et al. [11] investigated the homogeneous gas-phase kinetics based on the set of rate parameters presented by Catoire and Swihart [12] involving 92 reversible reactions among 32 species. Tan et al. [11] suggested AlOCl as the major rate-determining intermediate species in the gas phase. Nitodas and Sotirchos [10], though considering adsorbed AlOCl(ad), did not consider this species taking part in the homogeneous gas phase reactions. In the present work the simple overall reaction mechanism according to Eq. (9.1) was used. This was formulated as a Langmuir–Hinshelwood expression to describe the rate of Al_2O_3 deposition:

$$j(\mathrm{Al_2O_3}) = k \cdot \exp\left(-\frac{E_\mathrm{A}}{RT}\right) \cdot P^\mathrm{a}(\mathrm{CO_2}) \cdot P^\mathrm{b}(\mathrm{AlCl_3}) \cdot P^\mathrm{c}(\mathrm{H_2}) \cdot P^\mathrm{d}(\mathrm{HCl}) \quad (9.3)$$

The activation energy, E_A, and the reaction orders a to d were determined directly from experimental deposition rate data, whereas the pre-exponential factor, k, was adjusted during the PHOENICS-CVD simulation [8].

The CVD experiments were performed in a vertical tubular hot-wall reactor. The reactor consisted of a 1230 mm long mullite ($Al_6Si_2O_{13}$, "Pythagoras") tube with an inner diameter of 27 mm. A three-zone furnace giving an isothermal zone of 450 mm length was used. To measure the weight gain of the samples the furnace was equipped with a magnetic suspension balance. All measurements concerning reaction kinetics were conducted with sintered Al_2O_3 substrates with the dimension $50 \times 10 \times 0.63$ mm^3. Two series of measurements with the substrate either at the lower end or at the upper third of the isothermal zone were made. The precursor gas, consisting of a mixture of $AlCl_3$, HCl, CO_2, H_2, and Ar, was introduced at the bottom of the reactor at a total pressure of 100 hPa. The flow rate was varied from 5 to 40 l h^{-1} (STP), the substrate temperature from 900 to 1200 °C, the partial pressure of $AlCl_3$ from 0.25 to 3.5 hPa, the partial pressure of HCl from 0.1 to 24 hPa, the partial pressure of CO_2 from 1 to 20 hPa, and the partial pressure of H_2 from 1 to 90 hPa. In order to validate the predictions of the kinetic model chains of 40 substrates, covering the whole length of the isothermal zone, were coated. In this case the flow rate was 20 l h^{-1} (STP), the substrate temperature 1050 °C, the partial pressure of $AlCl_3$ 1.3 hPa, the partial pressure of HCl 1, 4, and 20 hPa, the partial pressure of CO_2 12 hPa, and the partial pressure of H_2 60 hPa.

The weight gain of the samples was always proportional to the deposition time, i.e. the deposition rate was time independent. At the lowest flow rate, 5 l h^{-1} (STP), dust formation in the gas phase was observed. Note that, since the reactor wall temperature is close to that of the substrate, the deposition rate at the reactor wall is nearly the same as at the substrate. The activation energy and reaction orders varied in a rather wide range, suggesting that the actual reaction mechanism is complex. The pre-exponential factor, k, was evaluated later during the PHOENICS-CVD simulation. Due to the large variation in parameter values, a single parameter set could not be chosen. It may be expected that the lower substrate position should yield more reliable parameters since the actual gas composition at that point is similar to that of the inlet gas.

Simulations of local deposition rates were performed using the general CFD (Computational Fluid Dynamics) code PHOENICS with a CVD module developed within the European Esprit project 7161 and implemented by Kleijn and Kuijlaars [13]. The mass flow included diffusion according to Wilke [14]. Thermodiffusion was not included since its contribution was found to be negligible. Heat transport by the mass flow, radiation, and conduction in the gas and solids was included. Absorption of radiation by the gas was not considered. It is possible to include detailed gas-phase and surface chemistry in the simulation, but for the reasons mentioned earlier only the overall reaction described by Eq. (9.3) is used. The experimental data concerning single substrates could be reasonably well described. However, when comparing simulated deposition rates with experimental deposition rates using chains of 40 substrates, the simulated deposition rates were found to be a factor 2–4 higher than the experimental deposition rates. To get a better agreement the pre-exponential factors were readjusted. The activation energy and the reaction orders were left unchanged. The simulation results are shown in Fig. 9.1. The overall agreement is reasonably good, but the dependence on the position in the reactor tube is quite different. In particular when the amount of HCl in the inlet gas is small, the experimental deposition rate decreases strongly from the lower to the upper end of the isothermal zone, whereas the simulated deposition rate shows a maximum followed by a moderate decrease. The first (lowest) substrate in the chain typically shows a very high deposition rate. The reason for this behavior may be the same that gives a higher deposition rate for single substrates than for substrate chains. The simulation method used does give reasonable results within the studied parameter space. However, details in the deposition rate are not described well since the chemical model used gives a too simplified picture of the actual reaction mechanisms. The homogeneous gas phase chemistry can probably be reasonably well described [11, 12], but the surface chemistry is not well understood at all so far.

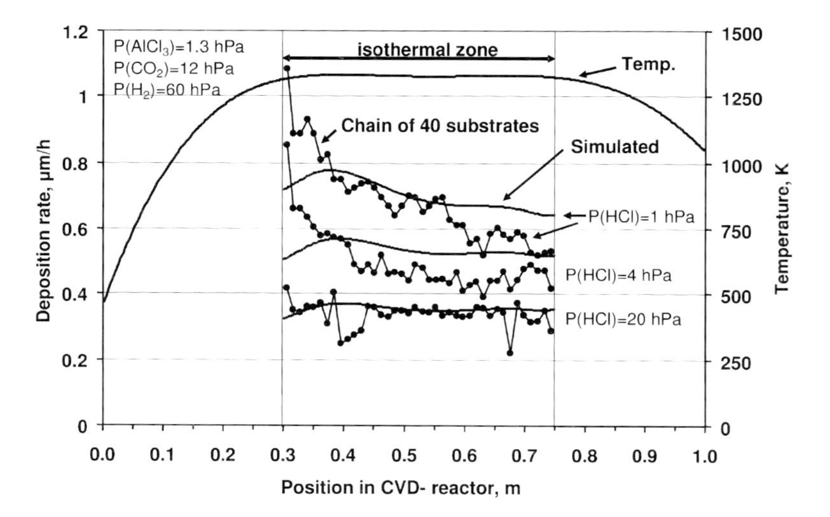

Fig. 9.1 Simulated and experimental deposition rates of α-Al_2O_3 in a hot-wall tube reactor with different amounts of HCl in the reactive gas.

9.3
Modeling and Simulation of the Magnetron Sputter Process

These days PVD is one of the commonly used coating technologies. Its principle is to evaporate solid material which after a transport phase will be deposited on a substrate. Within the SFB 370, magnetron sputtering was investigated using simulation techniques to allow predictions on the properties of the deposited coatings. The aim of the last period was to apply the knowledge gained earlier to perform a complete coating simulation with alloys like NiCoCrAlY onto complicated surfaces like a turbine blade.

Due to the variety of different physical phenomena within the sputtering process it is not efficient to describe and simulate it with one model. We will distinguish three separated parts, evaporation, particle transport through the plasma, and the film growth. All these parts are very different so multiple simulation techniques had to be applied.

In the magnetron sputter process the target material is evaporated by impacting ions, which transfer their kinetic energy to the target atoms. If the target atoms' kinetic energy exceeds the surface binding energy, they can leave the surface and travel to the substrate. The number of atoms leaving the surface per incident ion is called the sputtering yield Y and is strongly dependent on the impact energy and angle as well as the masses of the scattering partners. Since the late 1960s the mechanisms of the sputtering process have been studied theoretically. One of the first descriptions was given by Sigmund [15] followed by many researchers who investigated the sputtering for many decades [16, 17]. The evaporation and scattering can be considered as a statistical process consisting of single events which can by sufficiently approximated by analytical expressions. The simulation of the whole sputtering and transport process can be performed effectively using Monte Carlo methods.

The energy distribution of the material leaving the target surface was described using the modified Thompson distribution [17, 18]:

$$f(E)\,\mathrm{d}E = \frac{1-\sqrt{\dfrac{\bar{U}+E}{\Lambda E_\mathrm{i}}}}{E^2\left(1+\dfrac{\bar{U}}{E}\right)^3}\,\mathrm{d}E \tag{9.4}$$

where

$$\Lambda E_\mathrm{i} = \frac{4 m_\mathrm{g} m_\mathrm{t}}{(m_\mathrm{g}+m_\mathrm{t})^2} E_\mathrm{i} \tag{9.5}$$

is the maximum transferable energy and m_g and m_t are the atomic masses of the process gas and the target material, respectively.

For alloy targets like the one used in this project, the surface energy depends of the fraction of each component as well as their binding energy, which depends on the chemical binding energy of the surface atoms and also on all nonlinear properties of the surface like its roughness, etc. A good approximation for the surface binding energy is the heat of sublimation which is very well known for single-element targets. For metal alloys an approximation of the mean surface energy can be found as a linear combination of the surface energies U_i of each element, weighted with the mole fraction c_i in the alloy [19, 20]:

$$\bar{U} = \sum_i c_i U_i \tag{9.6}$$

During its travel from the target to the substrate the material collides with the process gas and loses part of its kinetic energy. Under the assumption that the process gas is sufficiently rarefied and the sputtered material is mostly neutral, which is a good approximation in conventional sputtering processes, the movement of the particles between two collisions can be regarded as force free. For the interaction between the plasma ions and the target particles, the hard-sphere potential was used [21–23].

For the sputtering yield an analytical formula, the revised Bohdansky formula

$$Y(E_i, \alpha = 0^\circ) = QS_n(\varepsilon)\left[1 - \left(\frac{E_{th}}{E_i}\right)^{2/3}\right]\left[1 - \left(\frac{E_{th}}{E_i}\right)\right]^2 \tag{9.7}$$

was applied which was fitted to available experimental data from the NIFS database SPUTY [24] and results from simulations performed using the SRIM package [25]. The formula describes the sputtering yield under normal incidence of the plasma ions ($\alpha = 0^\circ$) as the function of their kinetic energy E_i. The fitting parameters were the threshold energy E_{th} and the factor Q for the maximum of the yield curve. $S_n(\varepsilon)$ is the nuclear stopping cross-section for the stopping of the ions in the target material.

Due to the complicated curvature of the turbine blade it is problematic to achieve a uniform film thickness and chemical composition all over the coated surface. Several experiments and simulations were performed to verify the model. As a simple example the binary alloy CrAl was coated on a flat substrate. Figure 9.2 shows the radial film composition over the distance from the substrate center. The content of aluminum in the coating is slightly decreased compared to its content in the target owing to larger scattering angles of aluminum during the transport.

The influence of the substrate geometry, position, and orientation was also investigated experimentally and compared to equivalent simulations as shown in Fig. 9.3. The coated geometry contained a base plate on which two substrates were placed at an angle of 90° to each other (Fig. 9.3a). In Fig. 9.3b the measured and calculated values of the relative film thickness are exemplarily shown for the base plate at selected points [26, 27].

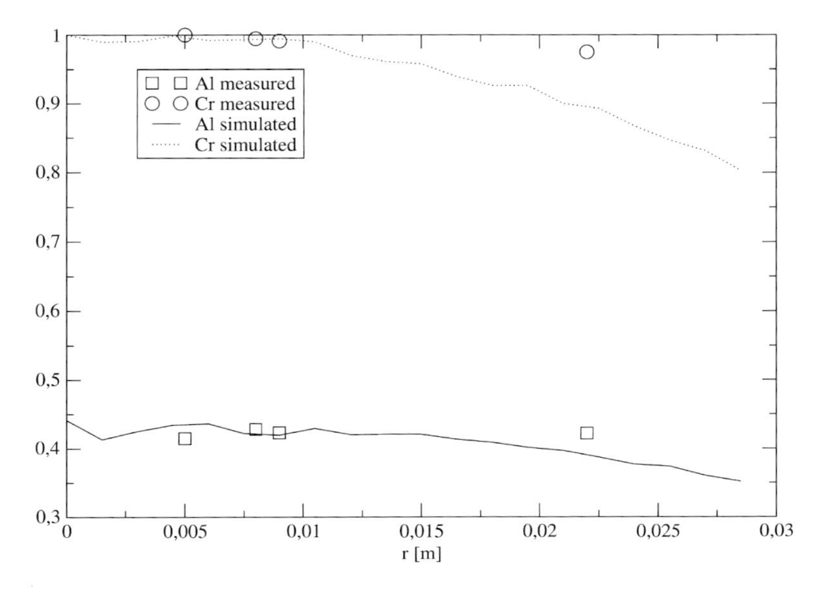

Fig. 9.2 Composition of the thin film deposited from a $Cr_{0.67}Al_{0.33}$ target.

Finally the real turbine blade geometry was imported as a 3D-CAD model and coating simulations were performed. Figure 9.4 shows the coating thickness in relative values from zero to one for a coating with a target parallel to the turbine blades top face. Shadowing effects due to the substrate curvature and several sharp edges are noticeable.

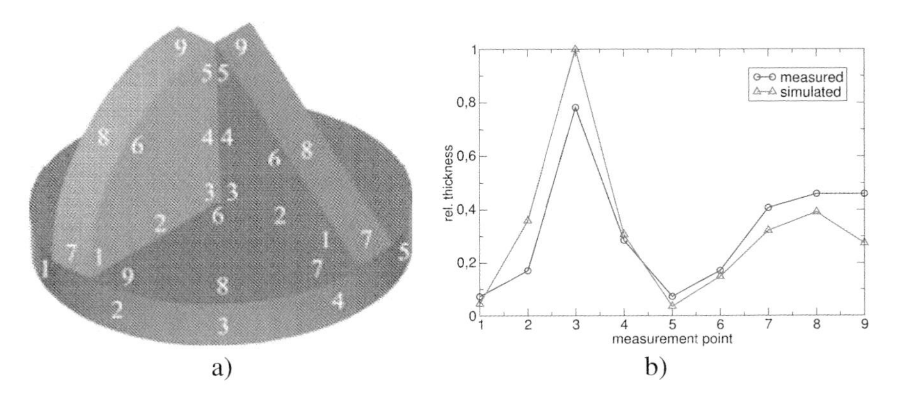

Fig. 9.3 Simulation and coating of a geometry composed of three different substrates. The values were measured and compared to the simulation results at the displayed measurement points.

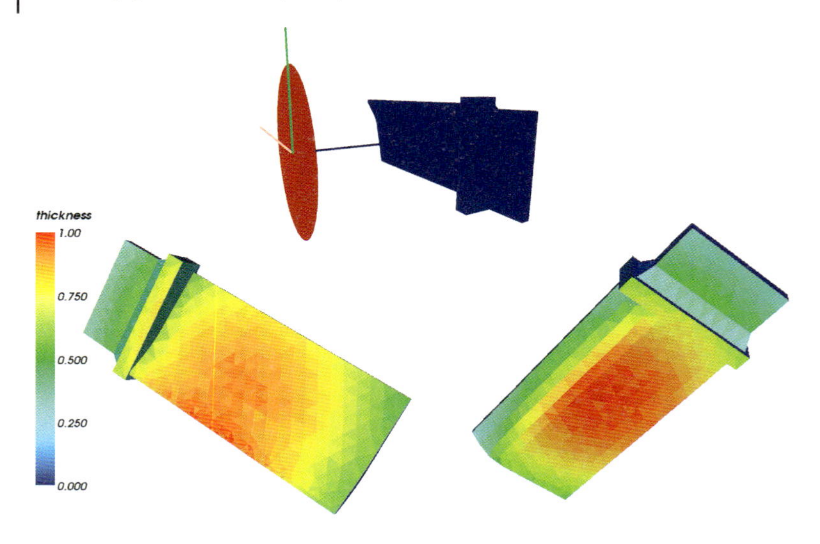

Fig. 9.4 Simulation results of the film distribution on a turbine blade which was placed parallel to the target. The film thickness is scaled to its maximal value.

9.4 Modeling and Simulation of Atmospheric Plasma Spraying and Thermal Barrier Coating Characterization

Atmospheric Plasma Spraying (APS) is a widely used technology for protective coatings against wear, heat, and corrosion. In the Collaborative Research Center "Integral Materials Modeling" (SFB370), the APS process was investigated, modeled, and simulated with respect to the deposition of thermal barrier coatings (TBCs) made of partially yttria-stabilized zirconia (PYSZ) for gas turbine applications.

The APS process is characterized by several physical phenomena. A consequence is that the process cannot be described by one simulation method. One way to avoid this problem is to subdivide the whole APS process into the subprocesses plasma torch, plasma free jet, sprayed powder particles characteristics, and coating formation [28, 29]. They can be considered as single processes, and suitable numerical methods are applied to simulate these subprocesses (Fig. 9.5).

9.4.1 Plasma Torch/Plasma Free Jet Simulation

A realistic CAD model of the plasma torch is shown in Fig. 9.6a. This CAD model was used as a basis for modeling and simulation of the torch plasma flow in a

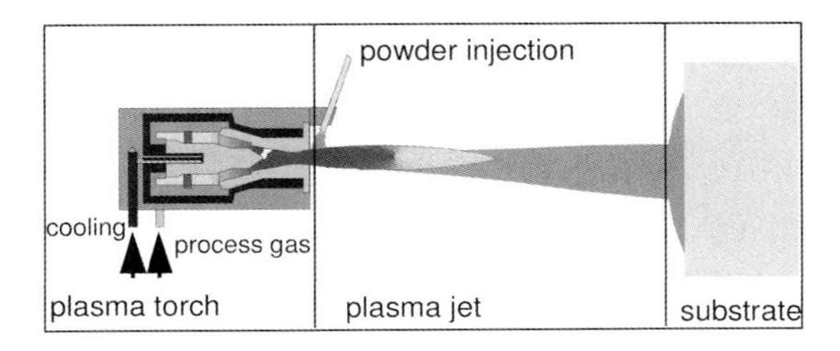

Fig. 9.5 APS and subprocesses in process simulation setup.

CFD tool (Fluent, Fluent Inc.) [30]. This was an important step in the realistic torch flow modeling in this project compared to formerly used simplified cylindrical torch models [31]. An electric arc is formed between the pin-shaped cathode and the anode which forms the nozzle wall of the plasma torch. The electrical arc can either be represented in a simplified approach by a volume heat source in the CFD model or the fully coupled solution of the flow balance equations with the electromagnetic field equations (magnetohydrodynamics) (Fig. 9.6b)

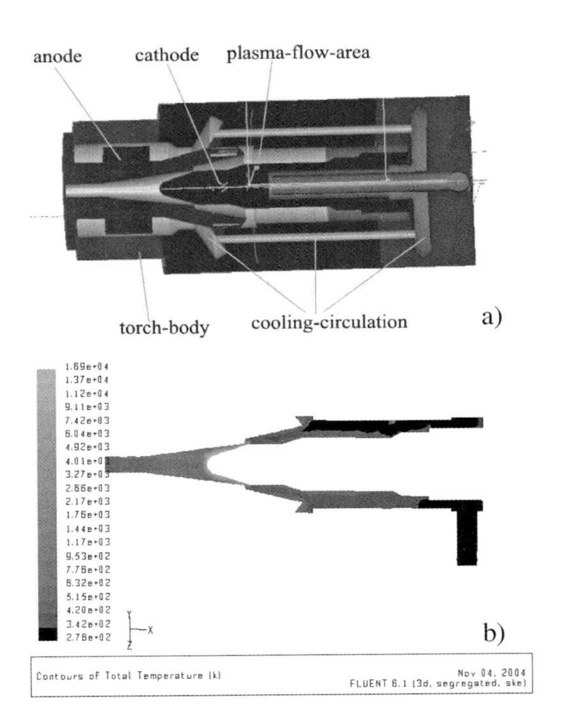

Fig. 9.6 (a) CAD model of the plasma torch. (b) Simulation/plasma flow area cross-section 35SLPM Ar/6SLPM H_2/20 kW.

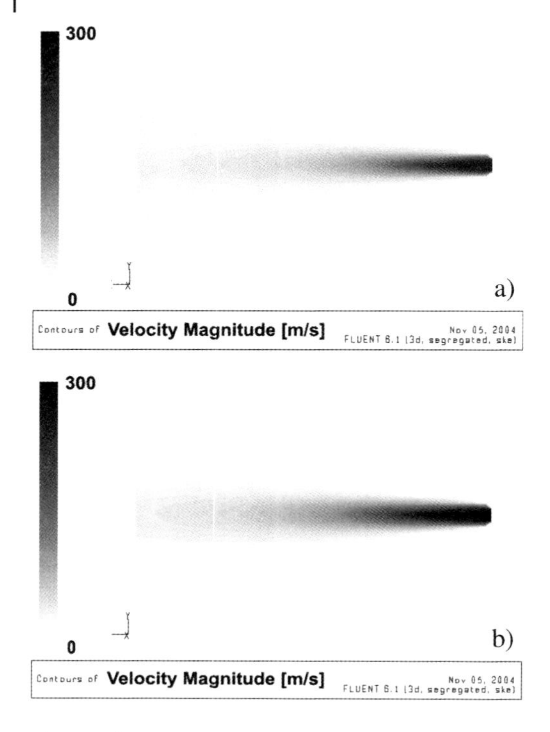

Fig. 9.7 Plasma jet: (a) 35SLPM Ar 6SLPM H_2 20 kW; (b) Ar 12SLPM H_2 20 kW.

[30]. The calculated plasma flow parameters at the nozzle exit define the inlet boundary conditions for the plasma free jet [28–30].

A cylindrical calculation domain was used for the plasma free jet simulation with free boundaries at the cylinder shell representing the plasma free jet behavior in the infinity of the ambient medium [31].

The simulation result in Fig. 9.7b deals with a higher hydrogen fraction than in Fig. 9.7a. The maximum velocity in the plasma jet is higher for the second example with the higher hydrogen fraction. This is due to the higher heating of the plasma in Fig. 9.7b and thus due to a higher expansion [30].

9.4.2
Powder Particles Characteristics

Powder particles are injected behind the nozzle together with an argon carrier gas. Particles are accelerated in the plasma flow due to the acting drag force. Furthermore, the particles are heated, melted, and partially evaporated in the plasma jet due to heat transfer from the hot plasma jet to the powder particles [28–33].

The equations of motion with the acting force on the right-hand side of the equation have to be solved to calculate the particle trajectories in the plasma jet:

$$\frac{du_P}{dt} = F_D(u - u_P) + \frac{g_x(\rho_P - \rho)}{\rho_P} + F_x, \quad F_D = \frac{18\mu}{\rho_P d_P^2}\frac{C_D \mathrm{Re}}{24} \tag{9.8}$$

where u_p is the particle velocity, $F_D(u - u_p)$ the drag force per unit particle mass, u the fluid phase velocity, ρ_p and ρ the particle and fluid density, g_x the gravity force, and F_x body forces per unit mass particle. In the term F_D, μ is the molecular viscosity of the fluid, d_P the particle diameter, Re the relative Reynolds number, and C_D the drag coefficient. Additionally, the heat conduction equation with a heat transfer at the surface of the spherical particles has to be solved to determine the temperature behavior of the particles with phenomena like melting and evaporation [29]. Two different ways were chosen for the simulation of the powder particle characteristics in the project. One way was the coupling between plasma free jet and powder particles simulation by a code developed under participation of IOT, the PLASMA2000 code [29, 31, 33]. As an alternative to this approach, the powder particle characteristics were calculated in a coupled multiphase model in the CFD software Fluent [30].

In Figs. 9.8 and 9.9, the calculated size and velocity distributions of PYSZ particles at the substrate surface are illustrated. The particles' diameters are distributed over a range of 5 to 45 μm, and they have an injection velocity of 15 m s^{-1} [29]. The velocity of particles is an important factor for the characteristics of a coating, e.g. the porosity or the deposition efficiency. But the maximum of the velocity is not correlated with the size of the particles. The fast particles travel close to the high-speed flow axis, and therefore get the best momentum transfer from the flow [29].

9.4.3
Coating Formation Simulation

One focus of the research work in the framework of SFB 370 in the last period was the coating formation modeling and simulation as well as the calculation

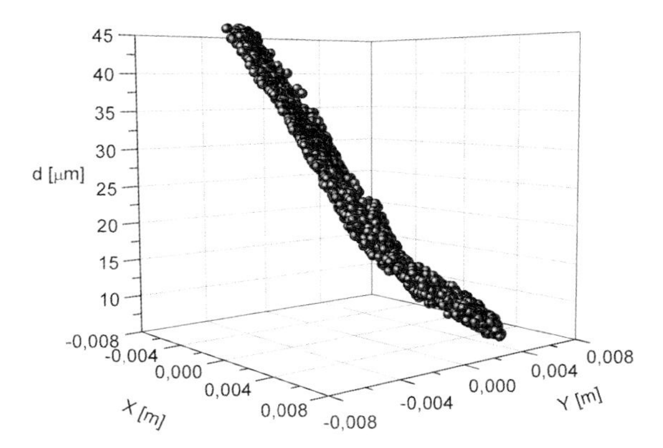

Fig. 9.8 Size distribution of $ZrO_2 + 7\%\ Y_2O_3$.

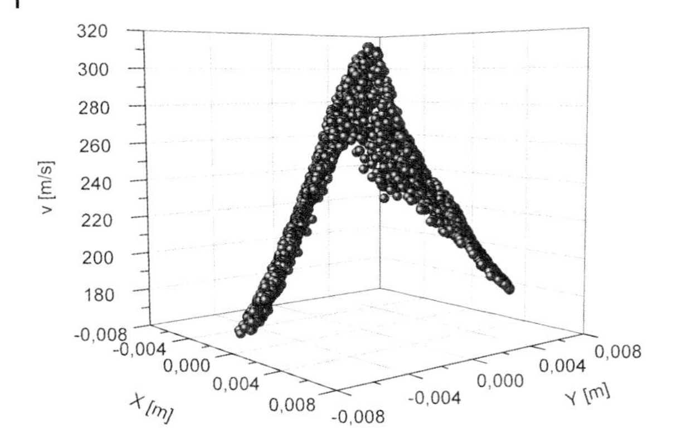

Fig. 9.9 Velocity distribution of $ZrO_2 + 7\% Y_2O_3$.

of coating properties depending on the TBC microstructure. The aim was to realize an integrated process simulation approach from production process (APS) parameters to product (PYSZ-TBC) properties. PYSZ means partially yttria-stabilized zirconia ($ZrO_2 + 7\% Y_2O_3$). Different model approaches with different physical assumptions and geometric scales were developed and applied.

One approach is the simulation of single powder particles hitting the substrate surface. The deformation, solidification, and cooling are simulated based on physical or analytical/empirical models. A lamellar, porous, microcracked microstructure is characteristic for APS-PYSZ-TBCs. The particles arriving at the substrate spread and form a splat. Some empirical approaches consider the form of a splat as a cylindrical disc [42]. The ratio of particle diameter prior to impact and the final splat (disc) height depends on the particle's Reynolds number in these approaches. The complex physical behavior of the ceramic particle cannot be taken into account, but properties like the coating porosity and local coating growth rates and thickness distributions can be investigated by this mesoscale approach which was also applied in the first phase of SFB 370 [43, 44].

The physical description of the particle impact is the most sophisticated modeling for the coating formation simulation [30]. The arriving particles are considered as a fluid with temperature dependent fluid properties (e.g. the viscosity) in a CFD and a volume of fluid (VOF) approach. The VOF model is a multiphase approach where two or more fluids (or phases) exist in a calculation domain and do not interpenetrate.

Examples for the impact simulation in the framework of SFB 370 of a PYSZ particle with a diameter of 40 μm, an initial velocity of 250 m s^{-1}, and an initial temperature of 3300 K are illustrated in Fig. 9.10 [30].

The physical single-particle setup is very suitable for the local investigation of the particle splashing and the microscopic structural coating growth if all physical phenomena are taken into account in the models. But only a few particles can be simulated with the available hardware resources in a reasonable calculation time. Therefore, this simulation approach is a simulation in the microscale domain.

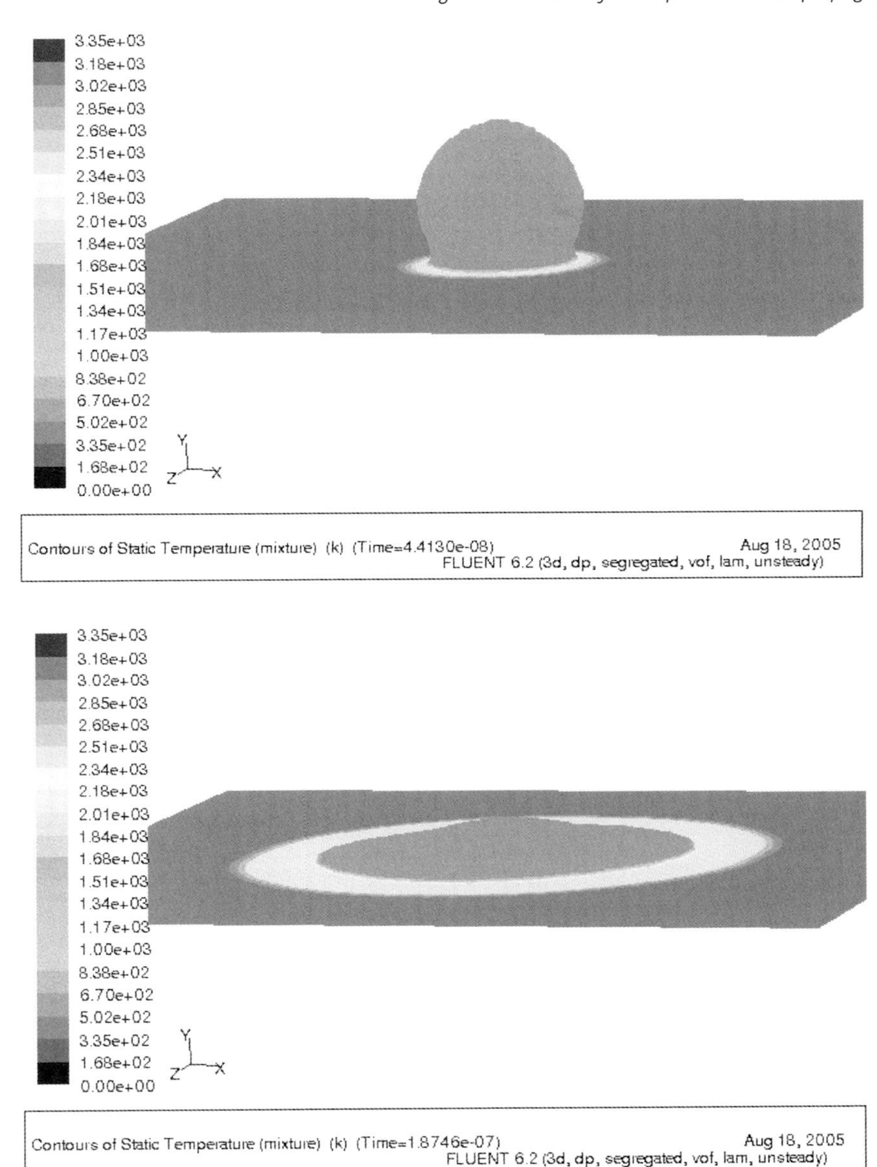

Fig. 9.10 PYSZ particle impact simulation: 44 ns (left), 0.19 μs (right).

A full coating build-up on complex components cannot be calculated by this method.

A macroscale coating formation modeling approach was developed in the framework of the SFB 370 to be applied in the investigation of the coating forma-

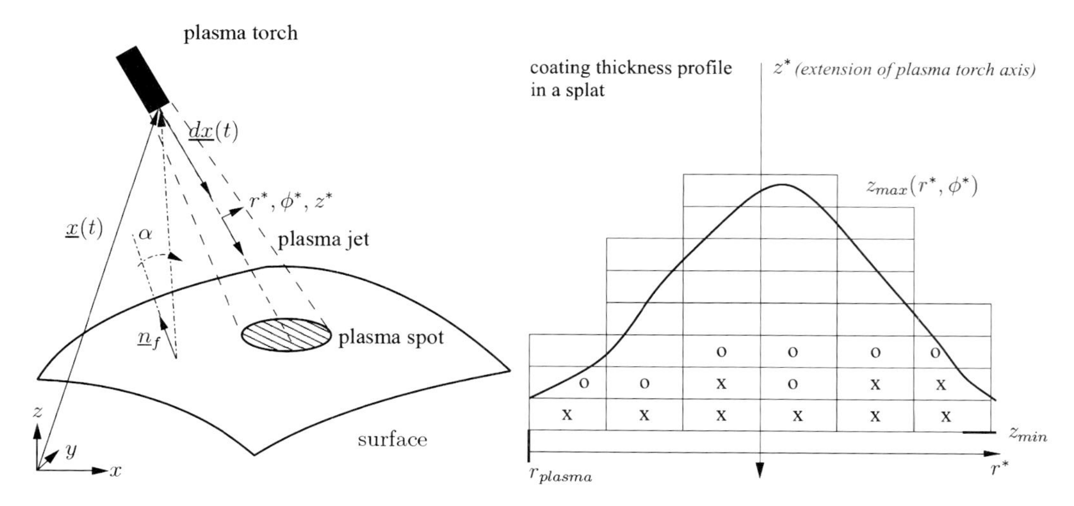

Fig. 9.11 Plasma torch movement with coordinates (left). Sketch of coating formation algorithm in the plasma spot (right).

tion during APS on arbitrary and complex components [20, 32, 34, 35]. The physical behavior (thermal, stress–strain state) of the coating/substrate system during the APS coating production process and the coating thickness distribution on arbitrary components depending on the process, especially the traverse path of the plasma torch, can be calculated. This simulation setup based on the finite element method (FEM) takes the time-dependence of the APS process into account. The plasma torch and, thus, the plasma spot moves relative to the substrate/component (Fig. 9.11). Due to the moving plasma spot, the splatwise coating formation is done layer by layer in this process. This simulation approach is aimed at modeling this process on the geometric scale of the substrate dimension.

The macroscale coating formation approach is based on a predefined meshed model of the coated component (including the discretized PYSZ-TBC layer). The TBC coating elements are deactivated at the beginning of the coating formation simulation. The elements are activated in such a way that they represent the powder particle distribution (coating thickness profile) in the plasma spot in front of the substrate. The x-marked elements in the cross-section of the splat in the plasma spot (right image of Fig. 9.11) are already activated FE elements while the o-marked elements are candidates for activation in the next step of the activation loop during one time increment. The realization of every plasma torch and substrate movement is possible in this algorithm described in [30, 32].

The powder particle distribution in front of the substrate (coating thickness profile) is calculated by modeling the powder particles' flight, heating, and melting during the plasma free jet simulation:

$$H(x) = \alpha \cdot h(x) \quad \alpha = (\mu \cdot \Delta t)/\rho \cdot v_{\mathrm{pl}} \cdot U \cdot \int_{-\infty}^{\infty} h(x)\,\mathrm{d}x \tag{9.9}$$

The coating thickness profile for the element activation in a plasma spot is obtained by a scaled (calculated) powder particle distribution. In Eq. (9.9), $H(x)$ is the coating thickness profile, $h(x)$ the powder particle distribution in front of the substrate, and α the scaling factor. This coating profiles perpendicular to the plasma torch movement are scaled dependent on the mass flow μ of the particles and the coating porosity U. Furthermore, the scaling factor α depends on the time increment Δt and the plasma torch velocity (relative to the substrate) v_{pl}.

An example of the calculated stress–strain state in a time increment of the coating formation simulation of a turbine blade is shown in Fig. 9.12. The development of microstructure dependent material models is described in the next section.

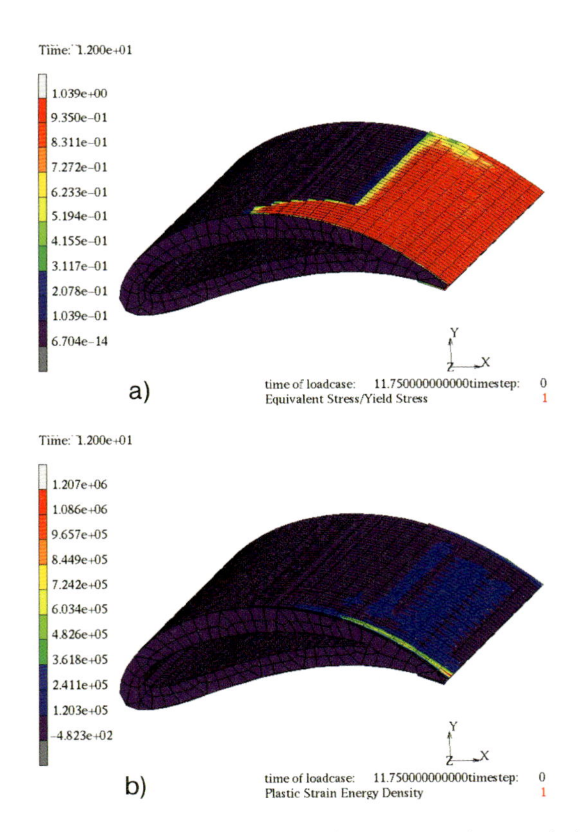

Fig. 9.12 FE macroscale coating formation simulation with developed material models: a) plastic strain energy; b) equ. stress/yield stress (FEM-tool: MARC2003/MSC) [30].

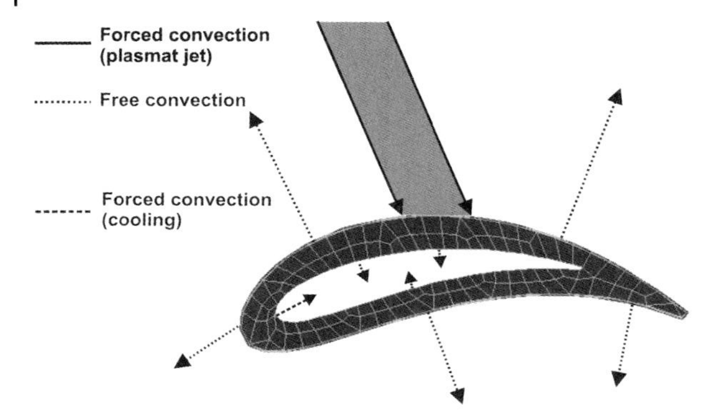

Fig. 9.13 Heat convection approach to turbine blade surface.

The developed macroscale approach is particularly useful for the investigation of process parameter and component geometry depending coating characteristics like the coating thickness distribution and the thermal and mechanical behavior of the coating/substrate system during the APS process.

Not only is the subprocess modeling and simulation important for the integrative modeling of the production process of atmospheric plasma spraying, but also the creation of interfaces between the subprocesses with a data transfer. The following example describes the modeling of the heat transfer to and from the coated substrate surface during the coating formation simulation. The heat transfer parameters are boundary conditions in the macroscale coating formation approach and were determined based on simulation results of the plasma free jet. The heat transfer boundary conditions have to be changed due to the moving jet as well as the boundary elements themselves due to the coating formation [36, 37]. The turbine blade is separated into different heat transfer zones in every time increment of the coating formation (Fig. 9.13). The heat transfer from and to the turbine blade surface is based on empirical-analytical free and forced convection approaches [36].

The equivalent von Mises stress values are calculated along a defined node path from the blade surface through the coating layers to the base alloy [36]. The influence of the plasma properties on the stress values, in this case the H_2 ratio in the Ar/H_2 process gas mixture, is illustrated in Fig. 9.14. The range of 10–90 mol% H_2 is used to investigate the dependency of the boundary conditions on the plasma properties, but in practice a maximum percentage of 40 mol% H_2 is applied to avoid damage in the plasma torch unit, especially damage of the torch electrodes, and also to limit particle evaporation.

The higher the H_2 ratio the higher the heat flux from the plasma jet to the substrate surface for the same plasma flow properties. One explanation for this result is an increase of the thermal conductivity of the gas/plasma.

One method is based on the physical equivalence homogenization [38]. The nonhomogeneous structure with the defined domain dimensions is compared to

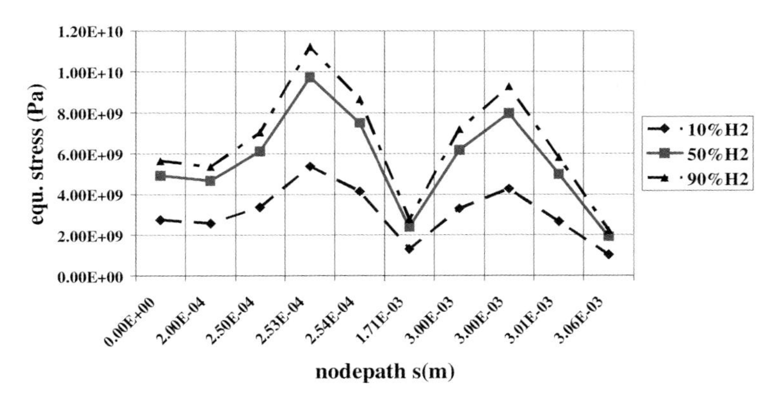

Fig. 9.14 Equivalent von Mises stress along a defined node path through the blade.

the homogeneous model with the same domain dimensions. The integral value of a physical parameter in both calculation domains is assumed to be equal. One physical parameter can be the strain energy for the mechanical behavior.

9.4.4 APS Coating Properties/Homogenization Methods

APS PYSZ coatings are characterized by a special lamellar, porous, cracked microstructure. Two homogenization methods were applied in this project to develop continuum materials models to be used in FE simulations. These material models and properties represent the defects in the microstructure as well as the bulk properties of the ceramic and the "void" phase [38]. The homogenization methods can be applied for the determination of the effective thermodynamic and mechanical behavior of the ceramic coatings.

In Fig. 9.15, a SEM micrograph of a PYSZ microstructure is shown. This micrograph is a basis for a black and white digitization (white: ceramic; black: pores) (Fig. 9.16). A FE model is built and a steady-state load case is applied to this FE model of the structure (Fig. 9.17). In the case of a mechanical homogenization, the strain energy is calculated. An effective stress–strain curve can be calculated based on different steady-state load cases (Fig. 9.18) [39].

The second homogenization method is based on the periodic structure theory [40, 41]. This method can be applied to a family of physical problems with the formulation

$$A^{\varepsilon} u_{\varepsilon} = f \quad \text{in } \Omega \tag{9.10}$$

where A^{ε} is a partial differential operator which depends on a small parameter ε. This parameter is of the order of magnitude of the structure period. One predic-

Fig. 9.15 SEM micrograph of a PYSZ microstructure.

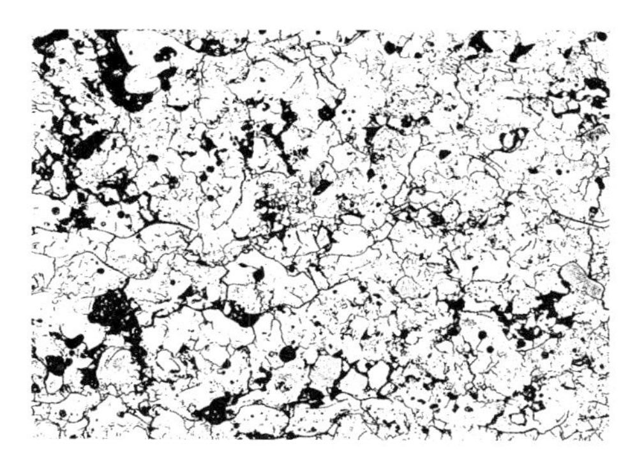

Fig. 9.16 Digitized SEM image for homogenization.

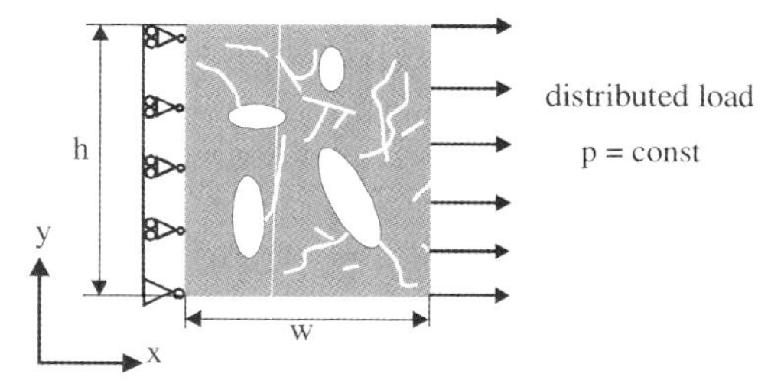

Fig. 9.17 Load case in mechanical homogenization.

tion for this homogenization method is that the period of the structure is much smaller than the dimension of the macroscopic calculation area Ω. The parameter u_ε is the periodic solution of Eq. (9.10) which fulfils the boundary conditions f.

An expansion of the solution vector u_ε is the basic approach of the homogenization of periodic structures. Another prediction for the application of this

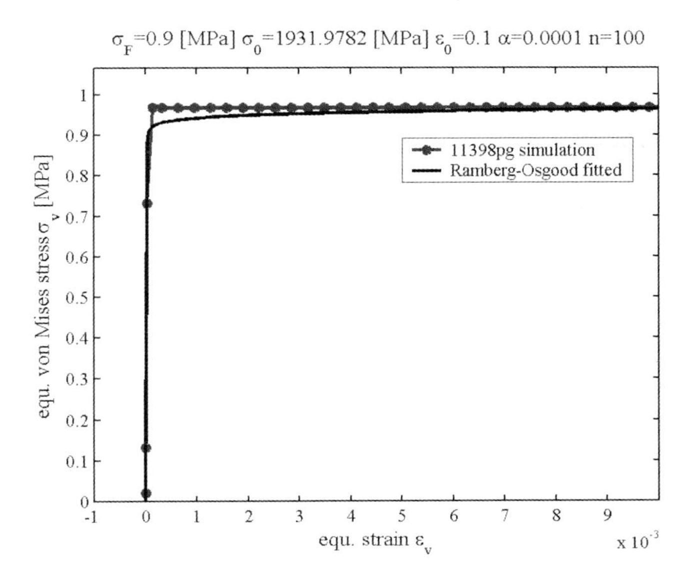

Fig. 9.18 Homogenized stress–strain curve fitted Ramberg–Osgood model.

method is the periodicity of the structure. The APS microstructure is not periodic. That is why a periodic assumption based on the spectral analysis of the digitized microstructure is applied to determine a dominant period or a group of dominant periods [41].

In Fig. 9.19, the power spectral density for a path in a digitized SEM image is illustrated. The peaks represent the periodic distribution of pores and defects of a special size. High peaks at a special period mean large defects with this periodic

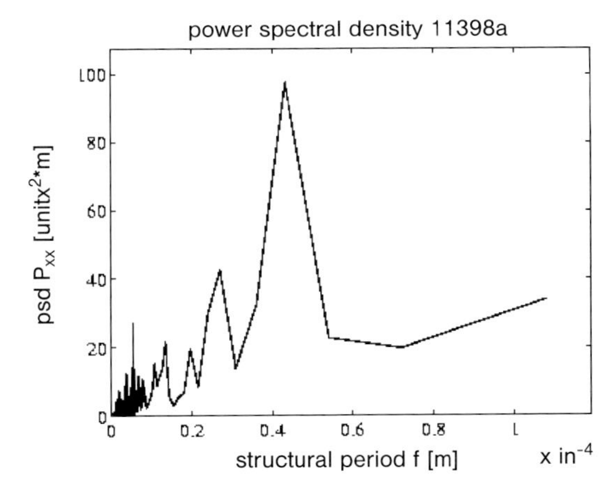

Fig. 9.19 PSD of a path in image 11398.

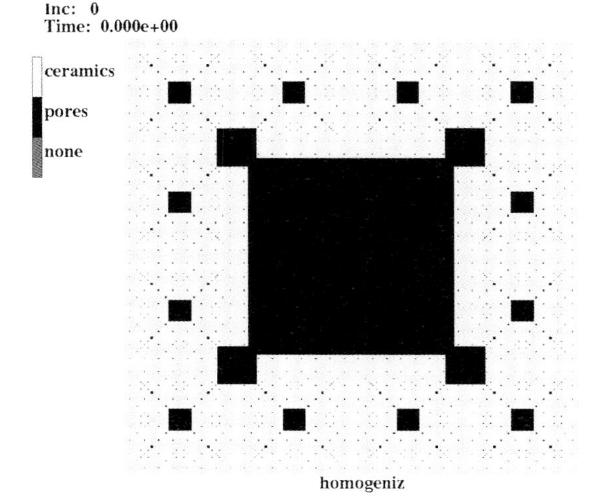

Fig. 9.20 Model for a multiperiod approach.

distribution. The PSD diagrams are used to define periodic models for the application of the periodic structure's homogenization method. If one-period models are used, the obtained effective Young's modulus values are too large [39]. That is why multiperiod models taking more dominant periods into account (Fig. 9.20) lead to more realistic effective material properties [45].

Acknowledgment

The authors gratefully acknowledge the financial support of the Deutsche Forschungsgemeinschaft (DFG) within the Collaborative Research Center (SFB) 370 "Integral Materials Modeling".

References

1 J. Müller, M. Schierling, E. Zimmermann, D. Neuschütz, *Surf. Coat. Technol.* 120–121, 1999, 16.
2 U. Eritt, G. von Hayn, E. Lugscheider, J. Müller, D. Neuschütz, *Mat.-wiss. u. Werkstofftech.* 33, 2002, 45.
3 J. Müller, Abscheidekinetik und Transporteigenschaften von CVD-Schichten aus α-Al_2O_3 als Diffusionsbarriere auf Nickelbasislegierungen, Ph.D. thesis, RWTH Aachen University, Shaker Verlag, Aachen, 2004.
4 J. Müller, D. Neuschütz, *Vacuum* 71, 2003, 247.
5 N. Bahlawane, S. Blittersdorf, K. Kohse-Höinghaus, B. Atakan, J. Müller, *J. Electrochem. Soc.* 151, 2004, C182.
6 C. Colombier, J. Peng, H. Altena, B. Lux, *Int. J. Refract. Hard Met.* 5, 1986, 82.

7 PHOENICS-CVD Ver. 3.4, CHAM, Concentration, Heat and Momentum Ltd, Wimbledon Village, London, UK, 2001.
8 J. Müller, D. Neuschütz, in *Chemical Vapor Deposition XVI and EUROCVD 14*, ed. M.D. Allendorf, F. Maury, F. Teyssandier, Electrochemical Society, Pennington, NJ, 2003, p. 186.
9 M. Schierling, E. Zimmermann, D. Neuschütz, *J. Phys. IV* 9, 1999, Pr8-85.
10 S.F. Nitodas, S.V. Sotirchos, *J. Electrochem. Soc.* 149, 2002, C130.
11 P. Tan, J. Müller, D. Neuschütz, *J. Electrochem. Soc.* 152, 2005, C288.
12 L. Catoire, M.T. Swihart, *J. Electrochem. Soc.* 149, 2002, C261.
13 C.R. Kleijn, K.J. Kuijlaars, *PHOENICS J. Computat. Fluid Dynamics Appl.* 8, 1995, 404.
14 C.R. Wilke, *Chem. Eng. Prog.* 46, 1950, 95.
15 P. Sigmund, *Phys. Rev.* 184, 1969, 383.
16 P. Sigmund, A. Olivia, G. Falcone, *Nucl. Instrum. Methods* 194, 1982, 541.
17 M. W. Thompson, *Philos. Mag.* 18, 1968, 377.
18 D. G. Coronell, E. W. Egan, G. Hamilton, A. Jain, R. Venkantraman, B. Weitzman, *Thin Solid Films* 333, 1998, 77.
19 P. K. Haff, Z. E. Switkowski, *Appl. Phys. Lett.* 29, 1976, 549.
20 G. W. Reynolds, F. R. Vozzo, R. G. Allas, P. A. Treado, J. M. Lambert, *Nucl. Instrum. Methods Phys. Res. B* 2, 1984, 804.
21 J. Herec, J. Sielanko, J. Wronski, *Vacuum* 63, 2001, 507.
22 K. Macak, P. Macak, U. Helmersson, *Comput. Phys. Commun.* 120, 1999, 238.
23 D. R. Gillen, W. G. Graham, A. Goelich, *Phys. Rev.* 194, 2002, 409.
24 SPUTY database, National Institute of Fusion Science (NIFS), Japan.
25 J. F. Ziegler, W. Eckstein, SRIM 2003, www.srim.org.
26 O. Knotek, E. Lugscheider, K. Bobzin, M. Maes, A. Krämer, in *Adhesion Aspects of Thin Films*, Vol. 2, ed. K.L. Mittal, VSP, Leiden, The Netherlands, 2005, p. 1.
27 E. Lugscheider, K. Bobzin, N. Papenfuß-Janzen, D. Parkot, *Surf. Coat. Technol.* 200, 2005, 913.
28 U. Eritt, Ph.D. thesis, RWTH Aachen University, Verlag Shaker, Werkstoffwissenschaftliche Schriftenreihe, Bd. 40, 2000.
29 N. Papenfuß-Janzen, E. Lugscheider, in *Proceedings of the Business and Industry Symposium, Simulation Series*, Vol. 34, No. 4, ed. M. Ades, L. Deschaine, Society for Modeling and Simulation International, 2002, p. 137.
30 E. Lugscheider, K. Bobzin, R. Nickel, in *2005 Business and Industry Symposium*, ed. M. Ades, T. Hang, Society for Modeling and Simulation International, 2005, p. 13.
31 E. Lugscheider, N. Papenfuß-Janzen, in *International Thermal Spray Conference*, ed. E. Lugscheider, C.C. Berndt, DVS Deutscher Verband für Schweißen, 1061, 2002, p. 42.
32 E. Lugscheider, R. Nickel, *Surf. Coat. Technol.* 174–175, 2003, 475.
33 E. Lugscheider, A. Fischer, D. Koch, N. Papenfuß, in *Thermal Spray: New Surfaces for a New Millennium*, ed. C.C. Berndt et al., ASM International, 1381, 2001, p. 751.
34 S. Kundas, T. Kashko, E. Lugscheider, G. von Hayn, A. Ilyuschenko, in Proceedings of the 15th International Plansee Seminar, Vol. 3, 2001, p. 360.
35 S. Kundas, T. Kashko, V. Hurevich, E. Lugscheider, N. Papenfuß-Janzen, in *International Thermal Spray Conference*, ed. E. Lugscheider, C.C. Berndt, DVS Deutscher Verband für Schweißen, 1061, 2002, p. 765.
36 E. Lugscheider, N. Papenfuß-Janzen, R. Nickel, *J. Phys. IV France* 120, 2004, 373.
37 E. Lugscheider, U. Eritt, V. Hurevich, S. Kundas, A. Kuzmenkov, *Proc. Natl Acad. Sci. Belarus, Ser. Phys.-Techn. Sci.* No. 1, 2000, 134.
38 E. Lugscheider, R. Nickel, T. Kashko, S. Kundas, in *2003 Business and Industry Symposium*, ed. M. Ades, L. Deschaine, Society for Modeling and Simulation International, 2003, p. 117.
39 E. Lugscheider, R. Nickel, S. Kundas, T. Kashko, in *Thermal Spray 2004:*

Advances in Technology and Application, ASM International, 1129, 2004, p. 311.

40 T. Kashko, S. Kundas, E. Lugscheider, R. Nickel, in *Proceedings of the 4th International Conference on Plasma Physics & Plasma Technology*, Vol. II, National Academy of Sciences of Belarus, Minsk, 2003, p. 475.

41 T. Kashko, S. Kundas, E. Lugscheider, R. Nickel, *Proc. Natl Acad. Sci. Belarus, Ser. Phys.-Techn. Sci.* No. 2, 2004, 69.

42 R. McPherson, *Thin Solid Films*, 83, 1981, 297.

43 E. Lugscheider, C. Barimani, P. Eckert, U. Eritt, *Comput. Mater. Sci.* 7, 1996, 109.

44 A. Dostanko, S. Kundas, S. Bordusov, M. Bosjakov, I. Svadkovsky, A. Illyuschenko, L. Anufriev, E. Lugscheider, in *Plasma Processes in Electronics Production*, Vol. 1, ed. A. Dostanko, FUAinform, Minsk, 2000.

45 E. Lugscheider, K. Bobzin, R. Nickel, T. Kashko, Advanced homogenization strategies in material modeling of thermally sprayed TBCs, Euromat 2005, 5–8 September 2005, Prague, Czech Republic (paper invited for publication in a special edition of *Advanced Engineering Materials* by the editor, paper in review).

10
Hot and Cold Rolling of Aluminum Sheet (TP B1)

L. Neumann, R. Kopp, G. Hirt, M. Crumbach, C. Schäfer,
V. Mohles, and G. Gottstein

Abstract

The objective of the study was the complete through-process modeling of crystallographic texture evolution during aluminum sheet production. The process was modeled from the hot rolling process through the final annealing of the samples by applying various models. The deformation texture (GIA) model was coupled to a statistical analytical recrystallization texture (StaRT) model. The nucleation spectra for recrystallization were modeled with ReNuc, a collection of several submodels describing different nucleation mechanisms. A finite element model incorporating dislocation density-based work hardening as well as texture served as a process model to describe the macroscopic production parameters based on microstructural information. The behavior so modeled was validated by comparison with experimental data. This comparison has confirmed the good quality of prediction of the models used.

10.1
Introduction

The crystallographic texture resulting from the thermomechanical processing of Al sheet plays a central role for subsequent deep-drawing operations of the sheet. The sheet texture will primarily induce anisotropic flow behavior which – in industrial processing – can be a cause for significantly increased scrapping. In the following, "T-Pack", a modeling tool for hot and cold forming, is presented. In addition to the process model, microstructural information, i.e. material texture and hardness, expressed in terms of dislocation densities, is input to the simulation. The anisotropic flow behavior during the finite element simulation of the forming step is predicted on the basis of microstructural information and the update thereof using texture and dislocation density-based flow stress models [1, 2]. In through-process simulations T-Pack delivers the microstructural input speci-

Integral Materials Modeling: Towards Physics-Based Through-Process Models
Edited by Günter Gottstein
Copyright © 2007 WILEY-VCH Verlag GmbH & Co. KGaA, Weinheim
ISBN: 978-3-527-31711-0

fied above which is necessary for the postprocessing by the grain interaction texture (GIA) model, statistical analytical recrystallization texture (StaRT), and hence for the prediction of texture-induced anisotropic flow behavior. Details have been published elsewhere [3].

10.2
Hot and Cold Rolling of Aluminum

First, the through-process modeling aspect of the forming simulations that follow the homogenization simulation is considered. Following this, the forming simulations themselves as well as the simulations between forming steps will be discussed.

Material forming operations like rolling are conventionally modeled in a continuum mechanical framework. The material characteristics are only considered in an integral manner. In recent years, new crystal plasticity models have been developed that describe the material behavior down to the atomistic scale. In the SFB project [1–3] the interdependencies between the levels are demonstrated, and integration has been accomplished. Since the material history itself can affect the plastic material behavior, the continuous tracing of the dislocation structure – and other microstructural variables, which describe the material response in a given state – is required through all process stages.

Thermomechanical models and material models can be linked in successive, decoupled, or fully coupled mode [4]. In a successive simulation a model would just receive its input from previous simulations but otherwise run independently. A typical example is the transfer of work hardening information to an annealing model. Nevertheless the model crucially depends on the quality of the input. For instance, since recrystallization which is modeled during annealing has no impact on the thermal model, a decoupled mode is sufficient here, but the full thermal history needs to be provided locally. A flow stress model, however, interacts with the forming model and therefore requires full integration. The models that are applied and coupled in this study are listed in Table 10.1.

Coupling is accomplished between all models via standardized interfaces, which enable the exchange of data on pretreated materials and allow the comparative application of different model combinations to the same processes or material state. In the following the fundamentals of each model and the coupling strategy applied are given.

10.2.1
Three-Internal-Variable Model (3IVM)

3IVM is a statistical model which describes the hardening and recovery behavior of metals in terms of three state variables that represent the microstructure. It was originally developed by Roters et al. [5] as a refinement of the Kocks–Mecking model that is based on only one state variable, the dislocation density. 3IVM ac-

Table 10.1 List of models used in this study.

Model	Phenomena covered by the model
3IVM	Flow stress description based on three dislocation density populations taking into account hardening and recovery.
FC-Taylor	Full constraints deformation texture model based on assumptions by Taylor used primarily for calculation of Taylor factor.
GIA	Relaxed constraints deformation texture model regarding eight grain aggregates with intergranular interactions while deformation of the aggregate as a whole is prescribed as in the full constraints Taylor model.
StaRT	Statistical recrystallization texture model.

counts for cellular dislocation substructures for materials with high stacking fault energy. This is the case for Al and most of its alloys. 3IVM is based on the densities ρ of three dislocation populations as microstructural state variables: the density of immobile dislocations stored in the cell interiors (ρ_i) and in the walls (ρ_w) as well as of the mobile dislocations (ρ_m), which accomplish the plastic flow. For these variables, evolution equations $d\rho_i/dt$, $d\rho_w/dt$, and $d\rho_m/dt$ have been formulated which account for dislocation interactions and reactions such as dipole and lock formation, instantaneous annihilation, or recovery by climb-controlled annihilation. In total this results in a system of three variables and three coupled differential equations, which 3IVM solves numerically in order to track the microstructural evolution. The flow stress is finally calculated from these dislocation densities, weighted by the volume fractions of cell interior and walls, complemented by the hardening contributions of solute atoms (Mg and Mn) and precipitates. Full model descriptions are published elsewhere [6–8].

10.2.2 Full Constraints Taylor Texture Model

Among a multitude of approaches to texture modeling developed in the past, the model developed by Taylor is well established and remains a reference [9]. The model's most important restriction is the enforced strain compatibility, meaning that each single grain undergoes the same distortion as the macroscopic sample. Models based on this assumption are referred to as "full constraints" models. Five independent slip systems are required to accomplish the deformation in each grain. The Taylor model chooses these systems from the ones available by minimizing the dissipation energy for this deformation. Hence it is possible to calculate the orientation change for a given macroscopic deformation and, therefore, the texture evolution of a polycrystal throughout the whole deformation process. The relation of the macroscopic stress to the resulting shear stress in the slip system is described by the Taylor factor. During a deformation simulation, the average Taylor factor is computed and transferred to the flow stress model 3IVM. Within the through-process model presented here, the full constraints Taylor model was found to be sufficient for a Taylor factor calculation for each grain. The exact texture evolution is calculated afterwards in postprocessing using the more accurate model GIA.

10.2.3 Grain Interaction Texture (GIA) Model

As texture itself is a key property of the finished sheet metal, the use of the full constraints Taylor model is not sufficient for an accurate deformation texture simulation. Therefore, a relaxed constraints texture model was developed that takes grain interaction into account [10, 11] and is referred to as GIA. The relaxed constraints are realized by considering aggregates of eight grains which – as a whole – must undergo the same distortion as the macroscopic sample. However, for each single grain of the aggregate, the components of the strain rate tensor are relaxed resulting in independent deformation of the single grains according to their kinematic hardness. A grain's kinematic hardness is reflected by the Taylor factor resulting from the grain orientation and the imposed deformation. Strain incompatibilities that result from the independent deformation of the aggregate's grains are compensated by the introduction of geometrically necessary dislocations. The associated energy of the geometrically necessary dislocations and the Taylor energies necessary for slip on the active slip systems of an eight grains aggregate add up to the total energy for the deformation of the cluster. Minimization of the total energy delivers the extent of relaxation, the active slip systems, and the corresponding amount of slip. This renders the resulting orientation change of each grain.

10.2.4 Recrystallization Nucleation (ReNuc)

The prediction of recrystallization textures requires knowledge of the textures of new nuclei. Nucleation of new grains is very difficult to investigate experimentally, but there is evidence [12–14, 19] that in Al alloys nucleation occurs by several distinct mechanisms which are related to specific features of the deformed microstructure. Such mechanisms in Al alloys are nucleation at shear bands, transition and Cube bands, strain-induced grain boundary migration, or particle-stimulated nucleation (PSN). The ReNuc model is a collection of several submodels describing the different nucleation mechanisms [17–19]. Three types of mechanisms are considered in the current study: random nucleation, grain boundary nucleation, and transition band nucleation. The frequency distribution of the nuclei resulting from the considered mechanisms can be deduced from special criteria applied to the output of the GIA model.

10.2.5
Statistical Recrystallization Texture Model (StaRT)

The production of Al sheet does not only include forming steps but also periods at elevated temperatures where primary static recrystallization may occur. For the simulation of the latter, the StaRT model was developed [15, 16]. This model predicts texture evolution during annealing after deformation and hence, represents an essential component of a through-process model. The input of StaRT is the deformation texture in form of a discrete set of orientations and their volume fractions. This information is provided by the ReNuc model. The model also considers usually the nucleation to be site saturated, i.e. all nuclei are generated at $t = 0$.

The growth rate of the nuclei is proportional to the grain boundary velocity which is the product of grain boundary mobility and the local driving force. The grain boundary mobility is strongly dependent on the misorientation between growing and deformed grain. The Taylor factor of the deformed matrix components is – in agreement with trends found in measurements – taken as a measure of the local driving force. Growth is assumed to be isotropic but impingement of nuclei will terminate the growth. The two extreme cases of possible growth scenarios are competitive and compromise growth, and the annealing texture development is affected by both.

10.2.6
Coupling GIA and 3IVM

The deformation texture model GIA and the strain hardening model 3IVM are coupled by alternately performing steps of each model and exchanging information on relevant variables. For the calculation of the Taylor energy in GIA the resolved shear stress is required. This is calculated with the 3IVM for every deformation increment in each grain of the GIA 8-grain aggregate. On the other hand, for solving the evolution equations in 3IVM, the slip rate $\dot{\gamma}$, the number of active slip systems N_{GLS}, and the temperature T are required. These data are provided by the GIA output of the previous time step. In that way, an incremental coupling is established between GIA and 3IVM on a grain scale [19].

10.2.7
Through-Process Model (TPM)

The through-process modeling scheme used for this study is realized as follows. For modeling of the forming stages, the finite element (FE) code Larstran/Shape [21] was used. The FE code calls the full constraints Taylor texture model in order to incrementally update the texture in an element if the accumulated plastic strain in that element has reached a user-defined value (typically $\varepsilon = 0.02$). The 3IVM is called up iteratively by the FE code. 3IVM delivers the flow stress of an element while relying on the texture information represented by the Taylor factor

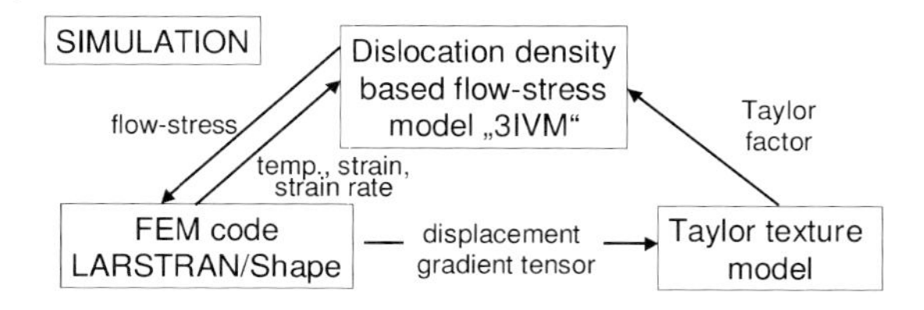

Fig. 10.1 Model interaction during simulation of forming processes with a FE-package.

of that element, the state variables, i.e. the dislocation density populations, and the values of temperature, strain, and strain rate. After the mechanical equilibrium has been reached in one increment, the Taylor factor and the dislocation densities are stored for this element. This model interaction is shown in Fig. 10.1.

Since texture itself is also of interest in through-process simulation (not only the influence of texture on flow stress evolution), the displacement gradient is traced for specific elements at different locations in the mesh of the FE simulation of the rolling process. For these elements, the texture can then be recalculated accurately with the GIA deformation texture model. A direct implementation of GIA into the FE simulation is currently not feasible due to GIA's high computational intensity, but also not necessary for calculation of the Taylor factor. The texture prediction delivered by GIA at the end of a rolling pass is then used as input for the primary static recrystallization model StaRT (see Fig. 10.2) [15, 16]. Along with the predicted temperature change during the inter-stand time (or thermal treatment, such as annealing), StaRT delivers a texture prediction that is then used as input for the FE simulation of the next rolling pass.

This through-process modeling strategy can be successfully applied to the process chain of Al sheet as shown in the following section.

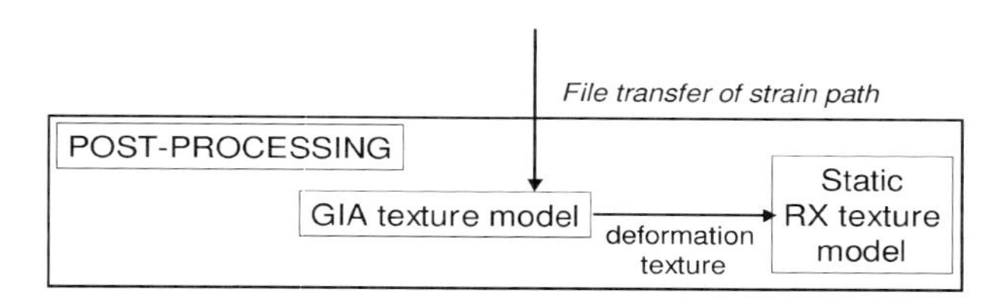

Fig. 10.2 Model interaction of computationally more intensive models for precise texture prediction.

10.2.8
Texture Predictions

As outlined before, the process chain is subdivided into hot rolling, cold rolling, and the intermediate heat treatments. For the simulation of the rolling process the FEM package Larstran/Shape was applied. After the FE simulation of each rolling pass the deformation textures were calculated with GIA-3IVM in post-processing, and the recrystallization textures were simulated with the ReNuc and StaRT models considering the deformation conditions and the microstructure data. The relevant information on the deformation conditions and the displacement gradient tensor were given as "path line" files by T-Pack. The process chain was modeled for two alloys to investigate the influence of chemical composition on the texture development. In the following the simulation results obtained for the commercial alloy AA3104 (1.1% Mg, 0.8% Mn, 0.5% Fe, 0.2% Si, 0.2% Cu, Cr, and Ti contents $\ll 0.1\%$) are compared with experimental data from x-ray texture measurements.

The material was homogenized for 24 h at $T = 500\ ^{\circ}\mathrm{C}$. Afterwards the material was hot rolled in three hot rolling steps at a temperature of about $T = 380\ ^{\circ}\mathrm{C}$. Between the passes the material was reheated which caused static recrystallization. The StaRT parameters used standard values, which can be found in [18]. Each of the three hot rolling passes had a logarithmic strain of $\varphi = 0.3 \ldots 0.4$. Since no interpass samples could be taken only the experimental texture after hot rolling and coiling was available. Subsequently, the material was subjected to industrial cold rolling to a reduction of 74%. The cold rolled texture was measured in the center sheet layer. Finally, after annealing for 30 min at $T = 360\ ^{\circ}\mathrm{C}$ the annealing texture was measured again in the center layer of the sheet.

10.2.9 Hot Rolling

The initial texture of the as-cast and homogenized ingot was assumed to be random at any position in the cast ingot. The conditions between the hot rolling passes were only simulated because no measured data was available. The rolling texture evolution was simulated for an element in the center plane of the sheet with the GIA model subsequent to FEM simulation. It was assumed that between the three hot rolling passes complete static recrystallization took place, which was modeled with StaRT by using standard parameters [18, 20]. After the third hot rolling pass the material was cooled down to room temperature in air at ambient conditions. Therefore, it was assumed that only recovery took place, which was taken into account by carrying out a static recovery process simulation with 3IVM. The simulated and the measured hot strip texture in the center layer are shown in Fig. 10.3.

10.2.10 Cold Rolling

Starting with the simulated hot strip texture, subsequent cold rolling was simulated for the sheet center plane. The resulting cold band texture is shown in Fig. 10.4.

Finally, the texture evolution during annealing of this cold band was addressed. The GIA and ReNuc models provided the nucleation spectra and stored energies

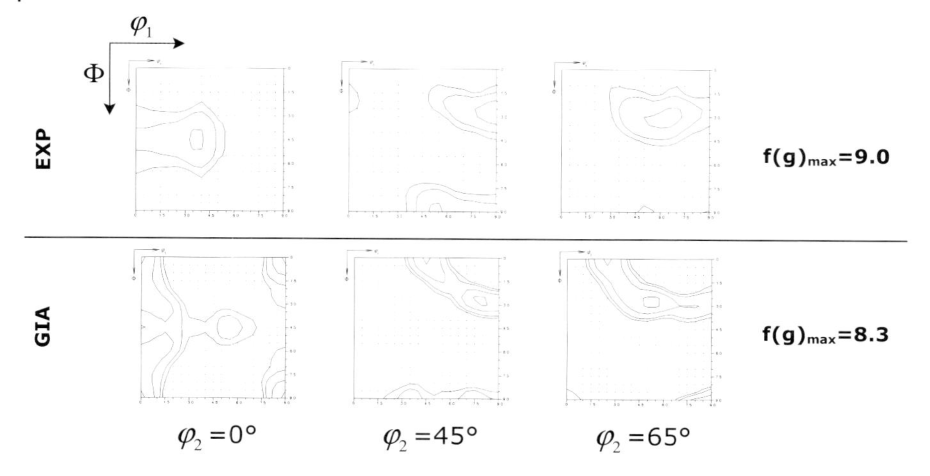

Fig. 10.3 ODF of (a) calculated and (b) measured deformation texture in the center layer of the commercial alloy after the third hot rolling pass. Contour levels: 1.2/2.0/4.0/7.0/12.0.

for the StaRT model. The fractions of the nuclei from different nucleation mechanisms in the ReNuc model were 70% random nucleation, 18% grain boundary nucleation and 12% transition band nucleation. Figure 10.5 shows the textures after final annealing treatment.

The simulated and measured through-process texture evolution is given in Fig. 10.6 and reveal good agreement for cold rolling and annealing.

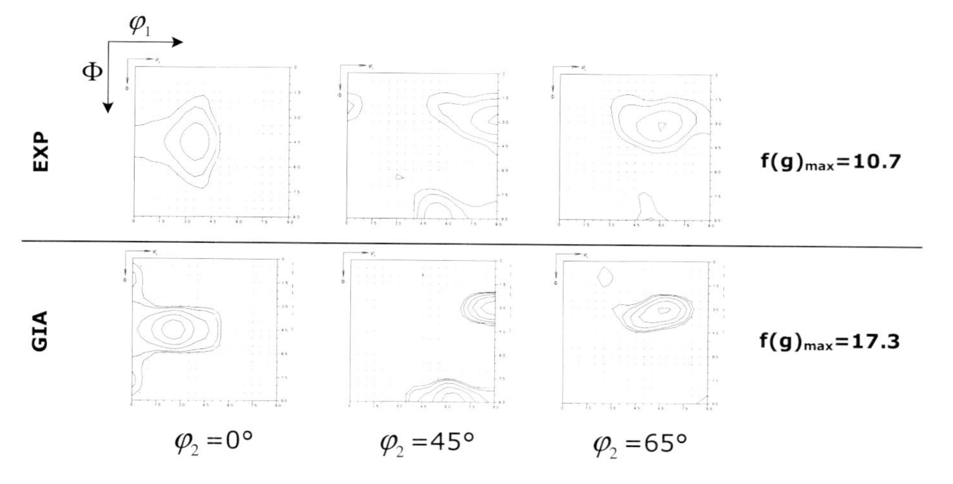

Fig. 10.4 ODF of (a) calculated and (b) measured deformation texture of the center layer of the commercial alloy after cold rolling. Contour levels: 1.2/2.0/4.0/7.0/12.0.

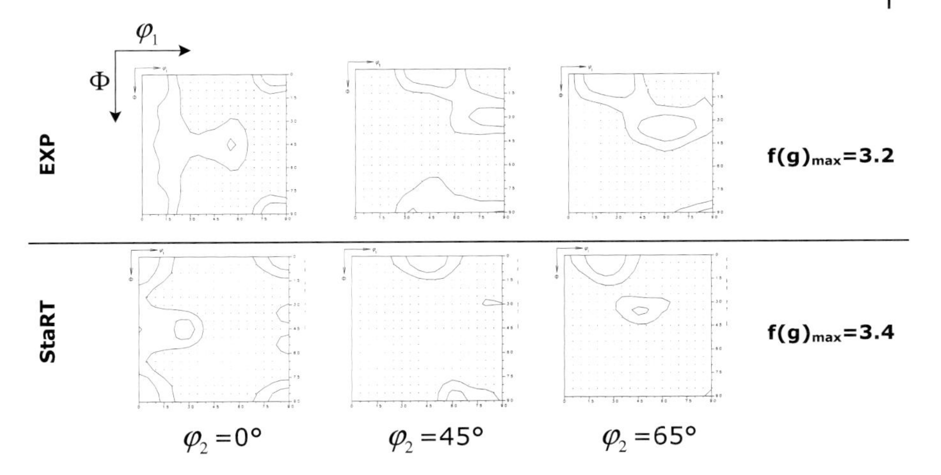

Fig. 10.5 ODFs of the (a) calculated and the (b) experimental measured recrystallization texture in the center layer of the commercial alloy after the recrystallization treatment. Contour levels: 1.2/2.0/4.0/7.0/12.0.

The final results of the through-process modeling exercise show good agreement with the experimental data. Only the experimentally observed P-orientation was not predicted. This is probably due to the neglecting of PSN. Also the intermediate steps show reasonable results even if some differences remain. Partly this can be attributed to the chosen initial texture, which was assumed to be random.

10.3 Database "StoRaDat" and Interfaces

During the previous phase of the program (2003–2005), the database "StoRaDat" was developed to offer an insight to active members and the broad public into the work of SFB 370. Tables that contain various data are stored as a database on a MySQL server at the Center for Computing and Communication of RWTH Aachen University. They can be modified through the Internet. Within the StoRaDat database, information concerning the subprojects is available. Furthermore, the subprojects are grouped within three main groups of subprojects (A "Transition to the Solid State", B "Thermomechanical Treatment", and C "Component and Material Properties"). Each subproject includes a list of simulations carried out therein, the participating institutes and staff members, as well as input and output data and the corresponding dependencies. Input and output data are variables necessary to carry out a specific simulation or resulting from the latter. The database is accessed via hypertext pre-processor files (php) ensuring that access to the database itself, which lies elsewhere, on the database server, is not possible for unauthorized users. The database is always online at http://

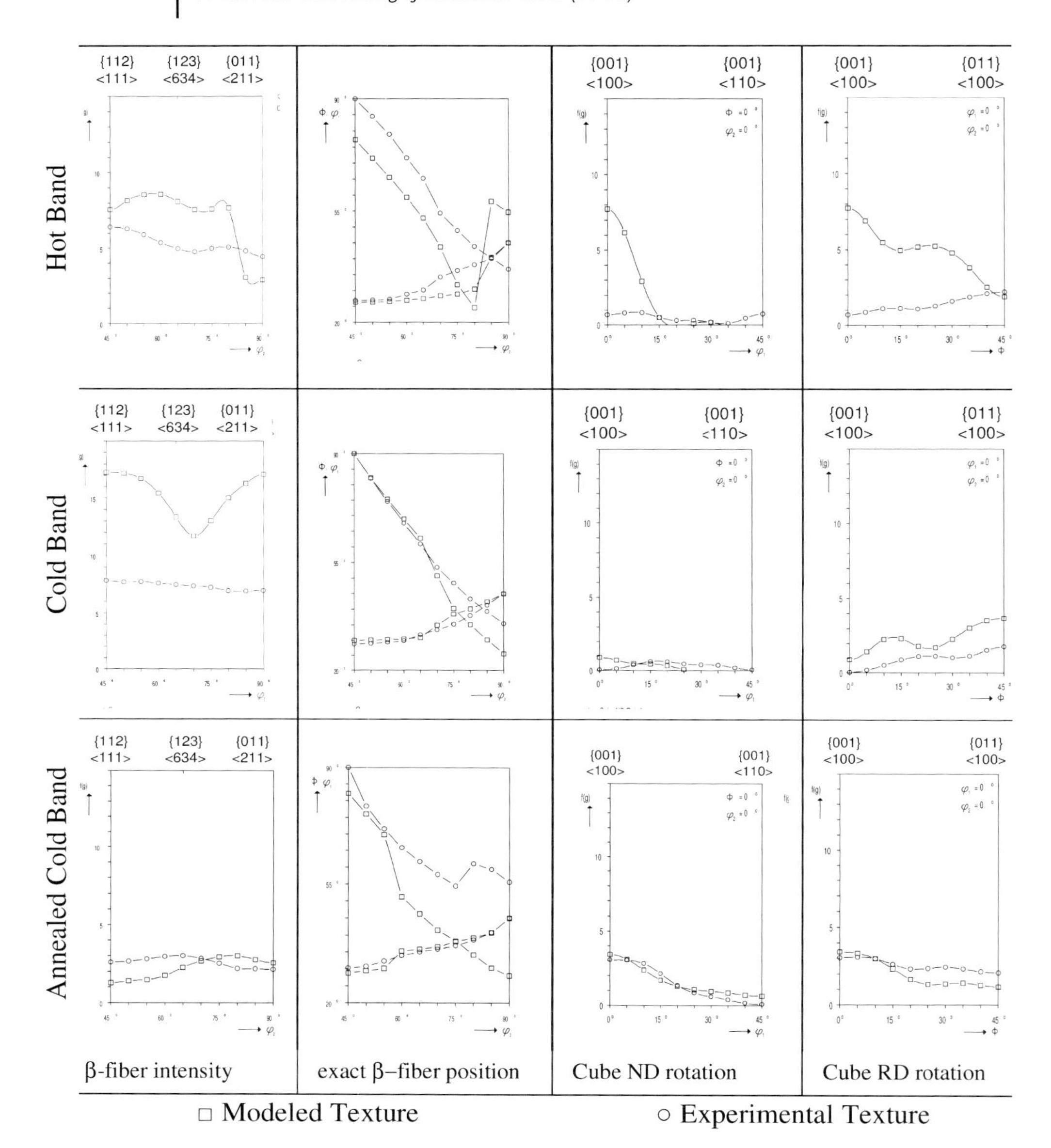

Fig. 10.6 Through-process texture simulation of the commercial alloy from hot rolling to cold rolling and final annealing.

www.ibf.rwth-aachen.de/storadat since the php files are stored on a web server. Datasets can be created, modified, and deleted after the authorization to do so has been verified by login. This also prevents unauthorized calling up of protected php files. The structure of StoRaDat is based on linkage of tables as shown

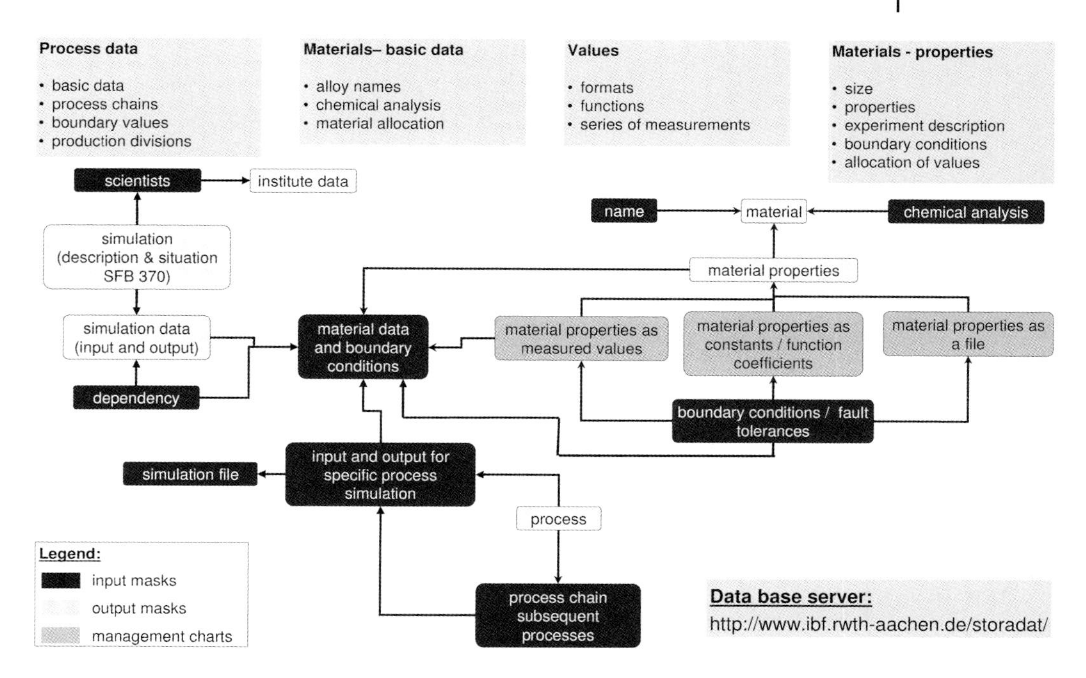

Fig. 10.7 Structure of the StoRaDat database; arrows indicate the direction of linkage.

in Fig. 10.7. Linkage – as indicated by the arrows – is realized with foreign keys. Furthermore, the tables are classified as output masks, input masks, or administration tables. Output masks transfer their data to other tables while input masks transfer them to php files and thus make them accessible via the Internet. Administration tables sort material properties into three categories before they are available for other tables.

Input of processes and corresponding results is realized by storing the link to the file on the database server. The files themselves are stored on a file server of the Center for Computing and Communication. Processes are different working steps within one simulation. Multiple working steps make up a process chain. The latter can be stored as a defined sequence of working steps in StoRaDat. Process chains can be modified later, if necessary. The results can be stored and accessed only by members of the project since they are mainly of the members' interest. General material data were also added to StoRaDat, making it possible to choose a material and display the corresponding material data.

New interfaces that are accessible to both the members of the project as well as the general public were added to StoRaDat. Within the interfaces, details of the corresponding simulations as well as their input and output data and the dependencies of the latter are stored. The structure of the Collaborative Research Center was modified in 2003, organizing it according to the processing routes of selected components. This change was integrated into StoRaDat.

10.4 Conclusions and Outlook

A through-process model was applied to hot and cold rolled aluminum sheet. The model was comprised of:

- the FE model Larstran/Shape,
- the GIA model for deformation texture development coupled with the hardening and recovery model 3IVM,
- the ReNuc model for recrystallization nucleation,
- the StaRT model for statistical modeling of recrystallization texture evolution.

The models were applied to hot and cold rolling and final annealing of an Al–1.1% Mg–0.8% Mn alloy and compared to corresponding experiments, to validate the prediction quality. The individual models (GIA, 3IVM, ReNuc, StaRT) were combined to a complete through-process texture model to simulate the texture for a commercial and a model alloy of similar composition throughout various consecutive processing steps.

The texture predictions are quite good albeit not perfect. It should be noted that the abundance of possible nucleation mechanisms in aluminum alloys still causes much room for uncertainties. Therefore, more precise recrystallization nucleation models need to be tested in future investigations.

Acknowledgments

The authors gratefully acknowledge the financial support of the Deutsche Forschungsgemeinschaft (DFG) within the Collaborative Research Center (SFB) 370 "Integral Materials Modeling".

The generous material donation by Hydro Aluminium Deutschland GmbH was of invaluable help for the success of the presented research. Furthermore, the authors thank Dr. H. Aretz, M. Schneider, M. Goerdeler, J. van Santen, M. Nutzmann, Ch. Wiedner, and A. Meyer for assistance and helpful discussions.

References

1 L. Neumann, H. Aretz, R. Kopp, M. Crumbach, M. Goerdeler, G. Gottstein, *AIP Conf. Proc.*, 712, 2004, 388.

2 L. Neumann, R. Kopp, M. Crumbach, M. Goerdeler, G. Gottstein, Proc. 2nd Int. Conf. on Multiscale Materials Modeling, Los Angeles, CA, p. 352.

3 L. Neumann, R. Kopp, A. Ludwig, M. Wu, A. Bührig-Polaczek, M. Schneider, M. Crumbach, G. Gottstein, *Modelling Simul. Mater. Sci. Eng.* 12, 2004, 19.

4 K.F. Karhausen, L. Neumann, R. Kopp, M. Goerdeler, G. Gottstein, *Aluminium* 80, 2004, 690.

5 F. Roters, D. Raabe, G. Gottstein, *Acta Mater.* 48, 2000, 4181.

6 M. Goerdeler, G. Gottstein, in *Recrystallization and Grain Growth*, ed.

G. Gottstein, D.A. Molodov, Springer-Verlag, Berlin, 2001, p. 987.
7 M. Goerdeler, M. Crumbach and G. Gottstein, *Mater. Sci. Forum* 408–412, 2002, 881.
8 M. Goerdeler, M. Crumbach, M. Schneider, G. Gottstein, L. Neumann, H. Aretz, R. Kopp, *Mater. Sci. Eng.* A387–389, 2004, 266.
9 G.I. Taylor, *J. Inst. Met.* 62, 1938, 307.
10 M. Crumbach, G. Pomana, P. Wagner, G. Gottstein, in *Recrystallization and Grain Growth*, ed. G. Gottstein, D.A. Molodov, Springer-Verlag, Berlin, 2001, p. 1053.
11 M. Crumbach, M. Goerdeler, G. Gottstein, L. Neumann, H. Aretz, R. Kopp, L. Löchte, 9th Int. Conf. on Aluminium Alloys, ed. J.F. Nie, A.J. Morton, B.C. Muddle, Institute of Materials Engineer-ing Australasia Ltd (CD), 2004, p. 684.
12 I.L. Dillamore, H. Katoh, *Metal Sci.* 8, 1974, 73.
13 G. Gottstein, Rekristallisation Metallischer Werkstoffe, DGM, 1984.
14 A. Duckham, R.D. Knutsen, O. Engler, *Acta Mater.* 50, 2002, 2881.
15 R. Sebald, G. Gottstein, *Acta Mater.* 50, 2002, 1587.
16 R. Sebald, G. Zhu, G. Gottstein, in *Recrystallization and Grain Growth*, ed. G. Gottstein, D.A. Molodov, Springer-Verlag, Berlin, 2001, p. 1027.
17 M. Crumbach, M. Goerdeler, G. Gottstein, *Mater. Sci. Forum* 408–412, 2002, 425.
18 M. Crumbach, M. Goerdeler, G. Gottstein, L. Neumann, H. Aretz, and R. Kopp, *Modelling. Simul. Mater. Sci. Eng.* 12, 2004, 1.
19 M. Crumbach, doctoral thesis, RWTH Aachen, 2006.
20 A. Abdurahman, diploma thesis, RWTH Aachen, IMM, 2005.
21 Larstran/Shape User's Manual, Revision F, LASSO Ingenieurgesellschaft mbH, Leinfelden-Echterdingen, 1994.

11
Modeling of the Hot Rolling Process of a C45 Steel (TP B1)

X. Li, R. Kopp, G. Hirt, B. Zeislmair, and W. Bleck

Abstract

In this study, models describing the precipitation kinetics during the hot rolling process for steel production were integrated in the FEM code LARSTRAN/SHAPE. Using the experiments "casting and rolling in one heat" the industrial hot rolling conditions for steel production were realized on a laboratory scale. The experiments were carried out on the construction steel C45. To specify the microstructural evolution during "casting and rolling in one heat", models describing the grain size development due to recrystallization and grain growth were set up. The hot rolling process was simulated with the integrated microstructure models. The experimental results and modeled results showed good agreement with respect to rolling force and grain size.

11.1
Introduction

One aim of the Collaborative Research Center (SFB) 370 is to achieve a through-process modeling of a material for process development and property control in the production line. It was the aim of this study to improve the accuracy of the modeling of steel hot rolling by better material characterization, so that in the long run the process layout can be more easily adjusted to obtain the desired product properties.

The typical processing steps for the production of a steel profile, which consists of casting, hot rolling, thermal treatment, and welding or bending, were taken into account in this study. The possible microstructure evolution during the hot rolling process for the production of a steel profile was considered including phase transformation, precipitation, work hardening, as well as softening effects due to recrystallization and grain growth, in order to enable the prediction of the mechanical properties of the steel profile at the end of the production.

Integral Materials Modeling: Towards Physics-Based Through-Process Models
Edited by Günter Gottstein
Copyright © 2007 WILEY-VCH Verlag GmbH & Co. KGaA, Weinheim
ISBN: 978-3-527-31711-0

Within this project phenomenological models describing the microstructure development and material behavior are applied, since more physically based relations are not available for the investigated materials.

In previous work, phase transformation models were developed for the material S460 [1]. In the current study, precipitation models were first developed for the material S420M Nb and validated by experiments. The models were integrated into the finite element method (FEM) program LARSTRAN/SHAPE. To show the flexibility of the model and the easy adaptation to given conditions, the hot rolling of a new material, the medium carbon steel C45, was studied in this work.

11.2 Experimental Procedure

11.2.1 Casting and Rolling in One Heat

To study and model the material properties during the hot forming steps, the laboratory processing concept "casting and rolling in one heat" has been developed within the Collaborative Research Center (SFB) 370. In the current project, this concept was further developed and optimized. As shown in Fig. 11.1, the studied steel slabs (about 40 mm × 100 mm × 400 mm) were directly packed in a hot transfer box after melting and hot charging. After the transport, the slabs were

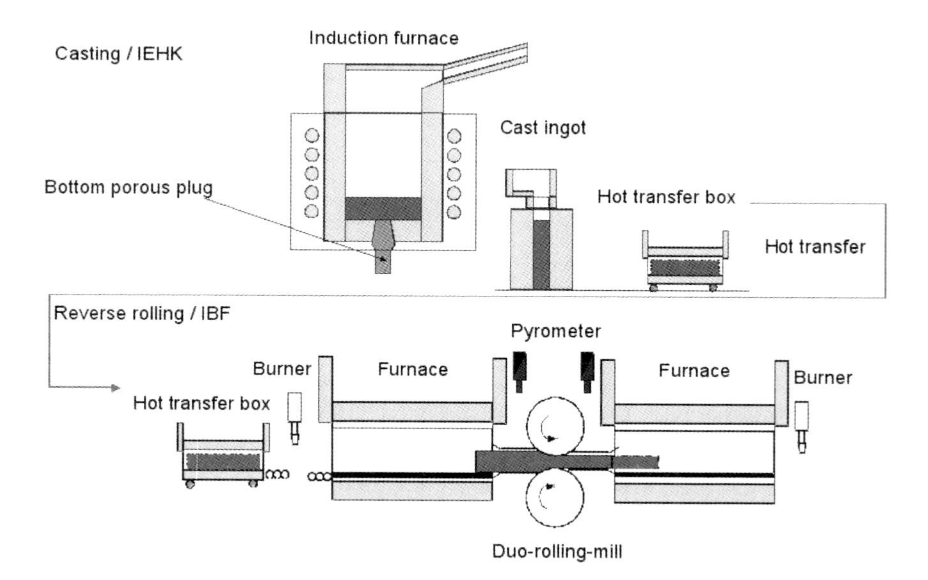

Fig. 11.1 Principle of casting and rolling in one heat.

Table 11.1 Hot rolling schedule for batch 33.

Pass	Entry thickness [mm]	Exit thickness [mm]	Time for thermal treatment [s]	Temperature at thermal treatment [°C]	Rolling velocity [mm s^{-1}]
1	41.96	27.78	190	1200	425
2	27.78	19.7	151	1200	430
3	19.7	14.28	637	1200	430
4	14.28	9.65	–	–	435

first handed over from the hot transfer box to an additional furnace. After a short homogenization treatment, the slabs were hot rolled down to a thickness of 2–3 mm according to process requirements. The heating furnaces in front of and behind the rolling mill enabled a controlled heating above A_3. All process parameters were monitored in the course of the process with a digital measuring system [3]. This experimental concept offered the possibility of a pilot-scale simulation of the industrial steel production steps, which were emulated by a sequence of single rolling passes, and the validation of the through-process simulation models.

Different rolling strategies were followed. As an example, Table 11.1 shows one of the hot rolling schedules, which is characterized by four consecutive rolling passes and three thermal treatments.

After the experiments a chemical analysis of the sheets was carried out (Table 11.2).

Table 11.2 Chemical composition (mass%) of the investigated steel C45.

Fe	C	P	S	Al	N	Cr	Cu	Mn	Mo	Ni	Si
Balance	0.47	0.015	0.028	0.054	0.011	0.22	0.22	0.74	0.03	0.19	0.32

11.2.2
Determination of Flow Curves

Hot upsetting tests were carried out for the steel C45 to determine the flow curves. Rastegaev cylindrical samples (10 mm diameter × 15 mm) were used. The tests were performed on a test machine with constant strain rates of 1 and 10 s^{-1}, which covered the strain rate field during the hot rolling process. The upsetting temperature ranged from 700 °C up to 1250 °C.

As an example, Fig. 11.2 shows the measured flow curves for a constant strain rate 10 s^{-1}. After the upsetting tests, the compressed samples were quenched

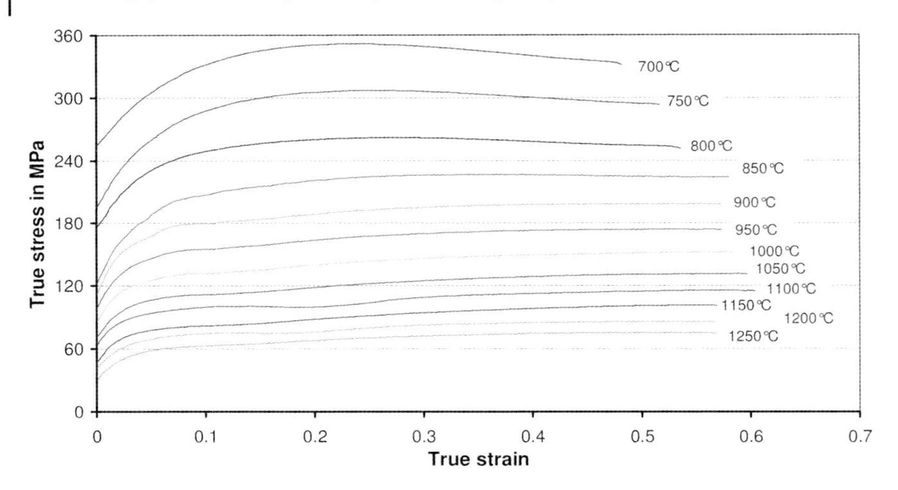

Fig. 11.2 Measured flow curves after the hot upsetting tests with constant strain rate of 10 s^{-1}.

immediately and later on studied metallographically. In this way, the dynamically recrystallized grain size was obtained.

11.3 Modeling of the Hot Rolling Process

11.3.1 Phenomenological Modeling of the Grain Size Development During Hot Rolling

Physics-based microstructure models were not available for the steels considered in this project. Therefore, the phenomenological models, as given in Table 11.3, were used in the present study. Based on the acquired flow curves and the metallographically measured grain size under certain forming conditions, the model parameters were determined according to Sellars [5–8] (see Table 11.3). The activation energy Q_{def} for hot forming was determined to be 363.881 kJ mol^{-1}.

The parameters in the static model, which describe the kinetics of static recrystallization, statically recrystallized grain size, and the grain size development due to grain growth (see Eqs. (11.1) to (11.3)), were derived from literature data [4].

$$t_{0.5} = 2.3 \times 10^{-15} \cdot \varepsilon^{-4} \cdot Z^{-0.18} \cdot d_0{}^2 \cdot \exp\left\{\frac{274000}{R \cdot T}\right\} \tag{11.1}$$

$$d_{SRX} = 87 \cdot Z^{-0.16} \cdot d_0{}^{0.48} \cdot \varepsilon^{-1} \tag{11.2}$$

$$d_{GG}{}^{10} = d_r{}^{10} + 5.3 \times 10^{32} \cdot t \cdot \exp\left\{\frac{400000}{R \cdot T}\right\} \tag{11.3}$$

Table 11.3 Dynamic model parameters for the studied material (see Appendix for meaning of symbols).

Equations	Parameters
$\varepsilon_{max} = a_1 \cdot {d_0}^{a_2} \cdot Z^{a_3}$ $\varepsilon_{crit} = a_4 \cdot \varepsilon_{max}$	$a_1 = 0.448$ $a_2 = 0$ $a_3 = -0.01$ $a_4 = 0.378$
$\varepsilon_{stat} = e_1 \cdot \varepsilon_{max} + e_2 \cdot {d_0}^{e_3} \cdot Z^{e_4}$	$e_1 = -0.072$ $e_2 = 1.333$ $e_3 = 0$ $e_4 = -0.0067$
$\frac{k_f}{k_{f\,max}} = \left[\frac{\varepsilon}{\varepsilon_{max}} \cdot \exp\left(1 - \frac{\varepsilon}{\varepsilon_{max}}\right)\right]^C$ with $C = C_1 \cdot [1 - \exp(C_2 \cdot (\ln Z)^{C_3})]$	$C_1 = -9089.29$ $C_2 = 0.812 \times 10^{-4}$ $C_3 = -0.395$
$X_{DRX} = 1 - \exp\left(d_1 \cdot \left(\frac{\varepsilon - \varepsilon_{crit}}{\varepsilon_{stat} - \varepsilon_{crit}}\right)^{d_2}\right)$	$d_1 = -4.81$ $d_2 = 1.892$
$\sinh(o_3 \cdot k_{f\,max}) = o_1 \cdot Z^{o_2}$	$o_1 = 0.4985 \times 10^{-2}$ $o_2 = 0.146$ $o_3 = 0.667 \times 10^{-2}$
$d_{DRX} = b_1 \cdot Z^{b_2}$	$b_1 = 252.14$ $b_2 = -0.083$

11.3.2
Modeling Precipitation During Hot Rolling

Under the hot rolling conditions, precipitation may take place, which has great influence on the material properties and, therefore, needs to be taken into account in the modeling of hot rolling.

Based on the work of Dutta and Sellars [9], new approaches for the phenomenological modeling of the relevant precipitation parameters as well as efficient ways for the experimental simulation of precipitation processes were developed [2]. As case studies, the precipitation start of niobium carbonitrides (Fig. 11.3a) and manganese sulfides (Fig. 11.3b) during hot rolling, the size distribution of iron carbides under variations of rolling and cooling parameters, and the size of manganese sulfide precipitates during solidification were simulated and validated experimentally. Furthermore, the experimental effort for the investigation of the precipitation processes was reduced by the use of creep tests [2].

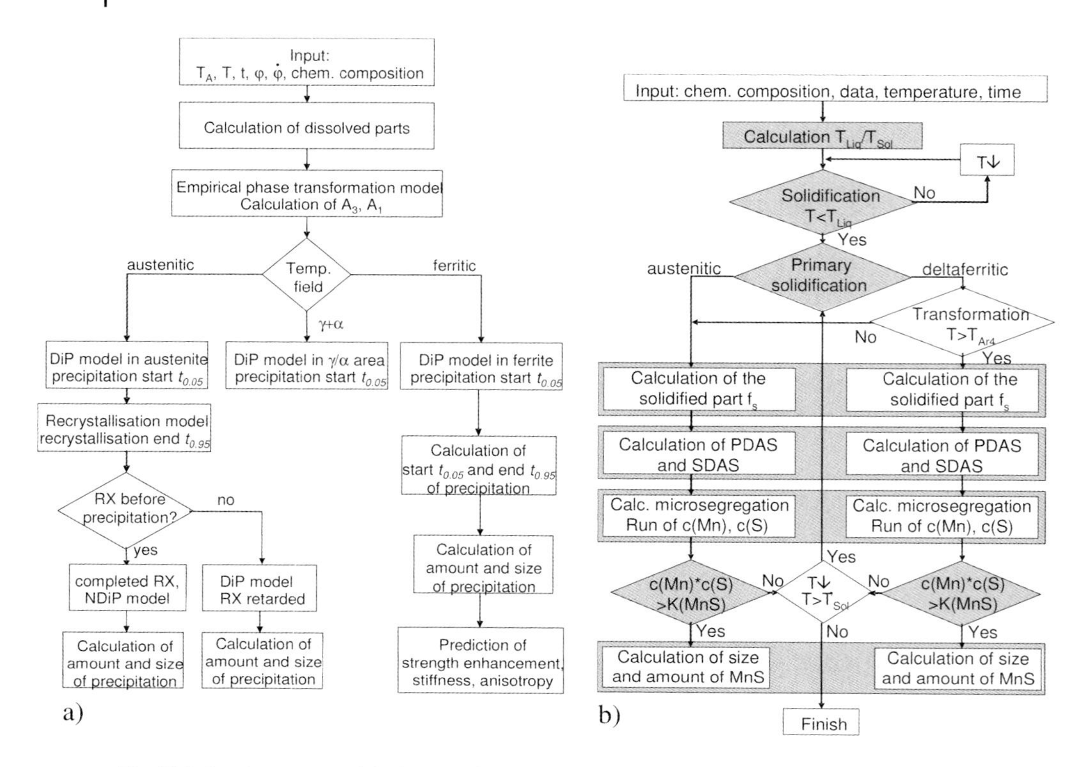

Fig. 11.3 Precipitation models for (a) NbC and (b) MnS [2].

To study the interaction between precipitation kinetics and recrystallization kinetics, the NbC precipitation model was implemented into the microstructure module STRUCSIM [10], which had been already integrated into the FE code LARSTRAN/SHAPE. The precipitation kinetics of MnS are not dependent on the forming conditions [2]. The flow chart in Fig. 11.4 shows the sequence of the microstructure simulation models that are integrated in LARSTRAN/SHAPE.

In previous work [2] it was shown that the occurrence of NbC precipitation depends strongly on the Nb concentration in the material. While the Nb concentration of the studied C45 is quite low, it can be assumed that NbC precipitation will not happen in the material during hot rolling. Thus, the effects of precipitation are not taken into account in the modeling of hot rolling of C45.

11.3.3
Microstructure Simulation of the Hot Rolling Process

A complete hot rolling process, consisting of four rolling passes and three thermal treatments, was simulated using the microstructure models integrated into

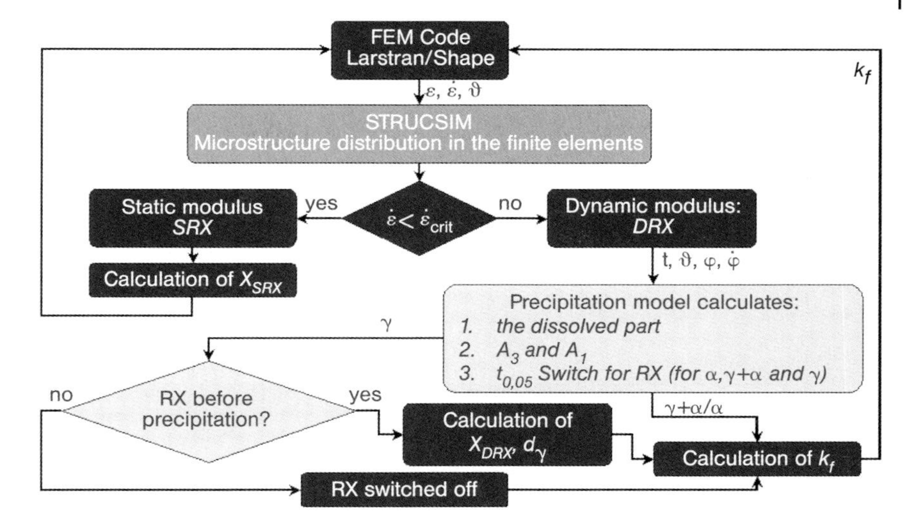

Fig. 11.4 Implementation of precipitation models in the microstructure modulus STRUCSIM.

LARSTRAN/SHAPE. In the FEM model (Fig. 11.5), the rolls were treated as rigid surfaces and a node at the upper right corner of the cross-section was chosen for the comparison with metallographic measurements.

The comparison of measured and simulated rolling forces is shown in Fig. 11.6. In the third rolling pass, there is a difference of about 10% between the simulated and measured forces. For the other three passes, the simulations agree well with the measurements.

Figure 11.7 shows the grain size development during the rolling process. The original grain size in the cast slab was measured to be 250 μm (ASTM 1), which

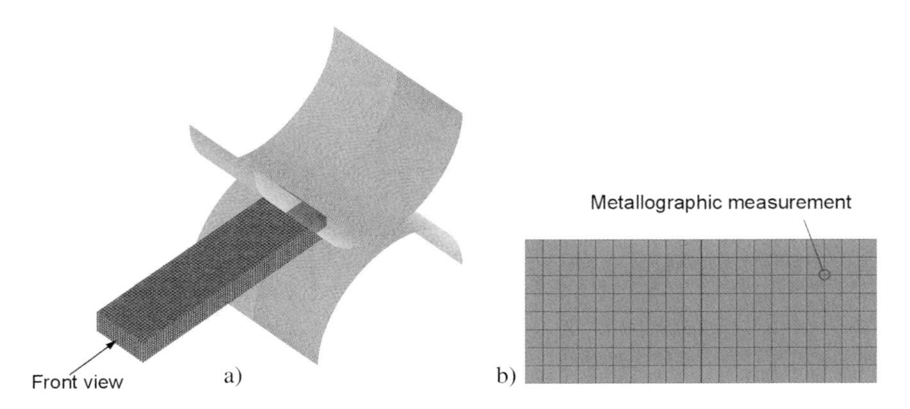

Fig. 11.5 (a) FEM model for the hot rolling and (b) front view of the workpiece model.

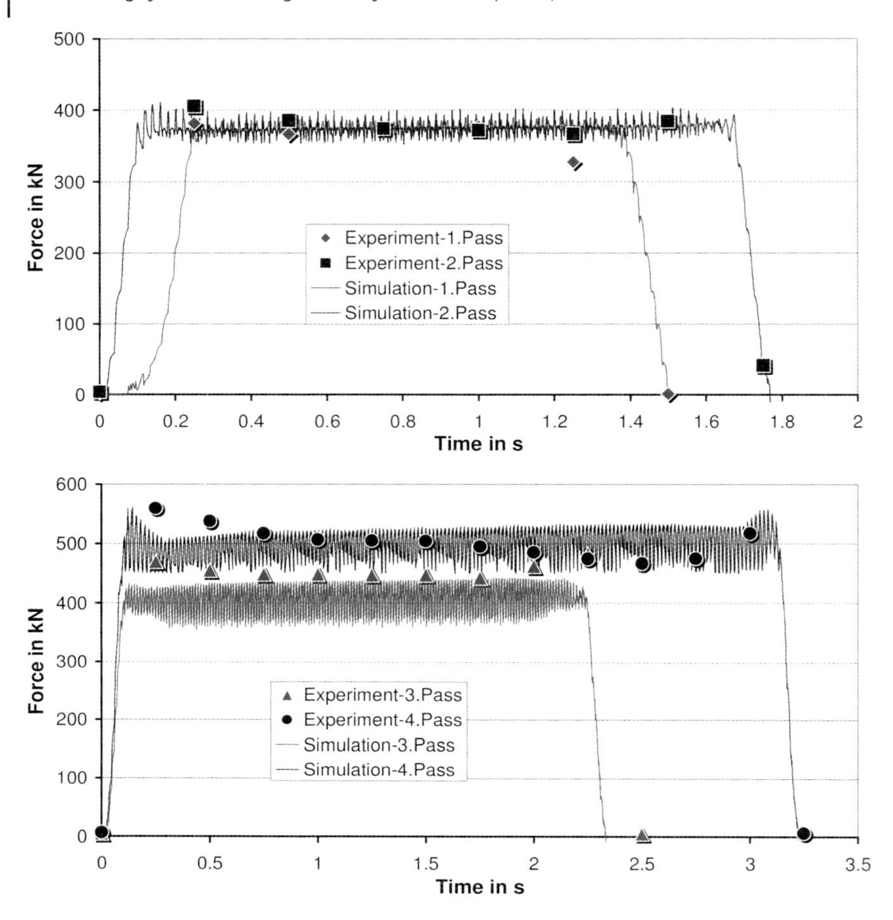

Fig. 11.6 Comparison of the simulated and measured rolling forces during the four passes.

was input to the FEM model. After each rolling step the average grain size dropped strongly due to dynamic recrystallization. In the subsequent thermal treatment between two rolling steps, the reduction of the average grain size continued in the beginning of the thermal treatment (interpass time) because of static recrystallization. This effect is particularly pronounced in the third thermal treatment because of its relatively long time interval. After static recrystallization the average grain size began to increase due to grain growth. The average austenite grain size at the end of the fourth rolling step was calculated to be 60 μm, which was close to the range of the metallographic measurements ASTM 3–5 (63–127 μm).

In both the comparisons of rolling force and grain size, good agreement between the modeling results and the measurements was found.

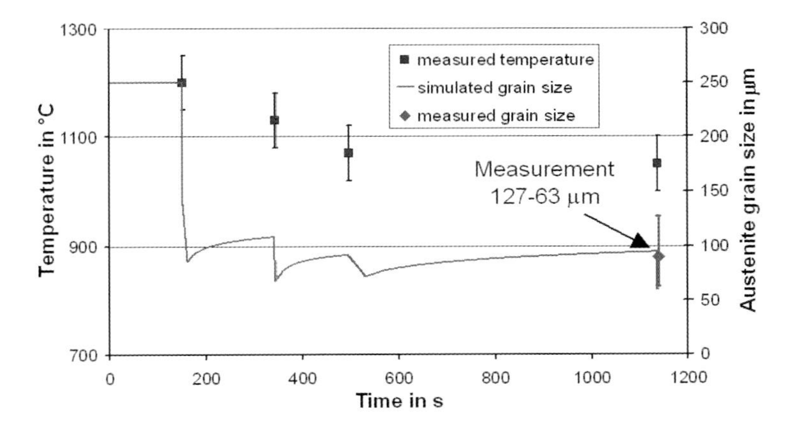

Fig. 11.7 Calculated grain size development and measured temperatures during the whole hot rolling process.

11.4 Summary

In this study, phenomenological models describing precipitation and recrystallization and grain growth during the hot rolling steps and the interpass time were implemented into the FEM code for through-process modeling. By means of process simulation (simulation of the processed casting and rolling in one heat) with models integrated into FEM, the accuracy, flexibility, and adaptability of the developed models were validated. A comparison of experimental results and simulations showed good agreement. The experimental approach to determine the precipitation parameters was improved, so that the number of required experiments could be strongly reduced. In the subsequent stage of the Collaborative Research Center (SFB) 370 the newly modified models and the results gained from experiments were applied to the last processing steps (bending or welding) to finish the process chain for the production of the steel profile and to predict the material properties over the whole process chain (see also project C7). In the future, the efficiency of the developed models and their sensitivity to the material composition need to be studied further.

Acknowledgment

The authors gratefully acknowledge the financial support of the Deutsche Forschungsgemeinschaft (DFG) within the Collaborative Research Center (SFB) 370 "Integral Materials Modeling".

Appendix

Symbols	Description
C [−]	Cingara parameter
d_0 [μm]	Original austenite grain size
d_{DRX} [μm]	Dynamically recrystallized grain size
d_{SRX} [μm]	Statically recrystallized grain size
d_{GG} [μm]	Grain size due to grain growth
ε [−]	Strain
ε_{crit} [−]	Strain, at which the dynamic recrystallization begins
$\varepsilon_{\max}$ [−]	Strain, at which flow curve reaches its maximum
ε_{stat} [−]	Strain, at which flow curve reaches its stationary area
$k_{f\max}$ [MPa]	Maximum flow stress
R [J K^{-1} mol^{-1}]	Gas constant
T [K]	Temperature
t [s]	Interval time
$t_{0.5}$ [s]	Time, to which 50% of the microstructure are statically recrystallized
X_{DRX} [%]	Dynamically recrystallized volume fraction
Z [−]	Zener–Hollomon parameter

References

1 W. Bleck, R. Diederichs, U. Lorenz, R. Kopp, L. Neumann, *Steel Res.* 74, 2003, 631.

2 R. Diederichs, Ph.D. thesis, RWTH Aachen, 2004.

3 R. Kopp, M. Nutzmann, O. Ziegelmayer, W. Bleck, R. Diederichs, *Steel Res.* 73, 2002, 321.

4 E. Ruibal, J.J. Urcola, M. Fuentes, *Z. Metallkd.* 76, 1985, 568.

5 C.M. Sellars, in *Hot Working and Forming Process*, ed. C.M. Sellars, G.J. Davies, TMS, London, 1979.

6 C.M. Sellars, Proc. 7th Risø Int. Symp. on Metallurgy and Material Science, Risø National Laboratory, Roskilde, Denmark, 1986.

7 C.M. Sellars, A.M. Irrisari, E.S. Puchi, Symp. on Microstructural Control in Aluminium Alloys, Met. Soc. AIME Meeting, 1985.

8 C.M. Sellars, J.A. Whiteman, *Metal Sci.* 13, 1979, 187.

9 B. Dutta, C.M. Sellars, *Mater. Sci. Technol.* 3, 1987, 187.

10 K. Karhausen, Ph.D. thesis, RWTH Aachen, 1994.

12
Simulation of Phase Changes During Thermal Treatments of Various Metal Alloys (TP B2)

M. Schneider, E. Jannot, V. Mohles, G. Gottstein, C. Walter, B. Hallstedt, J. M. Schneider, N. Warnken, I. Steinbach, F. Gerdemann, U. Prahl, and W. Bleck

Abstract

An overview is given on the physical models required and used to predict the phase distribution and precipitation behavior under given thermomechanical treatments relevant for industrial processing. Models for precipitation statistics, multicomponent solute diffusion and phase distribution evolution have been developed and applied to realistic heat treatments of aluminum sheet AA3104, of nickel-based superalloys, and of steel. Selected results are given in order to illustrate the usefulness and industrial applicability of the models.

12.1
Introduction

From a physical point of view heat treatment in industrial fabrication and processing of structural materials constitutes complex production steps because the material's state is far from thermal equilibrium. This means that the system reacts with strongly temperature-dependent kinetics – defined by local physical processes which are thermally activated – to thermodynamic forces. The latter depend on the local chemical composition, resulting from the kinetics, and vary strongly with temperature. To model the resulting complex behavior of phase distribution and precipitation, three approaches are used in the present chapter and applied to two materials in order to predict the outcome of realistic heat treatments. For aluminum sheet production, precipitation kinetics during homogenization (nucleation, growth, ripening) has been predicted using a statistical precipitation model without spatial resolution. Moreover, the spatial distribution of Fe and Mn solutes are predicted. For a nickel-based superalloy, the evolution of the γ'-phase has been considered by a phase-field model. For C45 steel, the isothermal proeutectoid ferrite precipitation has been investigated using a phase-

Integral Materials Modeling: Towards Physics-Based Through-Process Models
Edited by Günter Gottstein
Copyright © 2007 WILEY-VCH Verlag GmbH & Co. KGaA, Weinheim
ISBN: 978-3-527-31711-0

field model. The influence of different cooling cycles and of austenite conditioning has been modeled using a semiempirical transformation approach for C45 as well as for S460M steel.

12.2 Aluminum Sheet AA3104: Precipitation Kinetics and Solute Distribution During Homogenization

12.2.1 Challenge

In the course of aluminum sheet fabrication, homogenization is a necessary step to reduce the compositional heterogeneities introduced during casting. This thermal treatment (see Fig. 12.1) also permits one to design a specific precipitate density which will greatly influence the properties of the material during subsequent industrial processing.

The phase transformations occurring during the homogenization of alloy AA3104 are numerous and complex. The characteristic microstructure after homogenization is shown in Fig. 12.2. During solidification, primary phases of composition β-$Al_6(Mn,Fe)$ precipitated at the edge of the dendrites. These phases become thermodynamically unstable when reheated above 400 °C and transform partially into the stable α-phase $Al_{12}(Mn,Fe)_3Si_{1.8}$. At the same time, smaller precipitates, the dispersoids, copiously nucleate in the dendrite arms.

The aim of this study is to deliver predictions of these phenomena. In the following, precipitation kinetics and solute diffusion are simulated for an alloy AA3104 (with composition in weight fractions AlMg1Mn1Fe0.45Si0.2) and compared to experimental data.

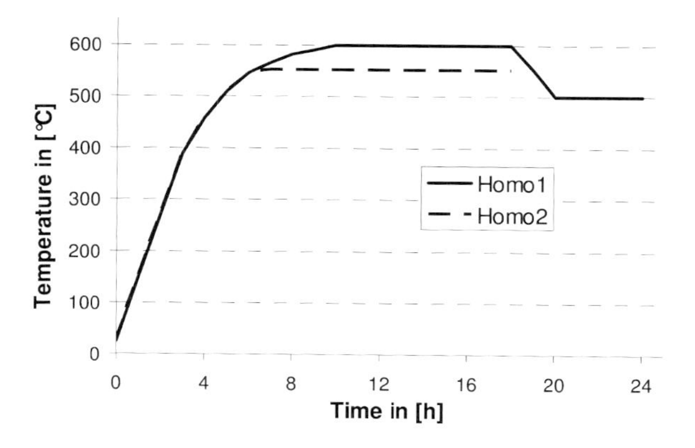

Fig. 12.1 Time–temperature program for the homogenization of AA3104.

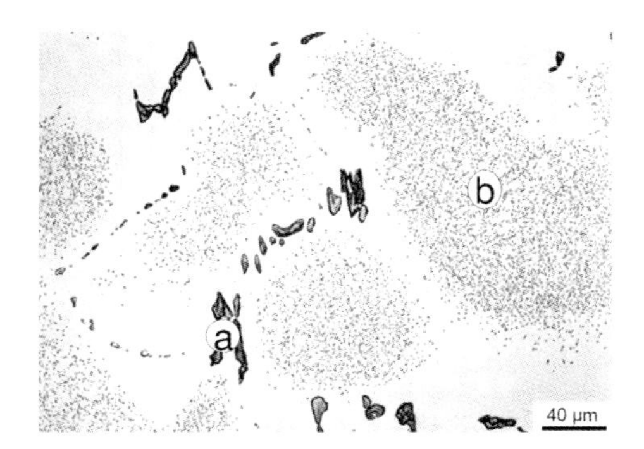

Fig. 12.2 Microstructure after homogenization: (a) primary phases; (b) dispersoids.

12.2.2 DICTRA Calculations

One-dimensional diffusion simulations were performed using the commercial software DICTRA. The DICTRA cell used models half a dendrite. The cell size is 20 µm. On the left, a thin layer (0.6 µm) of primary phase with the β-composition $Al_6(Mn,Fe)$ is present. The rest of the cell is occupied by the aluminum face-centered cubic (fcc) matrix. The right-hand end of the cell corresponds to the center of the dendrite. The concentration profiles in the matrix at the beginning of the simulation were taken from the simulations of the casting step, which were performed down to 550 °C. The homogenization simulation, thus, starts at 550 °C, i.e. after the heating period of 6 h (see Fig. 12.1). The heating period itself was not simulated. In the calculations, only diffusion in the fcc matrix was considered. Neither the transformation from $Al_6(Mn,Fe)$ to $Al_{12}(Mn,Fe)_3Si_{1.8}$ nor the presence of dispersoids were considered. Thermodynamic data were taken from a modified version of the COST II database [1–3] and diffusion data were taken from a mobility database developed by Prikhodovsky [3] for Al alloys.

12.2.3 Statistical Precipitation Model: ClaNG

The ClaNG model was used to predict the precipitation kinetics. This model is based on the classic nucleation and growth theory for precipitation. It follows the Kampmann and Wagner methodology to determine the evolution of the precipitate size distribution in the presence of a Becker–Döring nucleation rate.

The ClaNG model requires three kinds of input: the starting state of the material (alloy concentrations, already existing precipitate densities), the characteristics of the heat treatment (time–temperature curve given in Fig. 12.1), and physical parameters (physical constants, diffusion coefficients and the phase diagram). A specific feature of the tool is the determination of the state-dependent equilibrium at each time step. When the alloy concentration in the matrix or the temperature changes, the new thermodynamic equilibrium is calculated using the commercial Gibbs energy minimizer application ChemApp. The thermodynamic database ThermoTech TTAl serves as input to this application.

The output of the model comprises all the changes occurring during the thermal treatment: phase volume fractions (size distribution), equilibrium and instantaneous concentrations in the matrix and in the phases, as well as material properties (resistivity, thermoelectrical power).

A complete description of this model can be found in [4].

12.2.4
Evolution of the Primary Phases

The reality of the phase transformation from the β-$Al_6(Mn,Fe)$ to the α-$Al_{12}(Mn,Fe)_3Si_{1.8}$ phase is far too complex to be modeled accurately. However, the growth of the particles during the heat treatment can be estimated. Starting with precipitates of radius 2.00 μm, ClaNG simulations show that these particles remain virtually unchanged (2.04 μm) to the end of homogenization. Concentration profiles in the fcc matrix were simulated using DICTRA. Profiles for Mn and Fe are shown in Fig. 12.3. The alloy is practically completely homogenized after annealing for 8 h at 600 °C. Mg and Si are quite evenly distributed both before and after homogenization. Although the start compositions may be rather different from the real compositions after 6 h heating, this will probably have a rather limited influence on the final results. The amount of primary phase was found to increase slightly during homogenization, in agreement with the ClaNG simula-

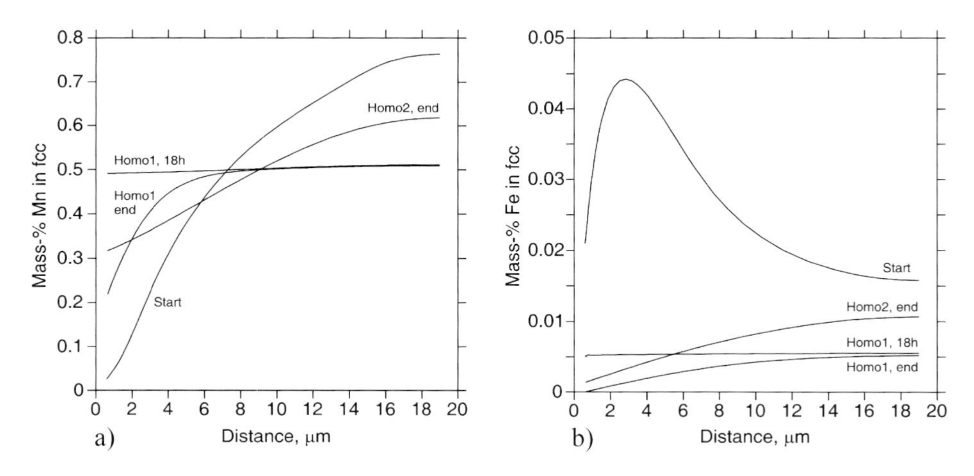

Fig. 12.3 Simulated concentration profiles in the fcc matrix using DICTRA: (a) Mn, (b) Fe.

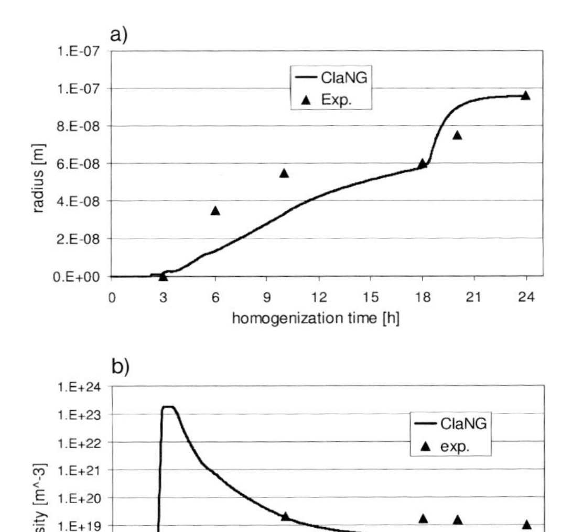

Fig. 12.4 Precipitation of dispersoids simulated with the ClaNG model, compared with experimental values: (a) average radius, (b) density of the particles.

tions. Observations by SEM confirm this minor change of size of the primary phases.

12.2.5 Precipitation of Dispersoids

This kind of phase transformation is a good example to test the validity of the classic nucleation and growth theory. The particle density and average radius obtained by simulation are compared with the values determined by TEM investigations in Fig. 12.4. The same order of magnitude is obtained by simulations for both quantities.

12.2.6 Conclusion on Al Sheet AA3104

By the combination of DICTRA and ClaNG simulations, most of the important aspects during homogenization, such as distribution of the alloying elements, evolution of primary phases, and formation of dispersoids, could be addressed.

The simulations were found to be in reasonable agreement with each other and also with experimental data.

12.3
Turbine Blades: Ni-Base Superalloys

12.3.1
Challenge

Turbine blades are commonly made from superalloys, a family of alloys based on nickel, cobalt, or iron, designed for high-temperature applications. Superalloys for single-crystal applications are based on Ni with major additions of Al, Cr, Ta, Ti, W, and other elements. The desired microstructure consists of finely dispersed γ' precipitates in a γ matrix. Volume fractions of 60–80% of γ' are found in modern single-crystal superalloys. The alloying elements Al, Ta, and Ti promote the formation of γ', while elements like W do the contrary. During solidification of the alloy, the γ' promotors segregate differently from the γ' inhibitors. This leads to a strongly inhomogeneous γ' distribution [5] and the formation of interdendritic γ'. Sophisticated heat treatments of cast parts aim at creating a segregation free microstructure and a well-defined γ' size distribution. In order to dissolve the interdendritic γ' and to dissolve the microsegregation pattern as fast as possible, high temperatures are needed for the heat treatment. But at the same time local remelting (incipient melting) has to be avoided.

12.3.2
Simulation Models and Experiments

The microstructure evolution during the heat treatment was simulated using the phase-field code MICRESS [9] coupled to thermodynamic calculations via the TQ-interface of ThermoCalc [6]. The phase-field method allows the simulation of spatially resolved microstructure evolution, controlled by constitution, diffusion, and stresses. Multicomponent diffusion was taken into account for the simulations. The necessary thermodynamic and kinetic data were obtained from CALPHAD calculations. In order to investigate the influence of finely dispersed γ' precipitates on the mesoscopic diffusion, results of phase-field simulations were fed into calculations according to the method of mathematical homogenization of periodic structures [10, 11]. Effective diffusion coefficients for microstructures consisting of γ and γ' were obtained this way.

In order to investigate the solution kinetics during the solutioning heat treatment, directionally solidified samples of a Ni–13.06 at% Al–10.49 at% Cr–2.66 Ta–2.92 at% W model superalloy were heat treated at 1285 and 1295 °C and quenched after different holding times (1–4 h). The samples were polished and etched, and the amount of residual interdendritic γ' was measured from the micrographs by means of image analysis.

12.3.3
Results

The solution heat treatment was simulated with MICRESS, using the NIST thermodynamical and kinetic databases [7, 8]. Microstructures obtained from solidification simulations were used as starting conditions. Figure 12.5 shows calculated results in comparison with experimental data. The amount of interdendritic γ' is plotted versus time for two temperatures. A temperature increase of 10 °C leads to a significantly increased dissolution kinetics. In Fig. 12.6 the incipient melting temperature calculated from the results of microstructure simulations is given with respect to time, for a multistep solution heat treatment. As local remelting

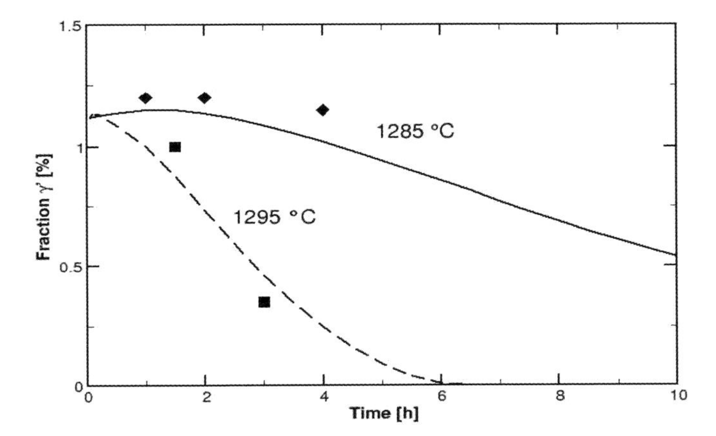

Fig. 12.5 Simulated (lines) and experimental fraction of interdendritic γ'.

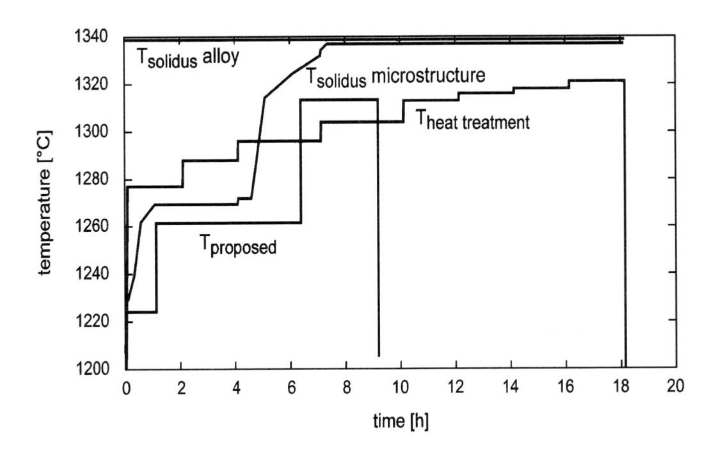

Fig. 12.6 Calculated incipient melting temperature for a given heat treatment.

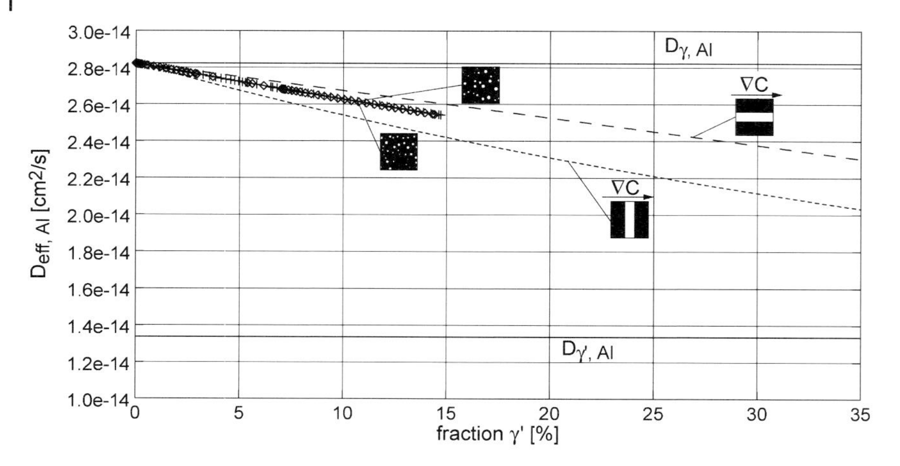

Fig. 12.7 Calculated effective diffusion coefficient of Al in dispersed and lamellar microstructures as a function of γ' fraction. The diffusion coefficients for pure γ and pure γ' are also given.

is predicted, the heat treatment was not suitable in this case. A new heat treatment was proposed, based on the simulation results.

Figure 12.7 shows effective diffusion coefficients calculated for γ–γ' microstructures, as a function of the γ' fraction. The cases of a lamellar structure with concentration gradients parallel and perpendicular, respectively, to the lamellae constitute the upper and lower limits for the calculation. It can be seen that in the case of dispersed structures, the effective diffusion coefficients depend on the γ'-fraction and stay within these limits.

Figure 12.8 shows microstructures in a Ni–Al–Cr alloy obtained by phase-field simulations. At the beginning of the simulations γ' precipitates from the matrix

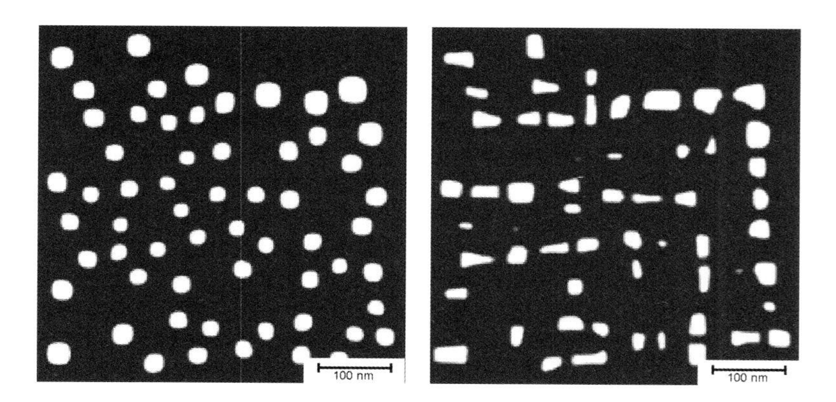

Fig. 12.8 Typical transition from spherical to cuboidal morphology of γ' precipitated under the influence of coherence stresses, calculated with the phase-field method.

and grows. The precipitates exhibit a spherical morphology. With progressing growth and subsequent ripening, the precipitates change their morphology from spherical to cuboidal due to coherency stresses and strains.

12.4 C45 and S460 Steel: Isothermal Phase Transformation and Austenite Conditioning

12.4.1 Challenge

The phase transformation of steel was investigated using two different approaches:

- The thermodynamics-based phase field approach was applied to the isothermal proeutectoid ferrite precipitation. Microstructural morphological information was taken into account. The main problem was the proper implementation of nucleation kinetics. The calculations were performed using the MICRESS code [9].
- A semiempirical approach based on isothermal input data was implemented in the SimZTU software. Using this approach, different cooling rates or cycles can be predicted, and the influence of austenite conditioning on the continuous cooling transformation can be predicted quantitatively.

12.4.2 Simulation Models and Experiments

The main problem for an isothermal proeutectoid ferrite transformation prediction is a proper description of the ferrite nucleation kinetics. Therefore, isothermal holding experiments were performed in order to measure nucleation densities and frequencies at triple junctions as a function of undercooling. Real microstructures and nucleation densities were transferred to phase-field calculations. Based on this input data, the ferritic phase transformation kinetics were calculated including solute partitioning.

For continuous cooling transformation calculations the JMAK (Johnson–Mehl–Avrami–Kolmogorov) approach was used. It is formulated for the description of isothermal transformation kinetics. In order to calculate CCT diagrams on the basis of the input data, an incremental method proposed in [12] was implemented using a stepwise isothermal approximation for continuous cooling paths. The approach is based on TTT diagrams, which are described by the 1 and 99% curve of transformed fraction of each microstructure. Concerning the austenite conditioning, it is assumed that it can be described effectively by the parameters grain size, effective grain boundary area per unit volume, and strain [13].

The validation of this approach was realized by a three pass hot rolling scheme. Here, the last pass was just above the transformation temperature A_{c1} such that at the cooled edges the transformation was accelerated. The simulation was performed by an FE calculation of the rolling process followed by a simulation of the austenite conditioned phase transformation [14].

12.4.3 Results

For the isothermal transformation it was found that the ferrite transformation proceeds in two steps (Fig. 12.9). In the first step, the transformation is controlled by carbon diffusion while other alloying elements still remain homogeneously distributed. This situation is termed LENP (local equilibrium non partitioned) and yields a semistable situation within technically reasonable times. For longer times, the transformation is diffusion controlled by the other alloying elements until these are adequately partitioned between ferrite and austenite. This equilibrium is referred to as LE (Local Equilibrium).

In Fig. 12.9, results are shown for C45 steel using an isothermal holding at 690 °C. A real microstructure had been used as input data for a two dimensional MICRESS calculation. The transformation is predicted to occur in two steps, where experimental data are available for up to 3 h holding times which marks the first semiequilibrium. For longer times the ferrite fraction reaches the equilibrium predicted by ThermoCalc.

Concerning deformation induced transformation, a simulation of a three pass rolling scheme for S460M steel was performed experimentally as well as numerically. Here, the last pass was just above the transformation temperature A_{c1} such that at the cooled edges the transformation was accelerated. The simulation was

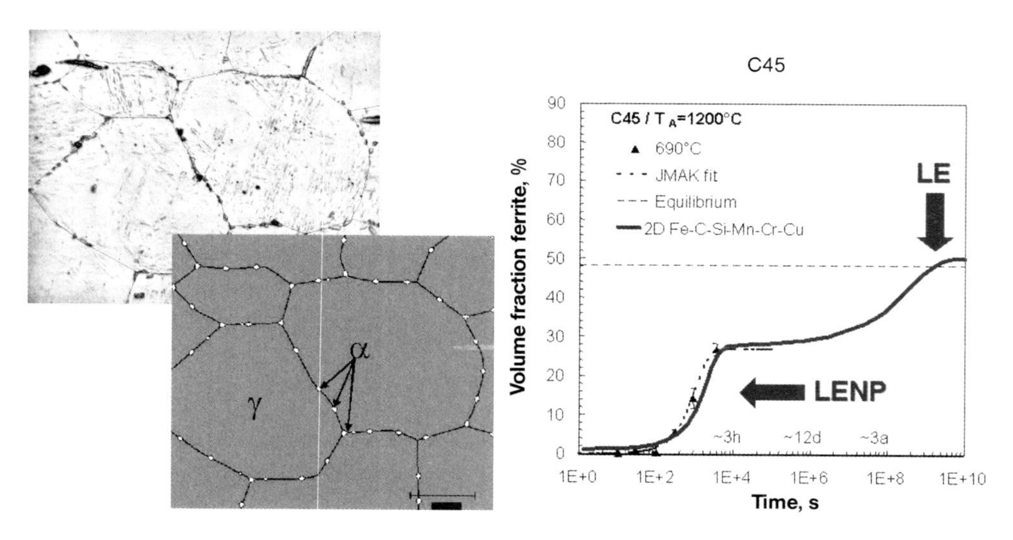

Fig. 12.9 Physical simulation of proeutectoid ferrite precipitation in C45 steel.

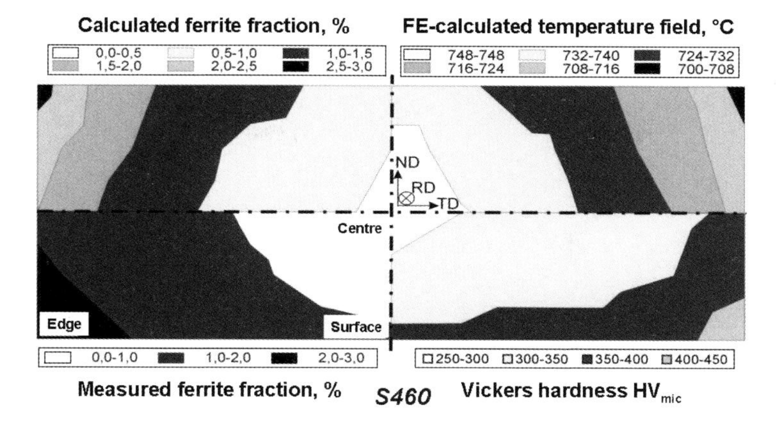

Fig. 12.10 Measured and calculated ferrite fraction after accelerated transformation due to cooled edges during hot rolling of S460M steel.

realized by an FE simulation of the rolling process followed by a simulation of austenite conditioned phase transformation.

In Fig. 12.10 tested and calculated results are compared. First the ferrite transformation was measured and calculated. A gradient in ferrite fraction can be observed from the edges to the center of the bar which was caused by the temperature gradient. The experimental results were measured at a specimen that had been cooled by water quenching which resulted in a hardness gradient due to decreasing quenching rates from the edge to the center.

Acknowledgment

The authors gratefully acknowledge the financial support of the Deutsche Forschungsgemeinschaft (DFG) within the Collaborative Research Center (SFB) 370 "Integral Materials Modeling".

References

1 I. Ansara, A.T. Dinsdale, M.H. Rand (eds.), *Thermochemical Database for Light Metal Alloys*, Vol. 2, COST 507 Final Report Round 2, European Communities, Luxembourg, 1998.
2 E. Balitchev, T. Jantzen, I. Hurtado, D. Neuschütz, *Calphad* 27, 2003, 275.
3 A. Prikhodovsky, Dissertation, RWTH Aachen, 2000.
4 M. Schneider, Dissertation, RWTH Aachen, 2006.
5 M. Durand-Charre, *The Microstructure of Superalloys*, 1997.
6 J.-O. Andersson, T. Helander, L. Höglund, P. Shi, B. Sundman, *Calphad* 26, 2002, 273.
7 U.R. Kattner, in *Calphad and Thermodynamics*, ed. P.E.A. Turchi, A. Gonis, R.D. Shull, TMS, Warrendale, PA, 2002, p. 147.
8 C.E. Campbell, W.J. Boettinger, U.R. Kattner, *Acta Mater.* 50, 2002, 775.

9 J. Eiken, B. Boettger, I. Steinbach, *Phys Rev E*, in press.
10 E. Sanches-Palencia, *Non Homogeneous Media and Vibration Theory*, Springer, Berlin, 1980.
11 A. Bensoussan, J.L. Lions, G. Papanicolaou, *Asymptotic Analysis for Periodic Structures*, North Holland, 1978.
12 H.-P. Hougardy, K. Yamazaki, *Steel Res.* 57, 1986, 466.
13 U. Lorenz, Dissertation RWTH Aachen, Shaker Verlag, 2004.
14 W. Bleck, R. Diederichs, U. Lorenz, R. Kopp, L. Neumann, *Steel Res.* 74, 2003, 631.

13
Deep Drawing Properties of Aluminum Sheet (TP C6)

L. Neumann, R. Kopp, G. Hirt, M. Crumbach, and G. Gottstein

Abstract

This chapter reports on recent work carried out on the through-process modeling of aluminum (Al) laboratory production. The process stage considered is cup drawing. Important microstructural phenomena were simulated for each preceding processing stage, and final results were passed on to the cup drawing simulation as input. The phenomena of preceding steps comprised diffusion, phase distribution, microsegregation, grain growth, strain hardening, deformation texture, and recrystallization texture. In order to validate the simulations, the production process was carried out on a laboratory scale. The phenomenon of interest in cup drawing is primarily the earing profile of the cup due to texture-induced anisotropy. Using simulated microstructural information on texture and dislocation densities, a phenomenological yield locus can be calibrated and updated for every finite element as texture evolves in the deep drawn blank.

13.1
Introduction

The crystallographic texture that is the result of the thermomechanical processing of Al sheet plays a central role for subsequent deep-drawing operations of the sheet. The sheet texture will primarily induce anisotropic flow behavior which – in industrial processing – can be a cause for significantly increased scrapping. In the following, "T-Pack" a modeling tool for cold forming will be presented. Apart from the process data such as tool geometries, microstructural information – material texture and hardness as expressed in terms of dislocation densities – is input to the simulation. The anisotropic flow behavior during the finite element simulation of the forming step is predicted on the basis of microstructural information and the update thereof using texture and dislocation density-based flow stress models. The setup of T-Pack and its application to through-process simulation is presented in Chapter 10. When used for through-process simulation, T-Pack delivers the microstructural input named above which is necessary for

Integral Materials Modeling: Towards Physics-Based Through-Process Models
Edited by Günter Gottstein
Copyright © 2007 WILEY-VCH Verlag GmbH & Co. KGaA, Weinheim
ISBN: 978-3-527-31711-0

the prediction of texture-induced anisotropic flow behavior. Results obtained for cup drawing simulations are presented in the following as well as an overview of the studies carried out on the modeling setup. For details the reader is referred to previously published results of these studies [1, 2].

13.2
Modeling Setup for Prediction of Texture-Induced Anisotropy

The extension of T-Pack to predict anisotropy owing to crystallographic texture and its development were developed by Neumann et al. [2–4] and are summarized in the following.

13.2.1
Interfacing Texture to Plastic Anisotropy

In previous work an eight-node solid finite element in updated Lagrange formulation was developed by Aretz [5]; it is specially designed for sheet metal forming applications. A significant variety of (currently 11) yield criteria is embedded into the finite element. An essential requirement for the element was the incorporation of a co-rotational material coordinate system in which orthotropy is approximately maintained and the yield criterion keeps its validity (see also Fig. 13.1). The approach that was chosen for T-Pack is the classic rigid body spin combined with an update for elements that have attained a preset value of accumulated strain.

Aretz showed that Hill48 delivers earing height predictions comparable in quality to "Yld91" [6], when it is calibrated with four directional yield stresses (see following section). Furthermore, since Hill48 is computationally less intensive than Yld91, it is preferred as long as four ears are expected.

Within the scope of this study, this modeling approach is used with a viscous material law, because it is computationally less intensive and more stable than elastoplastic material models. (This material law is used while hardening and softening are simulated with the models described above.) As long as elastic effects such as springback are not of interest, this approach appears to be justified.

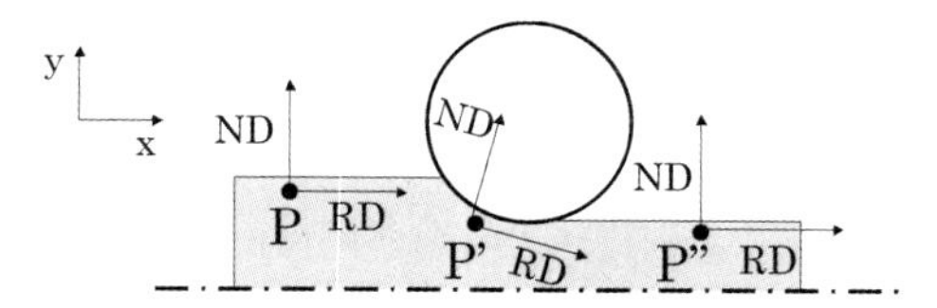

Fig. 13.1 Rotating specimen coordinate frame. This is much more relevant for the presented cup drawing than the two-dimensional rolling shown in the sketch. The implementation remains valid regardless of the forming process simulated.

13.2.2
Orthotropic Viscous Flow Approach

In the following, the most important constitutive equations for the viscoplastic potential approach used here are given [7]. The stress deviator $\underline{\underline{s}}$ with reference to the orthotropy axes is calculated according to:

$$\underline{\underline{s}} = 2\eta(\dot{\bar{\varepsilon}}) \cdot \underline{\underline{\dot{\varepsilon}}} \tag{13.1}$$

where $\eta(\dot{\bar{\varepsilon}})$ is the viscosity as function of the equivalent plastic strain rate.

The equivalent plastic strain rate $\dot{\bar{\varepsilon}}$ depends on the yield criterion applied. The yield function Hill48 is used in the present modeling setup. The power conjugate equivalent plastic strain rate for this yield criterion is given by:

$$\dot{\bar{\varepsilon}} = \sqrt{\frac{2}{3}(F'\dot{\varepsilon}_{11}^2 + G'\dot{\varepsilon}_{22}^2 + H'\dot{\varepsilon}_{33}^2 + 2L'\dot{\varepsilon}_{23}^2 + 2M'\dot{\varepsilon}_{31}^2 + 2N'\dot{\varepsilon}_{12}^2)} \tag{13.2}$$

The factors F', G', H', and N' can be associated with material parameters which have to be determined experimentally from the values of the following four directional yield stresses: Y_0 (yield stress in rolling direction), Y_{45} (yield stress in 45° direction), and Y_{90} (yield stress in transverse direction), as well as Y_b (yield stress in biaxial loading). Material data for L' and M' cannot be determined from mechanical tests since out-of-plane simple shear tests would be necessary. These would be very difficult to realize in thin sheet metals. Thus, $M' = L' = 1$ (isotropic case) is a simple and reasonable choice if no other information is available.

The deviatoric stress components with respect to the orthotropy axes finally become

$$s_{11} = 2\eta F'\dot{\varepsilon}_{11} \tag{13.3}$$

$$s_{22} = 2\eta G'\dot{\varepsilon}_{22} \tag{13.4}$$

$$s_{33} = 2\eta H'\dot{\varepsilon}_{33} \tag{13.5}$$

$$s_{12} = 2\eta N'\dot{\varepsilon}_{12} \tag{13.6}$$

$$s_{23} = 2\eta L'\dot{\varepsilon}_{23} \tag{13.7}$$

$$s_{31} = 2\eta M'\dot{\varepsilon}_{31} \tag{13.8}$$

The orthotropy axes rotate during deformation which is described by the total rotation tensor $\underline{\underline{R}}$ obtained from the polar decomposition of the total deformation gradient $\underline{\underline{F}}$:

$$\underline{\underline{R}} = \underline{\underline{F}} \cdot \underline{\underline{U}}^{-1} \tag{13.9}$$

where $\underline{\underline{U}}$ is the right (or material) stretch tensor. $\underline{\underline{F}}$ describes the deformation from the initial (at time $t = 0$) to the current configuration. Thus, $\underline{\underline{R}}$ describes the rotation of the initial orthotropy axes (at time $t = 0$) to the current configuration. This is expressed by

$$\underline{e}_i = \underline{\underline{R}} \cdot \underline{\hat{e}}_i \quad \text{with } i = 1, 2, 3 \tag{13.10}$$

Here, $\underline{e}_i$ represents the basis vectors of the orthotropy axes in the current configuration while $\underline{\hat{e}}_i$ corresponds to the initial configuration (at the time $t = 0$). $\underline{\hat{e}}_i$ are defined by the user. A coordinate transformation based on the relative orientation of $\underline{e}_i$ with respect to the global coordinate frame finally allows the calculation of the stress deviator $\underline{\underline{\tilde{s}}}$ in terms of the global coordinate axes.

13.2.3
Update of the Yield Locus

T-Pack offers an interactive update of the yield locus when all modules (full constraints Taylor model for texture, 3IVM for flow stress, and Hill48 for plastic anisotropy) are activated, and the parameters for the chosen yield locus can be calculated directly from the mechanical properties (see also Fig. 13.2).

After a preset summed up strain (usually of the value $\varepsilon_{plast} = 0.02$) for a given element has been reached, texture simulation is invoked for this element. This user setting avoids texture and anisotropy update in finite elements that are not in the forming zone, thus significantly reducing the computational effort neces-

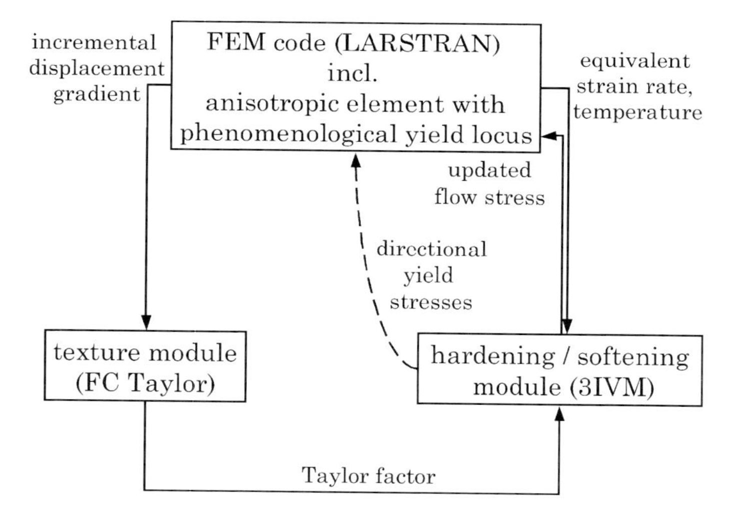

Fig. 13.2 This flow chart of T-Pack depicts the incorporation of texture and work hardening modules into FEM (solid lines). Input delivered for the yield locus update in the case of sheet metal forming simulations is marked by the dashed lines. (The hardening/softening module is called up iteratively while the texture module is called up incrementally).

sary for the simulation. Precision is not compromised up to the value of $\varepsilon_{plast} = 0.02$, where the orientation distribution is updated. Then, the finite element and its corresponding orientation distribution are subjected to the testing strain corresponding to the test for Y_0, Y_{45}, Y_{90}, and Y_b, which represent the yield stresses in rolling direction, in 45° direction, in transverse direction, and in biaxial loading, respectively, and used to calculate the yield stresses with the 3IVM. The resulting values are used to update the shape of the Hill48 yield locus in the current element. This procedure consists of the following calculation steps:

1. Appropriate rotation of the orientation distribution for tests Y_0, Y_{45}, Y_{90}, and Y_b.
2. Calculation of the Taylor factor $\overline{M}$ (for each test Y_0, Y_{45}, Y_{90}, and Y_b) after virtually subjecting the finite element to the appropriate displacement gradient tensor.
3. Calculation of Y_0, Y_{45}, Y_{90}, and Y_b with 3IVM and the previously calculated values for the Taylor factor $\overline{M}$.
4. Calibration of the yield locus Hill48 with the calculated values for Y_0, Y_{45}, Y_{90}, and Y_b.

13.3 Results and Discussion

A variety of studies were carried out during the development and testing phases of T-Pack, and for detailed information the reader is referred to Neumann [2]. Some results of these studies are given in the following; others are briefly summarized. The following are some of the aspects and their influence on earing profile predictions that were studied:

1. Earing profiles resulting from texture calculated in a through-process simulation.
2. Earing profiles resulting from texture measured in the deep-drawn sheet.
3. Comparison with established phenomenological yield functions used in the standard approach (Hill48, Yld91, Yld94).
4. Influence of the number of orientations with which the texture in T-Pack is discretized.
5. Influence of the hardening/softening law (3IVM or user-defined equation).
6. Influence of the value of the user-preset summed up strain ε_{plast}.
7. Earing profiles predicted for "ideal" textures consisting of orientations scattered around one orientation (brass, copper, Cube, Goss, S-orientation, and a mixture of S-orientation and Cube).

The evolution of texture (including possible static recrystallization) in through-process modeling was studied and reported on by Abdurahman [8]. The resulting textures were generally very weak – almost random – only showing slightly in-

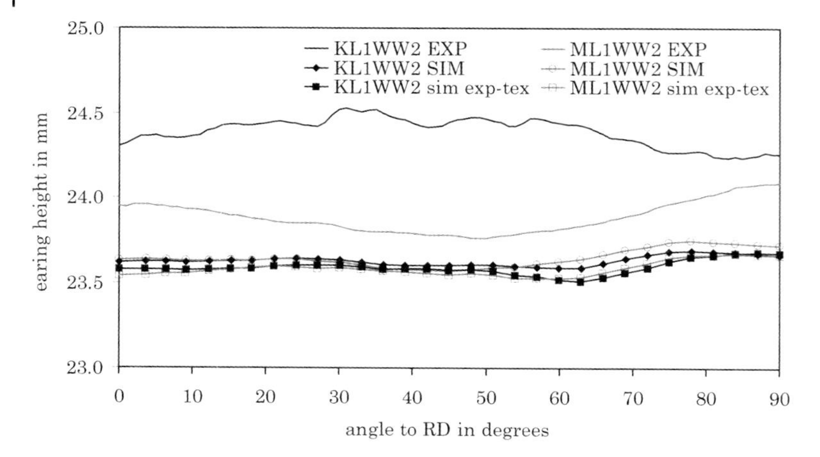

Fig. 13.3 Cup profiles of two slightly differently textured alloys: the commercial alloy AA3104 "KL" and the laboratory alloy "ML". Measurements are marked as "EXP" and simulations (that are results of the simulated through-process texture) as "SIM". Simulations that were carried out using the material measured texture are marked "sim exp-tex".

creased intensity in the Cube orientation. When comparing the commercial alloy AA3104 "KL" to the laboratory alloy "ML" (where both underwent homogenization 1 and were hot rolled at 380 °C) the only apparent difference is a somewhat stronger R-orientation in the commercial alloy AA3104 (equivalent to the S orientation but slightly shifted after recrystallization). The effect of this difference can be distinguished in the experimentally measured earing profiles (see Fig. 13.3). The slightly stronger Cube orientation in the lab alloy leads to weak ears in RD and TD with the latter being slightly higher. The commercial alloy AA3104 shows a very low and broad ear around 45° which is typical for textures with a slightly increased intensity around the β-fiber. T-Pack does not capture these differences. Nevertheless, T-Pack does predict extremely flat earing profiles, thus reproducing quite well the quantitative earing height (see Fig. 13.3). (Qualitatively, the prediction is more precise for the laboratory alloy than for the commercial alloy AA3104, because in both cases T-Pack predicts a small ear in TD.) This demonstrates an important feature: the quasi-isotropic behavior is excellently reproduced in the simulation while computation times remain within the order of magnitude as for simulations with the classical phenomenological approach. The absolute cup height prediction is a little too low, which may be due to the friction coefficient used. The friction coefficients were chosen as proposed by Meyer [9] for the lubricants used: $\mu_{punch} = \mu_{blankholder} = 0.1$ for deep-drawing plastic film, and $\mu_{die} = 0.03$ for deep-drawing oil.

13.4 Conclusions and Outlook

A modeling setup "T-Pack" for the through-process simulation of Al production has been developed. After homogenization, microstructural variables are traced throughout hot rolling, cold rolling, and annealing steps. Hardening, recovery, and static recrystallization are included in the modeling setup, and their influence on the dislocation substructure and texture evolution is accounted for. For sheet forming operations, T-Pack has been extended to account for texture-induced plastic anisotropy. Results of a cup drawing operation were presented and reference is given to further studies that have been conducted.

Further improvement in the modeling setup can be achieved in the modules themselves. For example, a more precise investigation and understanding of recovery and its implementation into 3IVM is desirable. In the finite element simulation, existing remeshing algorithms and adapted data transfer routines are necessary prerequisites for an application of T-Pack to geometrically more complex processes than flat rolling and deep-drawing.

Acknowledgments

The authors gratefully acknowledge the financial support of the Deutsche Forschungsgemeinschaft (DFG) within the Collaborative Research Center (SFB) 370 "Integral Materials Modeling".

Furthermore, the authors thank Dr. H. Aretz, M. Schneider, M. Goerdeler, J. van Santen, M. Nutzmann, Ch. Wiedner, and A. Meyer for assistance and helpful discussions.

References

1 M. Crumbach, Doctoral thesis, RWTH Aachen University, 2006.
2 L. Neumann, Doctoral thesis, RWTH Aachen University, in preparation.
3 L. Neumann, H. Aretz, R. Kopp, M. Crumbach, M. Goerdeler, G. Gottstein, *AIP Conf. Proc.* 712, 2004, 388.
4 L. Neumann, R. Kopp, H. Aretz, M. Crumbach, M. Goerdeler, G. Gottstein, *Mater. Sci. Forum* 495–497, 2005, 1657.
5 H. Aretz, Doctoral thesis, RWTH Aachen University, Shaker-Verlag, Aachen, 2003.
6 F. Barlat, D.J. Lege, J.C. Brem. *Int. J. Plasticity* 7, 1991, 693.
7 J.-L. Chenot, M. Bellet, in *Numerical Modelling of Material Deformation Processes*, ed. P. Hartley, I. Pillinger, C. Sturgess, Springer-Verlag, London, 1992, p. 179.
8 A. Abdurahman, M.Sc. thesis, Institut für Metallkunde und Metallphysik, RWTH Aachen Univer-sity, 2005.
9 A. Meyer, Diploma thesis, Institut für Bildsame Formgebung, RWTH Aachen University, 2004.

14
Simulation of Stress Response to Cyclic Thermal Loading in Thermal Barrier Composites for Gas Turbines (TP C8)

R. Herzog, P. Bednarz, E. Trunova, and L. Singheiser

Abstract

Finite element simulations of the stress response near the thermally grown oxide (TGO) in atmospheric plasma sprayed thermal barrier coatings (TBCs) were conducted corresponding to cyclic furnace tests. The deformation properties of bond coat (BC) and TBC as well as the oxidation kinetics were experimentally determined and implemented in the finite element code. The stress calculations showed two distinct features: (1) a fast development of large tensile stresses in the BC with a maximum directly at the BC/TGO interface below a roughness peak, which occurred during cooling, and (2) a development of large tensile stresses alongside the roughness peak. The simulation of crack formation at the BC/TGO interface using cohesive elements resulted in an early formation of a microcrack at the roughness peak corresponding to experimental observations.

14.1
Introduction

Finite element analysis (FEA) can contribute to the understanding of degradation and damage processes in thermal barrier coatings (TBCs) by providing a tool for analyzing the stress response in thermal barrier systems and for simulating the evolution of stresses and the formation and growth of cracks during cyclic and high-temperature loading. By doing so, FEA principally allows one to separate the parameters which affect the stress response and to assess their respective impact.

A large number of research groups have reported results from numerical simulations of the stress response in TBCs [1–12]. Some work comprised parametrical studies by implementing various but constant thickness values of the thermally grown oxide (TGO) scale, but growth stresses were neglected, because bond coat (BC) oxidation was not considered as a continuous process. Some studies included continued oxidation, but did not consider creep and stress relaxation. Others again addressed partly creep properties, but generally those proper-

Integral Materials Modeling: Towards Physics-Based Through-Process Models
Edited by Günter Gottstein
Copyright © 2007 WILEY-VCH Verlag GmbH & Co. KGaA, Weinheim
ISBN: 978-3-527-31711-0

ties were taken from literature and it was not possible to conclude whether the material data were representative for a real TBC composite or not. Thus, the demand for more realistic FE simulations for TBCs based on more realistic material properties for the coatings increased in recent years. The present work aims at a further step towards FE simulations for TBCs applied by atmospheric plasma spraying (APS) with improved significance. The simulations were conducted with material data, which had been determined predominantly on actual coatings. The numerical computations include stress calculations and the simulation of crack formation at the BC/TBC interface.

14.2 Experimental

The experimentally investigated thermal barrier composite consisted of the single-crystalline Ni alloy CMSX-4, a NiCoCrAlY bond coat applied by vacuum plasma spraying and an APS TBC of zirconia doped with about 7–8 wt% yttria. The thermal barrier composite was cyclically exposed to temperature changes between 60 °C (approx.) and 1050 °C with a dwell time of 2 h at 1050 °C. The material was tested until macroscopic spallation of the TBC was observed. The resulting failure mode is represented in Fig. 14.1. The failure crack path was located partly in the TBC and partly in the TGO.

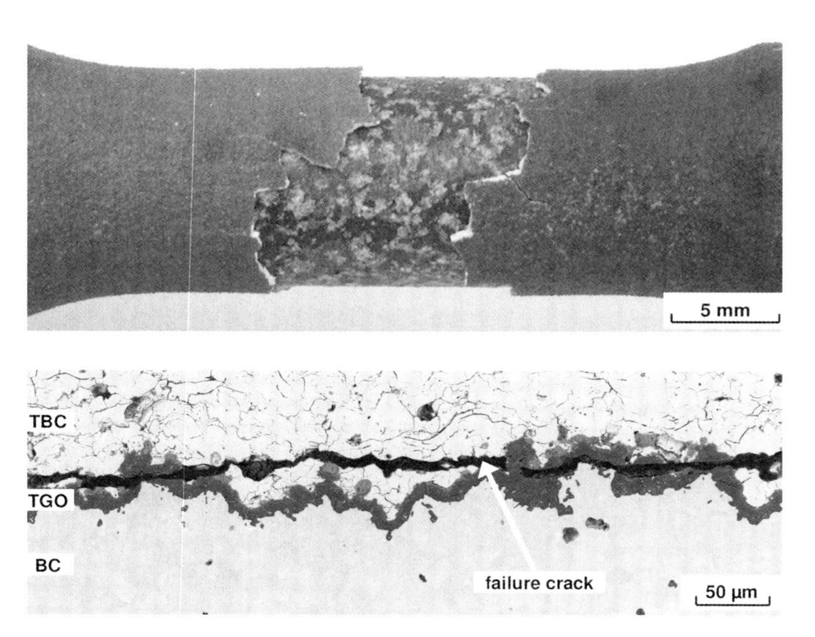

Fig. 14.1 Top: Observed failure of the plasma-sprayed TBC after cyclic thermal tests. Bottom: Cross-section directly next to the spalled area revealing a failure crack path which is located partly in the TBC and partly in the TGO.

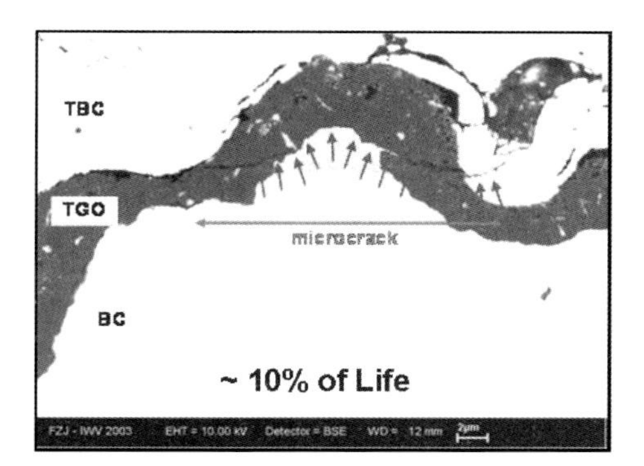

Fig. 14.2 Frequently observed crack pattern at early stages of exposure indicating weak points for crack formation and initial crack growth.

Additional experiments were conducted up to selected fractions of thermal fatigue life. Figure 14.2 shows examples of crack patterns observed at early stages of exposure (up to about 30% of time-to-failure). Frequently, microcracks were found at roughness peaks along the BC/TGO interface (Fig. 14.2). They partly crossed the TGO towards the TBC, and the crack tips were sometimes located within the TBC. Similar micrographs were taken from each sample, which had been exposed to a certain fraction of life. All microcracks associated with the TGO were counted and characterized with respect to their length. From these crack data the maximum crack length and the number (density) of microcracks were determined and plotted against the fraction of life (Fig. 14.3). Both curves reflect the evolution of damage at the metal/ceramic interface of the TBC. From about 10 to 50% of life more and more microcracks were formed along the inter-

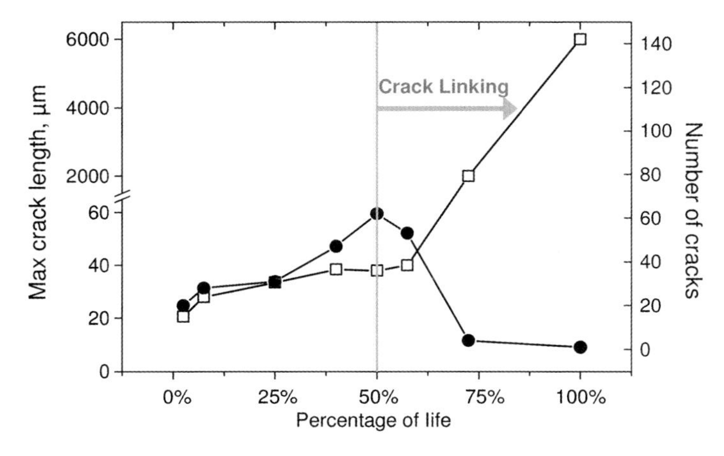

Fig. 14.3 Maximum crack length and number (density) of microcracks in or near the TGO plotted against the fraction of life. The data covers the range from early stages of exposure to macroscopic spalling.

face, but the maximum crack length was restricted to values below 50 μm. See also [14] for more detailed results about the evolution of damage in APS TBCs. The FEA described in the next section aims at simulating the local stress response and the crack formation near the TGO during the initial stage of exposure to obtain more information about the kinetics of the observed damage evolution.

14.3 Finite Element Simulation

14.3.1 Mesh and Boundary Conditions

A cylindrical geometry was chosen for the FE model, which corresponded to the specimen geometry used in the experiment. It consisted of four layers: base material, BC, TGO and TBC. The base material had an outer diameter of 10 mm. The thickness of the BC was 150 μm and that of the TBC 300 μm. The initial thickness of the TGO was 0.5 μm. The roughness profile at the BC/TBC interface was approximated as sinusoidal. The sinus was parameterized by an amplitude of 15 μm and a wavelength of 60 μm. The mesh consisted of four-node general plane strain (GPS) elements with reduced Gauss integration (type: CPEG4R of FE code ABAQUS). Geometry and mesh are shown in Fig. 14.4. The nodes on the edges of the segment (Fig. 14.4, right) were constrained as to the rotational symmetry and periodic boundary conditions.

14.3.2 Material Data and Bond Coat Oxidation

The base material (CMSX-4) was treated as an entirely elastic material, because of the low stresses which occurred during pure thermal cycling without external creep or fatigue loading. The BC was considered as elastic–viscoplastic, the TGO as elastoplastic, and the TBC as viscoelastic. All material properties were temperature dependent and experimentally determined on actual coatings [15, 16]. The data are summarized in [12, 18]. Creep of BC coat and TBC was generally considered for $T \geq 750\ ^{\circ}\mathrm{C}$. Primary and secondary creep stages were taken into account. The data were implemented using the following recently developed equation:

$$\dot{\varepsilon} = A' \cdot \sigma^{n'} \cdot e^{-\varepsilon/\varepsilon'} + A'' \cdot \sigma^{n''} \cdot e^{-\varepsilon/\varepsilon''} + A \cdot \sigma^{n} \tag{14.1}$$

where $\dot{\varepsilon}$ is the deformation rate, A, A', A'' are pre-factors, n, n', n'' are stress exponents, ε is the strain, ε' and ε'' are model parameters for the primary creep stage, and σ is the stress. The right-hand term covers steady-state creep; the first two cover primary creep. All model parameters are temperature dependent.

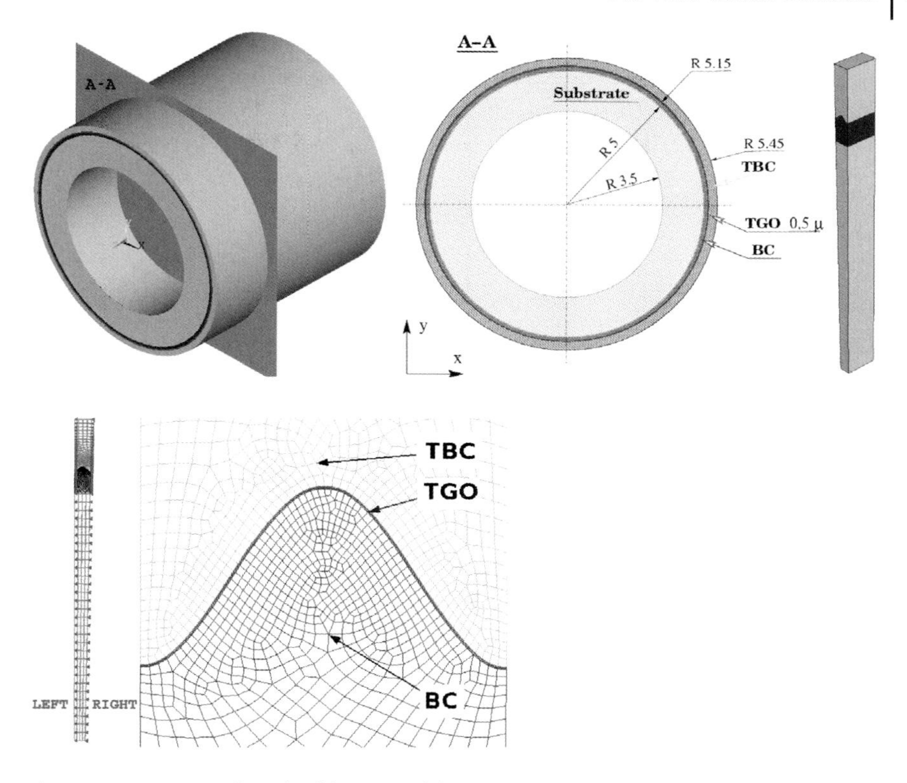

Fig. 14.4 Geometry and mesh of the FE model.

BC oxidation was simulated as a continuous volume increase of the alumina scale at high temperature using the swelling option in ABAQUS. It was modeled as an orthotropic swelling strain of the TGO, whereby lateral TGO growth (length increase) was considered as a constant fraction of thickness growth (generally 5%). The initial thickness of the TGO was defined as 0.5 µm. The oxidation kinetics of this thermal barrier composite were experimentally determined for three different temperatures (950, 1000, 1050 °C) [13] and were implemented for this temperature range using a parabolic time law.

14.3.3 Load Parameters

The simulated load cycle consisted of thermal cycling and high-temperature exposure corresponding to the experiments. Four steps can be distinguished: (1) heating from 20 to 1050 °C in 103 s (10 °C s^{-1}), (2) dwell time at 1050 °C for 2 h, (3) cooling from 1050 to 20 °C in 103 s (10 °C s^{-1}), (4) dwell time at low temperature (20 °C) for 15 min. A temperature of 200 °C was selected at which the TBC composite was initially stress free. It matches approximately with the material temperature during the APS process. The simulations comprised generally 160 load cycles, what amounted to about 30% of life. The TGO thickness was about 5.7 μm after the last cycle.

14.3.4
Simulated Stress Response

All numerical results presented in the following comprise the stress response near the TGO at room temperature after the 160th load cycle. Displayed stresses are radial (out-of-plane) stresses.

Figure 14.5 represents three different stress distributions, whereby the applied material properties were different for each case. The stress response at the left-hand side is a result from an entirely elastic calculation but with continuous TGO growth. Noticeable are rather high and localized compressive and tensile stresses. High tensile stresses (light gray regions) were present in the TBC and in the BC as well as in some smaller regions in the TGO (off-peak). Compressive stresses were developed primarily in the TGO and in the BC below roughness valleys, and also within smaller regions in the BC directly below the peak and in the TBC directly above the valley. The absolute stress values were quite high (>10 GPa and <−10 GPa) and probably not realistic.

The second stress distribution (Fig. 14.5, center) results from a calculation for which the plastic deformation properties of BC and TGO have been additionally taken into account in contrast to the first case. The main effects were an overall stress decrease and some redistribution of local stresses. The largest tensile stress was 1040 MPa and was located directly at the interface bond coat/TGO at the

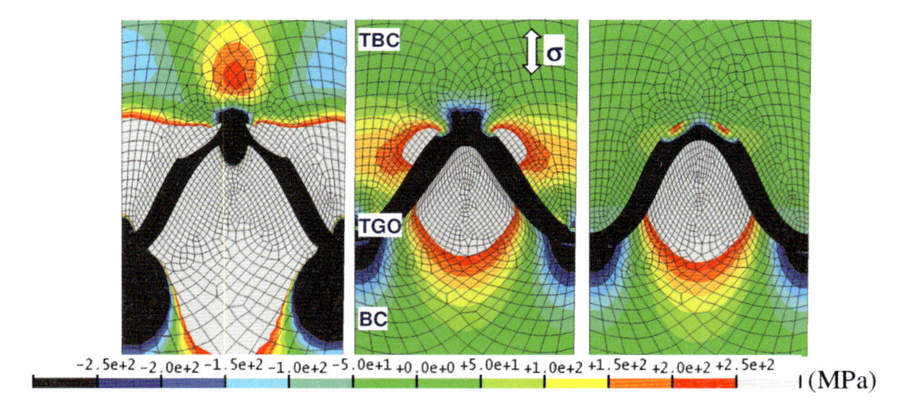

Fig. 14.5 Simulated stress response near the TGO after 160 load cycles at 20 °C with continued TGO growth (here: TGO thickness = 5.7 μm); case 1 (left): all materials elastic; case 2 (center): like case 1, but BC and TGO additionally plastic; case 3 (right): like case 2, but BC and TBC additionally with creep.

roughness peak. At this position, the stresses were changed from compression to tension compared to the first case. The maximum compressive stress occurred directly above the largest tensile stress in the TGO with approximately 2400 MPa. The third case (Fig. 14.5, right) comprised additionally creep in BC and TBC and thus the possibility of stress relaxation. By comparing the stress response at the center and at the right a decrease of the high tensile stresses alongside the roughness peak in the TBC, but also a shape change of the interface becomes apparent. The curvature at the peak became less sharp. This effect was directly due to stress relaxation, which relaxed the entire structure. In contrast, the tensile stresses at the BC/TGO interface were decreased only slightly by less than 10%. However, this region showed the largest tensile stresses.

The material properties of the last case were regarded as the most realistic parameter set and thus taken as a reference parameter set for further calculations. One of the first questions to be looked at was how the stress response is developing with increasing number of load cycles. In particular the stress distribution at 20 °C was of interest, because the largest stresses appeared at the lowest cycle temperature. Figure 14.6 displays the stress response after selected load cycles at 20 °C. Two remarkable features characterize the simulated stress response. At first, high tensile stresses occurred at the BC/TGO interface even after the first load cycle. Thus, early crack formation at this site appears quite likely depending of course on actual interface shape, material properties (deformation properties as well as resistance against crack formation at the interface), and load parameters. The corresponding cyclic furnace tests revealed crack formation at the BC/TGO interface within the first 10% of life (about 50 cycles). For comparison see also Fig. 14.2. Secondly, a coherent lateral region of tensile stresses was developing in the TBC at both sides of the roughness peak indicating higher loaded regions. According to the cyclic tests, the regions in between roughness peaks and over valleys showed remarkable cracking.

Starting from the reference simulation a couple of additional simulations with systematic modifications of certain material parameters, such as coefficient of thermal expansion (CTE), Young's modulus, and creep rate of the materials were carried out to analyze their influence. Here, only the influence of the stiffness variation of the TBC should be described exemplarily. Figure 14.7 displays the stress response associated with the reference parameter set (Fig. 14.7, center) as well as simulation results with a 50% higher (Fig. 14.7, left) and lower (Fig. 14.7, right) Young's modulus of the TBC (reference value: 17.5 GPa). The images show the stress distribution at 20 °C after the 160th cycle.

The stress plots indicate that an increase of the stiffness in the TBC increases the tensile stresses in the TBC as well as the compressive stresses in smaller zones directly above the roughness peak and the roughness valley. Particularly the lateral region of tensile stresses alongside the roughness peak in the TBC which are considered as lifetime relevant increased by the stiffness increase. The stresses in this zone directly at the boundary of the unit cell increased approximately linearly, the stresses directly at the interface TGO/TBC in the TBC are affected more than linearly. Beyond that, the tensile stresses at the BC/TGO inter-

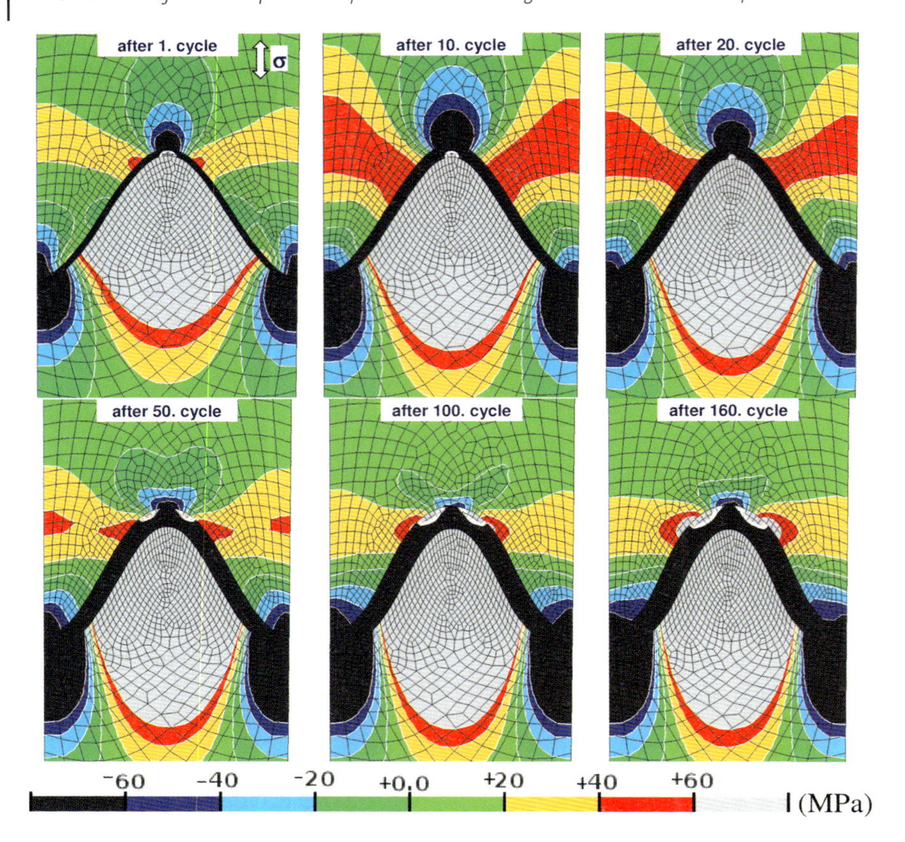

Fig. 14.6 Simulated stress response near the TGO after selected load cycles at 20 °C with continued TGO growth using the reference parameter set.

face increased also, but to a much lesser amount. On the other hand, when the stiffness was decreased the tensile stresses in the TBC were generally lowered.

In addition to the simulation of the stress response, simulations of crack formation and initial crack growth were carried out using cohesive elements [17]. Due to the fact that the highest local tensile stresses were developing at the BC/TGO interface, cohesive elements were implemented directly at the interface. Critical stress values of 600 MPa (normal) and 1200 MPa (shear) were applied for crack formation. For crack opening a critical strain energy release rate of 20 N m^{-1} was taken into account for normal and shear loading. Figure 14.8 represents the stress state after two selected cycle numbers taking into account the reference parameter set. After the 19th load cycle the maximum tensile stress exceeded the critical stress value of 600 MPa at the roughness peak, and a crack was formed during cooling (Fig. 14.8, left).

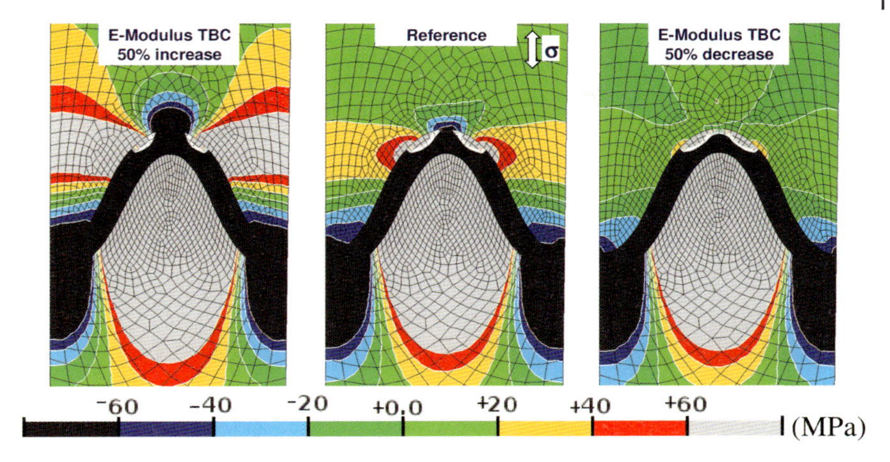

Fig. 14.7 Simulated stress response near the TGO after 160 load cycles at 20 °C with continued TGO growth; the Young's modulus of the TBC was modified by ±50% with respect to the reference parameter set.

The formation of the microcrack at the interface affected the stress field substantially. The tensile stress region directly below the crack was relaxed to a certain amount. In contrast, small regions with high tensile stresses occurred in the TGO close to the crack tips at both sides. After 160 cycles the crack was elongated downwards the roughness profile along the interface at both sides of the peak. However, in between the 19th and the 160th cycle the crack was not growing steadily. As a result of the interaction between thermal stresses, growth stresses, and stress relaxation the crack was after certain cycles partly closed and opened again. No tendency was found for the crack to propagate further downwards the roughness profile up to the 160th cycle. In contrast, the crack tip saw

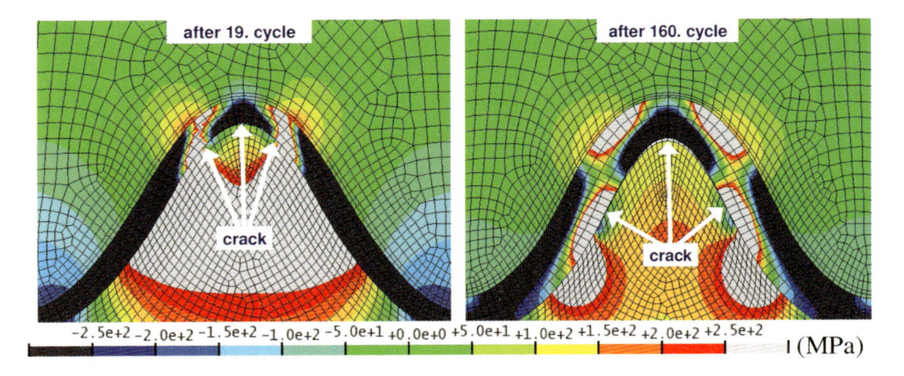

Fig. 14.8 Simulated stress response near the TGO and formation of a microcrack at the BC/TGO interface after 19 cycles and state after 160 cycles at 20 °C.

high tensile stresses in the adjacent TGO indicating a potential bending and a penetration of the crack into the TGO. This behavior would correspond to the frequently observed crack pattern, which is exemplarily shown in Fig. 14.2.

14.4 Conclusions

Simulations of the stress response in a plasma-sprayed thermal barrier system were presented and compared with corresponding cyclic thermal tests including a dwell time of 2 h at high temperature (cyclic furnace tests). The material properties of the MCrAlY bond coat and the ceramic TBC were determined in order to obtain more realistic deformation properties, particularly creep and thus stress relaxation properties as input data for the FE simulations. The simulation results showed that considering a continued oxidation process by simulating a continued TGO growth at high temperatures was required to cover the influence of growth stresses on the stress response. In general, the stress response in and near the TGO at the rough bond coat/TBC interface and its evolution during prolonged cyclic loading was the consequence of the complex interaction of thermally induced stresses, oxidation induced growth stresses and redistribution of stresses due to plastic deformation and even more due to stress relaxation as a result of local creep processes. Pure stress calculations showed two distinct features: (1) the fast development of high tensile stresses in the bond coat with a maximum value directly at the BC/TGO interface below a roughness peak, which occurred during the cooling stage and which were at maximum at the lowest cycle temperature, and (2) the development of a lateral region of larger tensile stresses alongside the roughness peak over roughness valleys. The simulation of crack formation at the BC/TGO interface using cohesive elements resulted in an early formation of a microcrack at the roughness peak (after the 19th cycle). Up to 160 cycles the crack was elongated downwards the roughness profile, but was also partly closed and opened again, and no tendency was observed for the crack to propagate further downwards. Instead, high tensile stresses in the adjacent TGO would suggest a penetration of the crack into the TGO in the upper half of the sinusoidal interface, but this process was not covered by the simulations, at this time. The results of the numericl simulations were in agreement with the experimental observations of crack patterns obtained within the first 30% of life from the corresponding cyclic furnace tests.

Acknowledgments

The authors gratefully acknowledge the financial support of the Deutsche Forschungsgemeinschaft (DFG) within the Collaborate Research Center (SFB) 370 "Integral Modeling of Materials".

Furthermore, the authors thank Priv.-Doz. Dr. R. Vaßen for manufacturing the coatings at the Institute of Materials and Processes in Energy Systems (IWV 1) of Research Centre Juelich.

References

1 G.C. Chang, W. Phucharoen, R.A. Miller, *Surf. Coat. Technol.* 30, 1987, 13.
2 R. Vaßen, G. Kerkhoff, D. Stöver, *Mater. Sci. Eng. A* 303, 2001, 100.
3 G. Kerkhoff, R. Vaßen, C. Funke, D. Stöver, Proceedings of the 6th Liege Conference on Materials for Advanced Power Engineering 1998, p. 1669.
4 M. Ahrens, R. Vaßen, D. Stöver, *Surf. Coat. Technol.* 161, 2002, 26.
5 A.M. Freborg, B.L. Ferguson, W.J. Brindley, G.J. Petrus, *Mater. Sci. Eng. A* 245, 1998, 182.
6 A.G. Evans, D.R. Mumm, J.W. Hutchinson, G.H. Meier, F.S. Pettit, *Prog. Mater. Sci.* 46, 2001, 505.
7 J. Cheng, E.H. Jordan, B. Barber, M. Gell, *Acta Mater.* 46, 1998, 5839.
8 J. Rösler, M. Bäker, M. Volgmann, *Acta Mater.* 49, 2001, 3659.
9 J. Rösler, M. Bäker, K. Aufzug, *Acta Mater.* 52, 2004, 4809.
10 E.P. Busso, J. Lin, S. Sakurai, *Acta Mater.* 49, 2001, 1529.
11 K. Sfar, J. Aktaa, D. Munz, *Mater. Sci. Eng. A* 333, 2002, 351.
12 P. Bednarz, R. Herzog, E. Trunova, R.W. Steinbrech, L. Singheiser, *Ceram. Eng. Sci. Proc.* 26, 2005, 55.
13 H. Echsler, Ph.D. thesis, RWTH Aachen University.
14 E. Trunova, Ph.D. thesis, *submitted to RWTH Aachen University*, Berichte des Forschungszentrums Jülich.
15 R. Herzog, E. Trunova, R.W. Steinbrech, E. Wessel, R. Vaßen, F. Schubert, L. Singheiser, Proceedings of the International Conference on Creep and Fracture in High Temperature Components: Design & Life Assessment Issues, Institution of Mechanical Engineers, London, UK, 2005.
16 P. Majerus, R.W. Steinbrech, R. Herzog, F. Schubert, Proceedings of the 7th Liege Conference, Liege, Belgium, Materials for Advanced Power Engineering 2002.
17 M. Cliez, J.-L. Chaboche, F. Feyel, S. Kruch, *Acta Mater.* 52, 2003, 1133.
18 R. Herzog, P. Bednarz, E. Trunova, V. Shemet, R.W. Steinbrech, F. Schubert, L. Singheiser, Proceedings of the 30th International Conference and Exposition on Advanced Ceramics and Composites, Cocoa Beach, 2006, to be published.

15
Through-Process Multiscale Models for the Prediction of Recrystallization Textures

D. Raabe

Abstract

This chapter reviews multiscale models for the prediction of crystallographic textures formed during static recovery and primary static recrystallization of metals. Two main approaches are presented, namely those that are based on the spatial and temporal discretization of the texture, interfaces, and energy and those which treat these phenomena in a spatially statistical fashion using the Avrami relationship between free and constrained growth.

15.1
Introduction to Recrystallization Models for Process Simulation

The development of models for recrystallization with the aim to predict crystallographic texture, microstructure, and properties during materials processing is a challenging task [1–4]. This opinion is due to the observation that recrystallization phenomena may be very sensitive with respect to small changes in the metallurgical state (e.g. purity, thermodynamics, texture, inhomogeneity, microstructure inheritance from preceding process steps), or external boundary conditions (time, temperature and strain fields, joint thermal and mechanical constraints). The sensitivity of recrystallization is due to the fact that most of the metallurgical mechanisms involved during recrystallization texture formation such as nucleation, grain boundary motion, or impurity drag effects are thermally activated. These mechanisms follow exponential functions the arguments of which may depend sensitively on some of the parameters listed above and interact with each other in a nonlinear fashion. Similarly, the microstructural state of the deformed material from which recrystallization proceeds is often not well known or not well reproduced in models. Although various types of homogenization models of crystal deformation are nowadays capable of providing a decent picture of the average

Integral Materials Modeling: Towards Physics-Based Through-Process Models
Edited by Günter Gottstein
Copyright © 2007 WILEY-VCH Verlag GmbH & Co. KGaA, Weinheim
ISBN: 978-3-527-31711-0

behavior of the material during a thermomechanical process, it is often the details and inhomogeneities in the deformed structure that strongly affect recrystallization so that even a good knowledge of averages does not generally solve the open questions pending in the field of recrystallization modeling. Moreover, in industrial manufacturing processes many of the influencing factors, be they of a metallurgical or of a processing nature, are usually neither well known nor sufficiently well defined to apply models which require a high degree of precision with regard to the input parameters. These remarks underline that the selection of an appropriate recrystallization multiscale model for the prediction of texture, microstructure, and properties for a given process must follow a clear concept as to what exactly is expected from such a model and what cannot be predicted by it in view of the points made above. This applies in particular to cases where a model for the simulation of textures is to be used for real manufacturing processes.

The main challenge of using multiscale process models for recrystallization, therefore, lies in selecting the right model for a well defined task, i.e. it must be agreed which microstructural property is to be simulated and what kind of properties should be subsequently calculated from these microstructure data. From that it is obvious that no model exists which could satisfy all questions that may arise in the context of recrystallization textures. Also one has to clearly separate between the aim of predicting recrystallization textures and microstructures on the one hand and the materials properties (functional or mechanical) on the other. While the first task may by pursued by formulating an appropriate recrystallization model within the limits addressed above, the second challenge falls into the wide realm of microstructure-property theory. This means that one should separate between the prediction of microstructure and properties from a given microstructure. Typically through-process modelers are interested in the materials properties rather than in the details of the microstructure. The modern attitude towards this discrepancy is the understanding that the description of properties requires the use of internal (i.e. of microstructurally motivated) parameters which can be coupled to microstructure–property laws. A typical example of that philosophy would be the grain size prediction via a recrystallization model and the subsequent application of the Hall–Petch law for the estimation of the yield strength. This chapter addresses exclusively the first question; more precisely, only models for primary static recrystallization will be tackled placing attention on the simulation of crystallographic textures. To be more specific the plan of this chapter is as follows. In the first part following this introduction two discrete recrystallization models are concisely discussed with respect to their capability to serve in through-process models for the simulation of textures. The second part presents a discussion of recrystallization models for texture prediction which treat these phenomena in a statistical form.

15.2
Models for Predicting Recrystallization Textures with Discretization of Space and Time

15.2.1
Introduction

The design of time and space discretized recrystallization models for predicting texture and microstructure in the course of materials processing which track kinetics and energy in a local fashion are of interest for two reasons. First, from a fundamental point of view it is desirable to understand better the dynamics and the topology of microstructures that arise from the interaction of large numbers of lattice defects which are characterized by a wide spectrum of intrinsic properties and interactions in spatially heterogeneous materials under complex engineering boundary conditions. For instance, in the field of recrystallization (and grain growth) the influence of local grain boundary characteristics (mobility, energy), local driving forces, and local crystallographic textures on the final microstructure is of particular interest. An important point of interest in that context, however, is the question how local such a models should be in its spatial discretization in order to really provide microstructural input that cannot be equivalently provided by statistical methods. In the worst case a problem in that field may be that spatially discrete recrystallization models may have the tendency to pretend a high degree of precision without actually providing it. In other words even in highly discretized recrystallization models the physics always lies in the details of the constitutive description of the kinetics and thermodynamics of the deformation structure and of the interfaces involved. The mere fact that a model is formulated in a discrete way does not automatically render it a sophisticated model *per se*. Second, from a practical point of view it makes sense to predict microstructure parameters such as the crystal size or the crystallographic texture which determine the mechanical and physical properties of materials subjected to industrial processes on a sound phenomenological basis. In this part particular attention is placed on models which are discrete in both space and time; i.e. on cellular automaton and Potts models because they are frequently used for the simulation of recrystallization phenomena.

15.2.2
Cellular Automaton Models of Recrystallization

Cellular automata are algorithms that describe the discrete spatial and temporal evolution of complex systems by applying local transformation rules to lattice cells which typically represent volume portions [5]. The state of each lattice site is characterized in terms of a set of internal state variables. For recrystallization models these can be lattice defect quantities (stored energy), crystal orientation, or precipitation density. Each site assumes one out of a finite set of possible discrete states. The opening state of the automaton is defined by mapping the initial distribution of the values of the chosen state variables onto the lattice.

The dynamical evolution of the automaton takes place through the application of deterministic or probabilistic transformation rules (switching rules) that act on the state of each lattice point. These rules determine the state of a lattice point as a function of its previous state and the state of the neighboring sites. The number, arrangement, and range of the neighbor sites used by the transformation rule for calculating a state switch determine the range of the interaction and the local shape of the areas which evolve. Cellular automata work in discrete time steps. After each time interval the values of the state variables are updated for all points in synchrony mapping the new (or unchanged) values assigned to them through the transformation rule. Owing to these features, cellular automata provide a discrete method of simulating the evolution of complex dynamical systems which contain large numbers of similar components on the basis of their local interactions. A more formal description of the basics of cellular automata for microstructure modeling is given in [5] and previous applications in the field of recrystallization and recovery are discussed in [6–11].

Cellular automata are – like all other continuum models that work above the discrete atomic scale – not intrinsically calibrated by a characteristic physical length or time scale. This means that a cellular automaton simulation of continuum systems requires the definition of elementary units and transformation rules that reflect the kinetics at the level addressed. If some of the transformation rules refer to different real time scales (e.g. recrystallization and recovery, bulk diffusion, and grain boundary diffusion) it is essential to achieve a correct common scaling of the entire system [1, 3, 11–16]. The requirement for an adjustment of time scaling among various rules is due to the fact that the transformation behavior of a cellular automaton is sometimes determined by no coupled Boolean routines rather than by local solutions of coupled differential equations.

The following examples on the use of cellular automata for predicting recrystallization textures are designed as automata with a probabilistic transformation rule [11–16]. Independent variables are time and space. The latter is discretized into equally shaped cells each of which is characterized in terms of the mechanical driving force (stored deformation energy) and the crystal orientation (texture). The starting data of such automata are usually derived from experiment (for instance from a microtexture map) or from plasticity theory (for instance from crystal plasticity finite element simulations). The initial state is typically defined in terms of the distribution of the crystal orientation and of the driving force. Grains or subgrains are mapped as regions of identical crystal orientation, but the driving force may vary inside these areas.

The kinetics of the automaton result from changes in the state of the cells (cell switches). They occur in accord with a switching rule (transformation rule) which determines the individual switching probability of each cell as a function of its previous state and the state of its neighbor cells. The switching rule is designed to map the phenomenology of primary static recrystallization. It reflects that the state of a non-recrystallized cell belonging to a deformed grain may change due to the expansion of a recrystallizing neighbor grain which grows according to the local driving force and boundary mobility. If such an expanding grain sweeps a non-recrystallized cell the stored dislocation energy of that cell drops to zero and a new orientation is assigned to it, namely that of the expanding neighbor grain. The mathematical formulation of the automaton used in this report can be found in [11–16]. It is derived from a probabilistic form of a linearized symmetric rate equation, which describes grain boundary motion in terms of isotropic single-atom diffusion processes perpendicular through a homogeneous planar grain boundary segment under the influence of a decrease in Gibbs energy. This means that the local progress in recrystallization can be formulated as a function of the local driving forces (stored deformation energy) and interface properties (grain boundary mobility). The most intricate point in such simulations consists of identifying an appropriate phenomenological rule for nucleation events.

Figure 15.1 shows the kinetics and some three-dimensional microstructures sketches of a recrystallizing aluminum sample which had an initial dislocation density of 10^{15} m^{-2} [11]. The simulation used site saturated nucleation conditions, i.e. the nuclei were at $t = 0$ s statistically distributed in physical space and orientation space. The grid size was $10 \times 10 \times 10$ μm^3. The cell size was 0.1 μm. All grain boundaries had the same mobility using an activation energy of the grain boundary mobility of 1.3 eV and a pre-exponential factor of the boundary mobility of $m_0 = 6.2 \cdot 10^{-6}$ m^3 N^{-1} s^{-1}. Small-angle grain boundaries had a mobility of zero. The temperature was 800 K. The time constant of the simulation was 0.35 s.

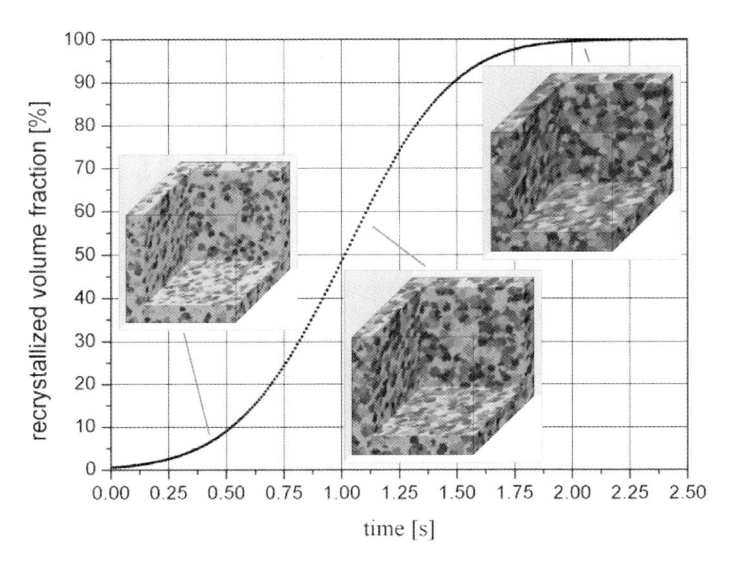

Fig. 15.1 Kinetics and microstructure of recrystallization in a plastically strained aluminum single crystal. The deformed crystal had a (011)[100] orientation and a uniform dislocation density of 10^{15} m^{-2}; site-saturated nucleation conditions at 800 K [11].

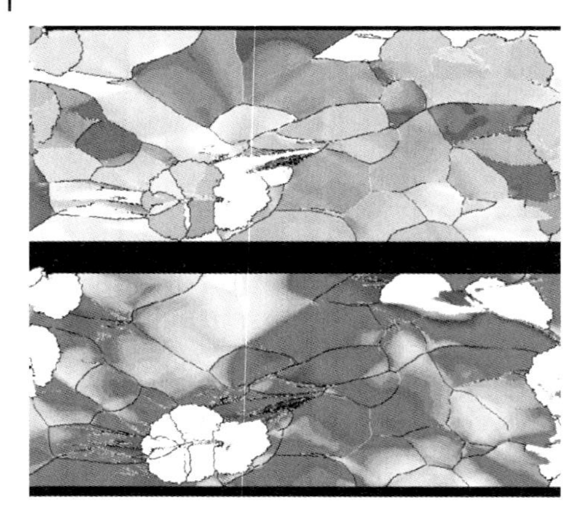

Fig. 15.2 Two-dimensional simulation of recrystallization in Al on the basis of crystal plasticity FE starting data. Bottom: change in dislocation density; top: texture in terms of the magnitude of the rotation angle with respect to the cube orientation (the rotation axis is not indicated). The white areas in the bottom figure indicates a stored dislocation density of zero, i.e. these zones are recrystallized. Parameters: 800 K; site-saturated nucleation in cells with 50% of the maximum dislocation density or above; growth misorientations above 15° at an activation energy of the grain boundary mobility of 1.46 eV and a pre-exponential factor of the grain boundary mobility of $m_0 = 8.3 \times 10^{-3}$ m^3 N^{-1} s^{-1}; details are given in [15].

Figure 15.2 shows an example of a coupling of a cellular automaton with a crystal plasticity finite element model for predicting recrystallization textures in aluminum [15]. The major advantage of such an approach, when compared to the example shown in Fig. 15.1, is that it considers the inherited material deformation heterogeneity as opposed to material homogeneity that was assumed in Fig. 15.1. This type of coupling the two models, therefore, seems more appropriate when aiming at the simulation of textures formed during materials processing. Nucleation in this coupled simulation works above the subgrain scale, i.e. it does not explicitly describe cell walls and subgrain coarsening phenomena. Instead, it incorporates nucleation on a more phenomenological basis using the kinetic and thermodynamic instability criteria known from classical recrystallization theory. The kinetic instability criterion means that a successful nucleation process leads to the formation of a mobile large angle grain boundary which can sweep the surrounding deformed matrix. The thermodynamic instability criterion means that the stored energy changes across the newly formed large angle grain boundary providing a net driving force pushing it forward into the deformed matter.

Nucleation in this simulation is performed in accord with these two aspects, i.e. nucleation sites must fulfill the kinetic and thermodynamic instability criteria. The nucleation submodel does not create new orientations. At the beginning of the simulation the thermodynamic criterion, i.e. the local value of the dislocation density was checked for all lattice points. If the dislocation density was larger than a critical value the cell was spontaneously recrystallized without an orientation change, i.e. a dislocation density of zero was assigned to it, and the original crystal orientation was preserved. In the next step the conventional cellular growth algorithm was used, i.e. the kinetic conditions for nucleation were checked by calculating the misorientations among all spontaneously recrystallized cells (preserving their original crystal orientation) and their immediate neighborhood considering the first, second, and third neighbor shell. If any such pair of cells revealed a misorientation above 15°, the cell flip of the unrecrystallized cell was calculated according to its actual transformation probability. In case of a successful cell flip the orientation of the first recrystallized neighbor cell was assigned to the flipped cell.

Figures 15.3 and 15.4 show a cellular automaton simulation of the recrystallization texture of a 75% cold rolled interstitial free (IF) sheet steel under consideration of Zener pinning [16]. The model is applied to experimentally obtained microtexture electron backscatter diffraction (EBSD) data. The simulation is discrete in time and space. Orientation is treated as a continuous variable in Euler space.

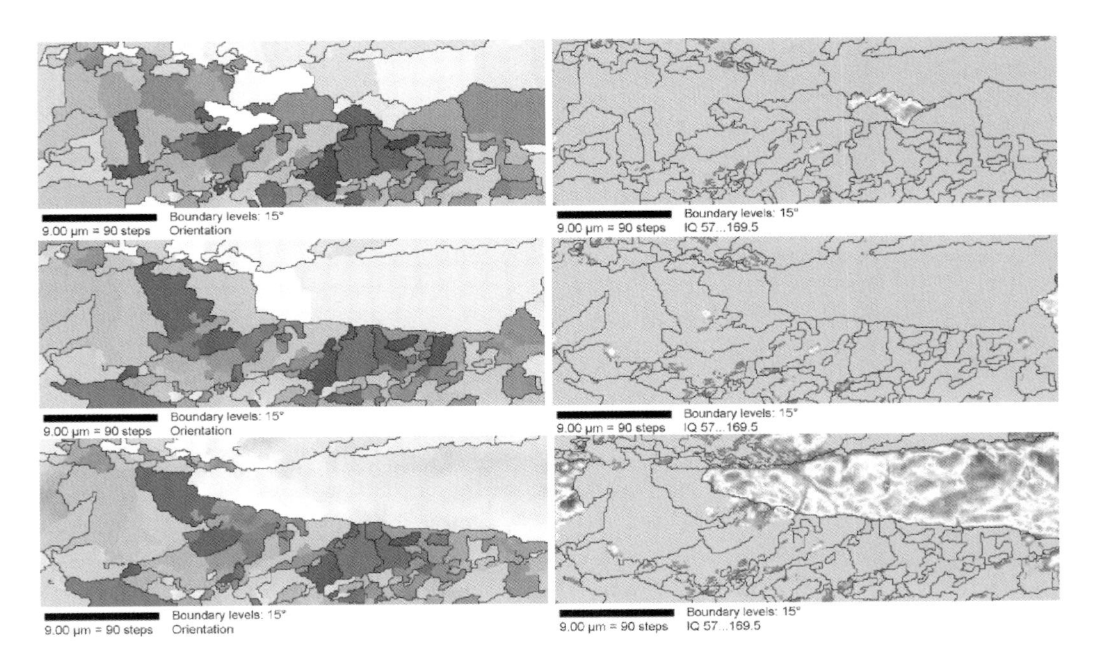

Fig. 15.3 Set of simulated microstructures for different values of the precipitated volume fractions, f, of foreign precipitates in a recrystallizing interstitial free (IF) steel (f: 28, 29, 30% from top to bottom row). The nucleation criterion was that all cells with a dislocation density equal and above 80% of the occurring maximum value undergo spontaneous recrystallization at $t = 0$ s (i.e. critical dislocation density $> 37.6 \times 10^{14}$ m^{-2}); classical Zener pinning. Left: texture map; right: dislocation density map. Details are given in [16].

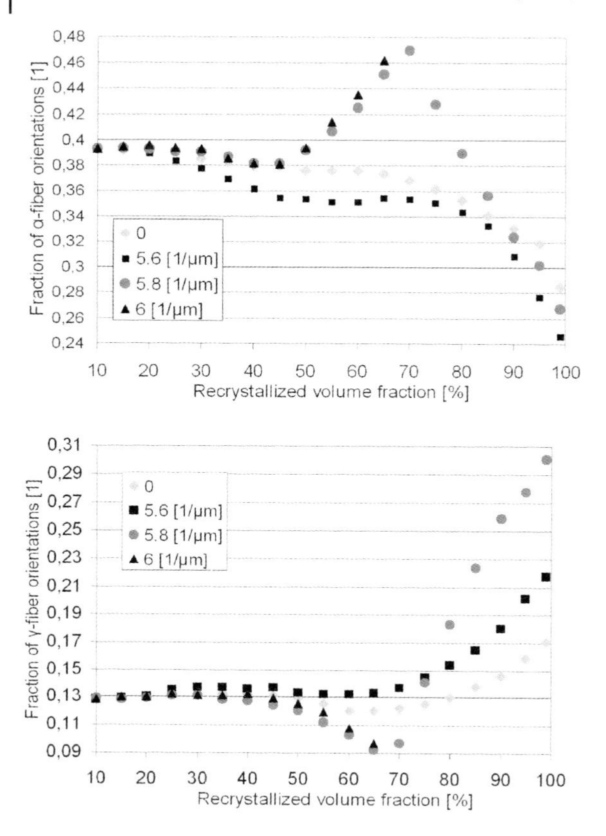

Fig. 15.4 Evolution of the crystallographic texture of the simulation shown in Fig. 15.3 for a recrystallizing IF steel in terms of the α-fiber and the γ-fiber during the simulated annealing under consideration of particle drag for different ratios of the volume fraction and particle radius [16]. The abbreviation f indicates the precipitated volume fractions of second phase particles.

The dislocation density distribution is approximated from the Kikuchi pattern quality of the EBSD data. It is used for calculating the driving forces. Different submodels for nucleation and for the influence of Zener-type particle pinning were tested. Figure 15.3 shows an approximation obtained by using the classic Zener formulation. Time and space calibration of the simulation is obtained by using experimental input data for the grain boundary mobility, the driving forces, and the length scale of the deformed microstructure as mapped by the EBSD experiments. The simulations predict the kinetics and the evolution of microstructure and texture during primary static recrystallization. Depending on the ratio of the precipitated volume fraction and the average radius of the particles the simulations reveal three different regimes for the influence of particle pinning on the resulting microstructures, kinetics and crystallographic textures.

15.2.3
Potts Monte Carlo Models of Recrystallization

The application of the Metropolis Monte Carlo method in microstructure simulation has gained momentum particularly through the extension of the Ising lattice model for modeling magnetic spin systems to the kinetic multistate Potts lattice model [1, 3, 17–24]. The original Ising model is in the form of a $\frac{1}{2}$ spin lattice model where the internal energy of a magnetic system is calculated as the sum of pair-interaction energies between the continuum units which are attached to the nodes of a regular lattice. The Potts model deviates from the Ising model by generalizing the spin and by using a different Hamiltonian. It replaces the Boolean spin variable where only two states are admissible (spin up, spin down) by a generalized variable which can assume one out of a larger spectrum of discrete possible ground states, and accounts only for the interaction between *dissimilar* neighbors. The introduction of such a spectrum of different possible spins enables one to represent domains discretely by regions of identical state (spin). For instance, in microstructure simulation such domains can be interpreted as areas of similarly oriented crystalline matter. Each of these spin orientation variables can be equipped with a set of characteristic state variable values quantifying the lattice energy, the dislocation density, the Taylor factor, or any other orientation-dependent constitutive quantity of interest. Lattice regions which consist of domains with identical spin or state are in such models translated as crystal grains. The values of the state variable enter the Hamiltonian of the Potts model. The most characteristic property of the energy operator for coarsening models is that it defines the interaction energy between nodes with like spins to be zero, and between nodes with unlike spins to be one. This rule makes it possible to identify interfaces and to quantify their energy as a function of the abutting domains.

The Potts model is very versatile for describing coarsening phenomena. It takes a quasi-microscopic metallurgical view of grain growth or ripening, where the crystal interior is composed of lattice points (e.g. atom clusters) with identical energy (e.g. orientation) and the grain boundaries are the interfaces between different types of such domains. As in a real ripening scenario, interface curvature leads to increased wall energy on the convex side and thus to wall migration entailing local shrinkage. The discrete simulation steps in the Potts model, by which the system proceeds towards thermodynamic equilibrium, are typically calculated by randomly switching lattice sites and weighting the resulting interfacial energy changes in terms of Metropolis Monte Carlo sampling. Microstructure simulations using the Potts model have been devoted to recrystallization and grain growth [1, 3, 17–24]. Figures 15.5 and 15.6 show two examples of Monte Carlo (MC) models for predicting recrystallization textures. The example presented in Fig. 15.5 is a Potts MC simulation of subgrain coarsening applied to orientation data of a low carbon steel obtained from TEM orientation measurements inside a single grain. Details on this simulation are given in [25, 26]. Figure 15.6 shows Potts MC simulations of grain and subgrain coarsening applied to

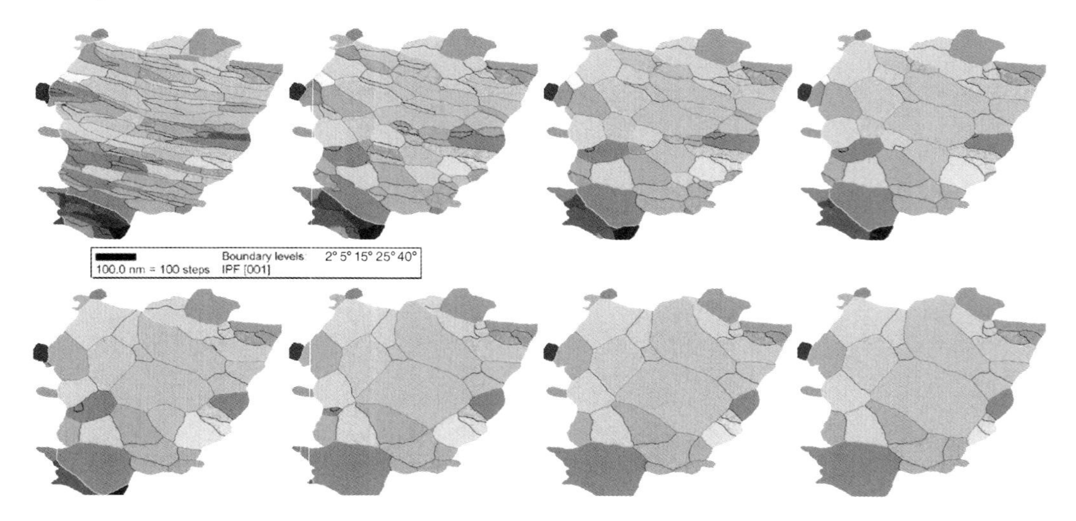

Fig. 15.5 Potts simulation of subgrain coarsening applied to orientation data of a low-carbon steel obtained from TEM orientation measurements inside a single deformed grain. Details on this simulation are given in [25, 26]. The gray scale indicates the grain orientation in terms of one Miller triple.

Fig. 15.6 Potts Monte Carlo simulation of grain and subgrain coarsening phenomena applied to subgrain orientation data of a low-carbon steel obtained from EBSD orientation measurements inside a small group of deformed grains. Details on the simulation are given in [25–27]. The gray scale indicates the affiliation with one of the main texture components occurring in such steels.

subgrain orientation data of a low carbon steel obtained from high-resolution EBSD orientation measurements obtained inside a small group of deformed grains [25, 26].

15.3 Statistical Models for Predicting Recrystallization Textures

15.3.1 The Sebald–Gottstein Model

A very versatile statistical model to predict texture changes during primary static recrystallization on the basis of nucleation and crystal growth is the approach of Sebald and Gottstein [28, 29]. In this model the texture of the deformed material is discretized in terms of a large set of discrete orientations which approximate a given orientation distribution function. The authors have investigated several nucleation mechanisms with respect to the orientation and misorientation distribution they create at the incipient stages of recrystallization. The information provided by the submodels for nucleation is the orientation-dependent density of nuclei within the deformed crystals with a given crystallographic orientation, i.e. the nucleation sub-models initiate the orientation and misorientation distributions of the nuclei. The authors have investigated different types of nucleation processes such as random nucleation which creates a random nucleus texture and a random misorientation distribution; nucleation at shear bands which forms a random nucleus texture with a nonrandom misorientation distribution; and nucleation due to preexisting nuclei which creates a nucleus texture similar to the deformation texture in conjunction with a narrow misorientation distribution.

The growth rate of the nuclei corresponds in the model to their grain boundary velocity which is given by the product of the grain boundary mobility and the local driving force. The mobility of a grain boundary depends on the misorientation between the growing and the deformed grain. The authors distinguish three categories of grain boundaries, namely low-angle, high-angle, and special boundaries. Low-angle grain boundaries are assumed to be essentially immobile. In the case of aluminum it is generally accepted that grain boundaries with $40^\circ\langle 111\rangle$ misorientation may show particularly high mobilities so that such interfaces are treated as special boundaries. This is realized in the simulations of Gottstein and Sebald by assigning higher mobilities to such grain boundaries. The mobility of an average high-angle grain boundary is typically set to 20% of the maximum occurring mobility. The driving force for primary static recrystallization is the difference of the stored energy density between the deformed matrix and the nucleus which in such models typically approximated in terms of the Taylor factor. Growth of the newly formed nuclei is assumed to be isotropic, but it ceases when the nuclei impinge. This means that a growing nucleus can only grow into the non-recrystallized volume fraction. This portion of the material can be calculated according to Avrami who gave a relation between the increase in the recrystallized volume fraction for unconstrained growth (so-called expanded volume fraction) and the actual increase under consideration of impingement for a random spatial distribution. In the model of Sebald and Gottstein this relation is used to confine the growth of the nuclei to the non-recrystallized volume fraction. Details of the method and its mathematical formulation are given in the original works of Sebald and Gottstein [28–30].

15.3.2
A New Texture Component-Based Avrami Model

In this section a texture component-based Avrami model is introduced. The model is based on discretizing both the deformation texture and the texture of the nuclei in the form a small set of Gauss-shaped spherical texture components. The texture components are characterized in terms of their center orientation, volume portion, and scatter width in orientation space. The approach consists of reducing the number of deformed and newly formed recrystallized orientations to only a small number of discrete components. The components which reproduce the deformation texture can be approximated from experimentally obtained pole figures or from single orientation data sets. A fixed set of texture components which may occur as potential nucleation components is then defined which can be fitted from known typical recrystallization textures for the material investigated. For instance in the case of rolled body centered cubic steels the most prominent potential recrystallization texture components are placed on the ⟨111⟩// ND texture fiber (ND is the normal direction of the sheet) [27, 30–38]. This fitting procedure which is required to empirically identify potential recrystallization components does only include the center orientation of the recrystallization texture components but not their scatter width or volume portion. The latter two properties are predicted in the course of the actual simulation.

All nucleation texture components have the possibility to grow into all existing deformation texture components following their individual kinetics for nucleation, mobility, and driving force in each of the different deformation texture components. The kinetic parameters are essentially defined by the activation energy for nucleation, the activation energy for grain boundary mobility, and the stored deformation energy in the various deformation texture components relative to the nucleation texture components that grow inside them.

The kinetic parameters enter for each of the growing components, relative to the deformation components affected, a differential form of the Avrami–Johnson–Mehl–Kolmogorov kinetic equation under the assumption that the set of growing texture components sweeps the deformed texture components homogeneously. It can be shown that this form is equivalent to using a discrete analytical form for the free expansion of the nuclei which is subsequently corrected by tracking the change in the deformed volume for calculating the constrained (or true) expanded volume.

Figures 15.7 and 15.8 show some exemplary results from this statistical texture component-based simulation of primary static recrystallization of an interstitial

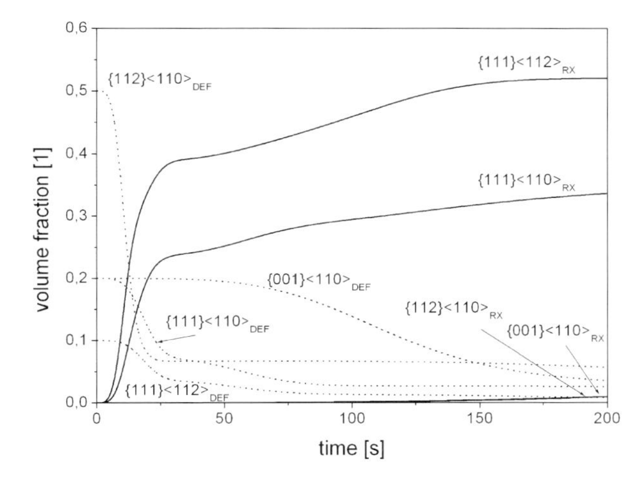

Fig. 15.7 Simulation by use of a texture component-based Avrami model at 900 °C for a rolled IF steel. DEF indicates the volume fraction of the respective texture component in the deformed microstructure; RX indicates the volume fraction of the texture component which is developed during primary recrystallization.

free steel (IF steel) at 900 °C. The index DEF indicates the volume fraction of the respective texture component in the deformed microstructure and the index RX indicates the volume fraction of the recrystallized texture component. Particularly the kinetics of the 45° ND rotated cube component is so slow that the simulated specimen does not undergo complete recrystallization after 200 s.

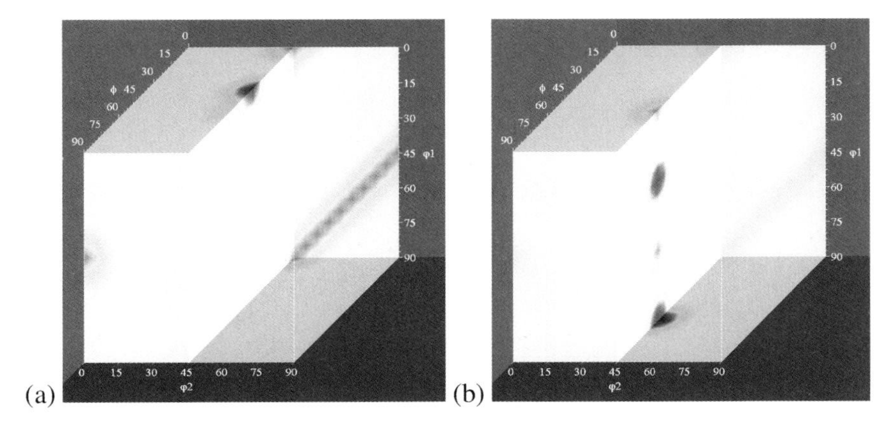

Fig. 15.8 Simulation at 900 °C for an IF steel (see kinetics in Fig. 15.7); orientation distribution functions. (a) Starting texture of the deformed specimen; (b) texture of the simulated recrystallization texture after 200 s. The dark areas show maxima in the orientation distribution functions.

15.4
Input to Recrystallization Models for Texture Prediction

15.4.1
Incorporation of Stored Deformation Energy into Recrystallization Models

Various approaches have been suggested to incorporate the stored deformation energy in recrystallization models for the prediction of crystallographic textures. One important method lies in approximating the stored deformation energy from EBSD data [16, 38, 39]. In this method one uses the Kikuchi pattern quality of the EBSD data with the aim to relate it to the stored dislocations density. The approach is based on the assumption that a high magnitude of the image quality corresponds to a small value of the stored deformation energy (Fig. 15.9). A second common method lies in using the Taylor model. In this approach it is typically assumed that the stored deformation energy for each deformed grain with a given orientation can be related to the Taylor factor. The Taylor factor is the sum of the crystallographic shear for an imposed von Mises strain step as calculated via a polycrystal homogenization models. Two different approaches occur in current recrystallization models when using this method.

One approach is based on the understanding that orientations which have a small Taylor factor correspond to areas with a high deformation potential, and thus to areas with a high stored deformation energy. An alternative view assumes that orientations which have a large Taylor factor correspond to areas with large accumulated strains, and thus to areas with a high stored deformation energy. In either case the Taylor approximation of the stored energy has two drawbacks. First, the method cannot properly capture the inhomogeneity of the deformation.

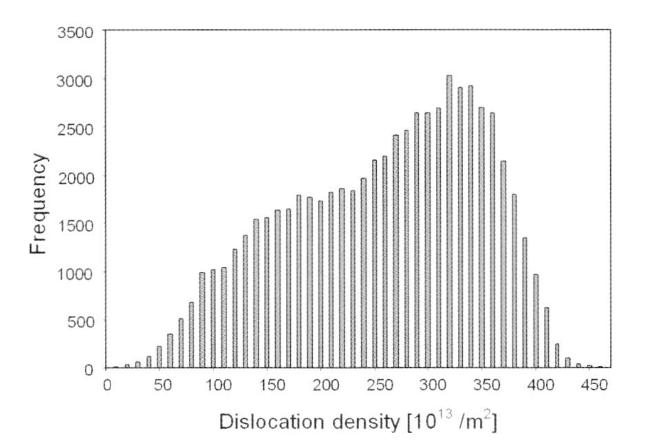

Fig. 15.9 Distribution of the dislocation density in a portion of a deformed steel sample as approximated from the Kikuchi pattern quality of its EBSD data. The maximum occurring dislocation density is equal to 47×10^{14} m^{-2} [16].

Second, the Taylor factor alone only provides a measure of the current deformation state, but it does not consider the deformation history. It is likely though that the accumulation of the stored energy during a deformation path is determined not only by the *final* Taylor factor but by the path that the Taylor factor takes during grain rotation in the course of deformation. A third method for including the stored energy is based on the subgrain substructure. When assuming that the dislocation substructure can be described in form of subgrains with a given radius and a certain interface energy the stored energy can be approximated as the sum of the entire inner surface energy. A fourth and more direct approach for obtaining the stored deformation energy consists in extracting the stored deformation energy directly from a crystal plasticity finite element deformation provided it uses a suited constitutive model [15]. A fifth method for determining local values for the stored deformation energy consists in directly measuring local strains via digital image correlation at the grain scale [40–42].

15.4.2 Grain Boundary Input Parameters into Recrystallization Models for Texture Prediction

Typically it is pertinent in recrystallization models to differentiate between low-angle grain boundaries which have a small mobility and large-angle grain boundaries which have a high mobility [28–30]. Additionally it makes sense to define special boundaries which may have a very large or low mobility [43–46]. Typical examples are twin boundaries which have a low mobility or highly mobile special grain boundaries such as the $40^\circ\langle 111\rangle$ grain boundaries in face-centered cubic metals or the $27^\circ\langle 110\rangle$ boundaries in body-centered cubic metals.

References

1 D. Raabe, *Computational Materials Science*, Wiley-VCH, Weinheim, 1998.
2 G. Gottstein, V. Marx, R. Sebald, *Jpn Inst. Metals Proc.* 13, 1999, 15.
3 D. Raabe, F. Roters, F. Barlat, L.-Q. Chen (eds.), *Continuum Scale Simulation of Engineering Materials*, Wiley-VCH, Weinheim, 2004.
4 F.J. Humphreys, M. Hatherly, *Recrystallization and Related Annealing Phenomena*, Pergamon Press, Oxford, 1995.
5 J. von Neumann, *Collected works of John von Neumann*, Vol. 5, ed. A.W. Burks, Pergamon Press, New York, 1963.
6 H.W. Hesselbarth, I.R. Göbel, *Acta Metall.* 39, 1991, 2135.
7 R.K. Sheldon, D.C. Dunand, *Acta Mater.* 44, 1996, 4571.
8 C.H.J. Davies, *Scripta Mater.* 33, 1995, 1139.
9 V. Marx, F. Reher, G. Gottstein, *Acta Mater.* 47, 1998, 1219.
10 C.H.J. Davies, L. Hong, *Scripta Mater.* 40, 1999, 1145.
11 D. Raabe, *Philos. Mag. A* 79, 1999, 2339.
12 D. Raabe, *Annual Rev. Mater. Res.* 32, 2002, 53.
13 D. Raabe, *Adv. Eng. Mater.* 3, 2001, 745.
14 K.G.F. Janssens, *Mod. Simul. Mater. Sci. Eng.* 11, 2003, 157.
15 D. Raabe, R. Becker, *Mod. Simul. Mater. Sci. Eng.* 8, 2000, 445.
16 D. Raabe, L. Hantcherli, *Comp. Mater. Sci.* 34, 2005, 299.
17 D.J. Srolovitz, G.S. Grest, M.P. Anderson, *Acta Metall.* 34, 1986, 1833.

18 R.D. Doherty, D.J. Srolovitz, A.D. Rollet, M.P. Anderson, *Scripta Metall.* 21, 1987, 675.
19 D. Rollett, D.J. Srolovitz, M.P. Anderson, R.D. Doherty, *Acta Metall.* 40, 1992, 3475.
20 P. Sahni, D. Srolovitz, G. Grest, M. Anderson, S. Safran, *Phys. Rev. B* 28, 1983, 2705.
21 D. Raabe, *Acta Mater.* 48, 2000, 1617.
22 E.A. Holm, C.C. Battaile, *JOM* 9, 2001, 20.
23 D. Rollett, D. Raabe, *Comput. Mater. Sci.* 21, 2001, 69.
24 S.G. Radhakrishnan, H. Weiland, P. Baggethun, *Mod. Simul. Mater. Sci. Eng.* 8, 2000, 737.
25 I. Thomas, doctoral dissertation thesis, Max-Planck-Institut für Eisenforschung and RWTH Aachen, 2006, in preparation.
26 S. Zaefferer, *Mat. Sci. Forum* 495–497, 2005, 3.
27 I. Thomas, S. Zaefferer, F. Friedel, D. Raabe, *Adv. Eng. Mater.* 5, 2003, 566.
28 R. Sebald, G. Gottstein, *Acta Mater.* 50, 2002, 1587.
29 R. Sebald, G. Gottstein, in Proc. ICOTOM 12, Montreal, 1999, Vol. 1, p. 292.
30 R. Sebald, doctoral dissertation thesis, RWTH Aachen Germany, 2001.
31 M. Hölscher, D. Raabe, K. Lücke, *Steel Res.* 62, 1991, 567.
32 W.B. Hutchinson, *Int. Mater. Rev.* 29, 1984, 25.
33 D. Raabe, K. Lücke, *Mater. Sci. Technol.* 9, 1993, 302.
34 D. Raabe, K. Lücke, *Scripta Metall.* 27, 1992, 1533.
35 C. Klinkenberg, D. Raabe, K. Lücke, *Steel Res.* 63, 1992, 227.
36 D. Raabe, *Steel Res.* 66, 1995, 222.
37 P. Juntunen, D. Raabe, P. Karjalainen, T. Kopio, G. Bolle, *Metall. Mater. Trans A* 32, 2001, 1989.
38 S.-H. Choi, *Acta Mater.* 51, 2003, 1775.
39 S.-H. Choi, Y.-S. Jin, *Mater. Sci. Eng. A* 371, 2004, 149.
40 D. Raabe, M. Sachtleber, Z. Zhao, F. Roters, S. Zaefferer, *Acta Mater.* 49, 2001, 3433.
41 J.-C. Kuo, S. Zaefferer, Z. Zhao, M. Winning, D. Raabe, *Adv. Eng. Mater.* 5, 2003, 563.
42 S. Zaefferer, J.-C. Kuo, Z. Zhao, M. Winning, D. Raabe, *Acta Mater.* 51, 2003, 4719.
43 G. Gottstein, L.S. Shvindlerman, *Grain Boundary Migration in Metals: Thermodynamics, Kinetics, Applications*, CRC Press, Boca Raton, FL, 1999.
44 L.S. Shvindlerman, G. Gottstein, Jpn Inst. Metals, 13, 1999, 431.
45 M. Upmanyu, D.J. Srolovitz, L.S. Shvindlerman, G. Gottstein, *Acta Mater.* 47, 1999, 3901.
46 M. Furtkamp, G. Gottstein, D.A. Molodov, V.N. Semenov, L.S. Shvindlerman, *Acta Mater.* 46, 1998, 4103.

16
Analytic Interatomic Potentials for Atomic-Scale Simulations of Metals and Metal Compounds: A Brief Overview

K. Albe, P. Erhart, and M. Müller

Abstract

Atomic-scale simulations of extended systems by means of molecular dynamics or Monte Carlo methods are only feasible if analytic interatomic potentials are used that are computationally efficient. Despite of the tremendous progress in developing realistic and transferable interatomic potentials they still can suffer from severe shortcomings, and there is an ongoing demand for more refined methods that are able to treat reliably a wide variety of atomic configurations. This chapter gives a brief overview of the various model potentials for modeling metals and discusses recent extensions of analytical bond-order potentials for modeling metal compounds.

16.1
Introduction

Over the last decades atomic-scale simulations have evolved as an important tool in condensed matter physics, chemistry, and materials science. Techniques such as molecular dynamics (MD), molecular statics or Monte Carlo (MC) simulations enable a detailed investigation of materials processes and phenomena with atomic resolution. These methods are applied, for example, in the study of point defects, dislocations, grain boundaries, disordered phases, liquids, and amorphous materials. Other applications include modeling of materials processes such as condensation, grain growth, bulk and surface diffusion, thin-film growth, cluster deposition, nanoindentation, sintering, crack growth, and many more. The outcome of atomistic simulations is, however, largely dependent on a realistic description of the interatomic interaction and a reliable model is required as input for either one of these methods.

First-principles techniques, such as density-functional theory (DFT), for calculating the quantum mechanical ground states of condensed matter systems have gained increasing importance in materials research and developed to be a

Integral Materials Modeling: Towards Physics-Based Through-Process Models
Edited by Günter Gottstein
Copyright © 2007 WILEY-VCH Verlag GmbH & Co. KGaA, Weinheim
ISBN: 978-3-527-31711-0

standard tool (see references in [1]), but are computationally expensive and limited to small system sizes. Tight-binding (TB) methods provide an approximate solution of the Schrödinger equation. They work by replacing the exact many-body Hamiltonian with a parameterized matrix and by writing the eigenstates in an atomic-like basis set (see [2] for details). Independent of the specific approach all TB methods have in common that the energy terms, which are not included in the single-electron eigenstates can be approximated by a sum of pair terms. The configurational energy then can be written as sum of the repulsive ϕ_{rep} and band energy ϕ_{band}, which is evaluated by integrating the electronic density of states $\rho(E)$:

$$\Phi = \phi_{\text{rep}} + \phi_{\text{band}} = \sum_{i \neq j} \phi_{ij} + \int_{-\infty}^{E_f} E\rho(E)\,\mathrm{d}E \tag{16.1}$$

TB methods are computationally more efficient than DFT calculations, and on present computers it is possible to carry out MD simulations for of up to a few thousand atoms over several hundred picoseconds.

Many phenomena of interest, however, require one to simulate a much larger number of atoms on extended time scales. Examples are fracture dynamics, dislocation motion, surface phase transitions, and so on. In this regime, atomistic modeling is only possible if the electronic degrees of freedom are removed and interatomic potentials are introduced, which are computationally more efficient and provide a real-space description of bonding energetics. In the hierarchy of modeling, atomistic methods based on such interatomic potentials close the gap between electronic structure calculations and coarse-grained modeling techniques such as dislocation dynamics, cellular automata, and continuum models. They are the method of choice for problems that are computationally not tractable with quantum mechanical schemes, but simple enough to be treated with interatomic potentials. With expanding computing capabilities the range of problems that can be handled by fully quantum-mechanical methods will grow. Similarly, we can expect that in the foreseeable future problems become treatable by atomic scale simulations that can nowadays only be accessed by continuum methods.

In the following, we will give a brief overview of the most established analytic potentials with focus on metals. At the end a specific type of bond-order potentials will be discussed as implemented by our group for several face-centered cubic (fcc) and body-centered cubic (bcc) d-transition metals and their compounds.

16.2 Overview of Established Potential Schemes

In the past a large variety of analytic expression for interatomic potentials has been proposed. Following the categorization of Carlsson [3], we will briefly discuss the various potential schemes and discuss the most prominent examples.

Pair Potentials are mathematical expressions of the form

$$\Phi = \phi_0 + \frac{1}{2}\sum_{i,j} \phi_2(\vec{R}_i, \vec{R}_j) \tag{16.2}$$

where ϕ_0 is a reference energy and ϕ_2 is an effective pair potential with a very small number of adjustable parameters. Because of their simplicity, pair potentials are widely used and have provided many useful insights especially for such problems where it is more important to statistically sample the configuration space rather than to precisely describe the energetics of the system of interest. Pair potentials stabilize close-packed lattices, but they fail to describe open structures. Therefore they are mostly used for densely packed metals and glasses [4]. However, there are a number of inherent shortcomings of pair potential models (see [3] for a complete overview). In cubic materials, for example, they only satisfy the Cauchy relation $C_{12} = C_{44}$, which is violated in most transition metals and semiconductors. Vacancy formation energies are overestimated and as large as cohesive energies, while lattice relaxations around defects and at surfaces are poorly described.

If the potential energy is not a direct sum of pair-like or three-body terms, but a function F of a quantity, that can be an electron density or sum of hopping integrals, then this type of potentials can be classified as **Pair Functional**:

$$\Phi = \frac{1}{2}\sum_{i,j} \phi_2(\vec{R}_i, \vec{R}_j) + \sum_i F\left(\sum_j g_2(\vec{R}_i, \vec{R}_j)\right) \tag{16.3}$$

Pair functionals allow one to mimic the effect of variation in electron density by explicitly describing the local environment of a particular atomic site. In principle two closely related approaches have been established in the last two decades: the second moment tight-binding (SMTB) approximation and the embedded atom method (EAM). The starting point for the tight-binding analysis is the physical expectation that the energetics of d-transition metals are dominated by the width, shape and occupation of the d-band. In most cases, it suffices to consider the width of the d-band only, which is closely related to the second moment μ_2^i of the projected density of states on site i [5]. As the energy of the occupied d-states is proportional to the width of the d-band, the band structure energy (Eq. 16.1) can be written in the second moment approximation as

$$\phi_{\text{band}} = -A\sum_i (\mu_2^i)^{1/2} = -A\sum_i \sqrt{\sum_j h_{ij}^2} \tag{16.4}$$

where h_{ij} is the distance dependent hopping parameter from site i to j. This simple expression for the band energy laid the ground for a number of SMTB potentials, like those proposed by Finnis and Sinclair [6] and Cleri and Rosato [7].

EAM potentials are formally very similar, but they build upon the basic idea that the bonding energy is gained by embedding an atom into a background charge density due to all of the other atoms. In the mid-1980s Daw and others [8, 9] proposed a functional form, where the band energy reads like

$$\phi_{\text{band}} = -A \sum_i F_i(\rho_i) \tag{16.5}$$

with an electron density

$$\rho_i = \sum_j \varphi_j(R_{ij}) \tag{16.6}$$

This has become the most widely applied empirical potential for describing pure metals and alloys. The only difference to the SMTB potential (Eq. 16.4) is the different choice of the *embedding function* F and the treatment of alloys. In the EAM formalism the interaction of the atom with the electron gas is independent of the origin of the gas, but is only given by the electron density at its site. Thus, the embedding function F_i used for the pure metals can be identically applied for alloys. SMTB potentials [7], in contrast, assume that the hopping integrals depend on the types of the pair atoms. Both potential types, however, require repulsive contributions that depend on the type of pair interaction.

Pair functionals have two significant advantages over simple pair potentials. They can describe materials with non-vanishing Cauchy pressure $1/2(c_{12} - c_{44})$ and the vacancy formation energy can differ from the cohesive energy. From the functional form (Eq. 16.3) it becomes obvious, however, that pair functionals cannot be applied to materials in which angular bonding is known to play a crucial role. Even in d-transition metals d-orbitals exhibit a certain degree of angularity that cannot always be described by pair-like potentials. Typical examples are the elastic constants of Pt. All established pair potentials predict an anisotropy ratio $c/c' \gg 2$ of the two cubic shears $c = c_{44}$ and $c' = 1/2(c_{11} - c_{12})$, while the experimental value is about $c/c' = 1.48$ [10].

The use of angular forces to model properties of open structures goes back to Keating [11] who used higher order expansions to describe small deviations from the equilibrium structure. The class of **Cluster Potentials** includes three-body (ϕ_3) and higher order terms that contain information about bond angles and therefore cover a broader range of configurations. They have the general form

$$\Phi = \phi_0 + \frac{1}{2}\sum_{i,j} \phi_2(\vec{R}_i, \vec{R}_j) + \frac{1}{3!}\sum_{i,j,k} \phi_3(\vec{R}_i, \vec{R}_j, \vec{R}_k) + \cdots \tag{16.7}$$

Most cluster potentials include decaying radial terms, so-called cut-off functions, and therefore restrict the interactions to a finite distance. Stillinger and Weber (SW) were the first to propose a cluster potential for silicon which successfully reproduces the equilibrium properties and yields the diamond structure as the most stable conformation [12]. In their model, the energy is given as a sum of a two- and three-body term

$$\phi_3 = g(R_{ij})g(R_{ik})\left(\cos\theta_{ijk} + \frac{1}{3}\right)^2$$

which is composed of radial cut-off functions g and an angular contribution that vanishes for the tetrahedral angle and therefore stabilizes the diamond lattice. Although the angularity is introduced *ad hoc*, the SW potential yields a fair description of the configurations close to the equilibrium structure but exhibits deficiencies when it comes to far-from-equilibrium configurations. Its overall performance is remarkable though and it has become one of the two most widely employed potentials for silicon.

Finally, if higher expansion terms are included in the embedding function one ends up with the more generalized type of **Cluster Functionals**:

$$\Phi = \frac{1}{2}\sum_{i,j}\phi_2(\vec{R}_i, \vec{R}_j) + \sum_i F\left(\sum_j \rho_2(\vec{R}_i, \vec{R}_j), \sum_{j,k}\rho_3(\vec{R}_i, \vec{R}_j, \vec{R}_k), \ldots\right) \tag{16.8}$$

The Tersoff bond-order potential (BOP) is probably the most prominent example, which is describing covalently bonded systems [13]. It is based on the concept of bond order developed by Pauling and Abell. The energy is expressed as a sum over bond energies which are composed of a repulsive V_R and an attractive branch V_A. The latter is moderated by a bond-order term b_{ij} which includes three-body interactions:

$$\Phi = \sum_{i>j} f_{\mathrm{cut}}(R_{ij})\left(V_R(R_{ij}) - \frac{b_{ij} + b_{ji}}{2} V_A(R_{ij})\right) \tag{16.9}$$

TB models with higher moments were introduced by Carlsson [14] as well as Xu and Adams [15] for modeling the transition metals V, W, and Mo. Similarly, Moriarty's generalized pseudopotential theory (GPT) for d-transition metals is an expansion of the total energy with the inclusion of higher moments up to the fourth order [16]. Although the inclusion of higher order terms in semiempirical d-band models provides higher accuracy, because of the better treatment of angularity, these models are computationally expensive and calculating derivatives of fourth-moment terms is a nontrivial task.

Horsefield et al. [17] made the connection between the Tersoff-type potential and higher moment TB methods by introducing an inter-site formulation of the bond order. They showed that Tersoff's bond-order potential is a second-moment approximation which includes a correct treatment of angular contributions due to the presence of many-body contributions.

The modified embedded atom scheme (MEAM) was introduced by Baskes [18]. It utilizes the pair functional of the embedded atom method

$$\Phi = \frac{1}{2}\sum_{i,j} \phi_2(\vec{R}_i, \vec{R}_j) - A\sum_i F(\rho_i) \tag{16.10}$$

and incorporates the angularity in the electron density

$$\rho_i = \sum_j \varphi(R_{ij}) - B\sum_{j,k\neq i}(1 - \cos\theta_{ijk})\varphi(R_{ik})\varphi(R_{jk}) \tag{16.11}$$

where A and B are adjustable parameters, and F is the embedding function. Parameterizations of the MEAM scheme have been published for fcc [18] and hexagonal metals [19]. Recently, extended MEAM potentials which include second-next neighbor interactions were proposed by Lee et al. [20].

Both the Tersoff–Abell and MEAM approaches describe the environmental dependence of the effective pair interactions. In the bond-order scheme, for example, the term $b_{ij} \propto 1/\sqrt{Z}$ is decreasing with increasing coordination Z and is thus reducing the bond strength. The MEAM scheme works in a similar way, but assumes a different functional form for the embedding functional.

16.3 Analytic Bond-Order Potentials (BOP)

Modeling of technologically relevant processes also requires one to consider materials consisting of several components. In the past, (M)EAM potentials have been applied to describe metal alloys [9] and multicomponent glasses [4] by adjusting the pair parameters to the equation-of-states of all relevant binary subsystems. There is also a rich literature on ionic potentials for metal oxides (e.g. [21, 22]). Considerably less effort, however, has been put into the development of potentials for metallic-covalent systems like metal carbides and silicides, although these build another class of technologically important materials. Obviously, the inclusion of angular terms is of vital importance for modeling such systems. For this reason, the MEAM scheme has been applied by Baskes et al. for studying Mo and Ni silicides [23, 24].

A different route was recently taken by Albe and coworkers. They made use of the fact that the SMTB approximation (Eq. 16.4) and the analytic bond-order scheme (Eq. 16.9) are formally equivalent methods, as pointed out by Brenner [25], and proposed to describe transition metals by means of a Tersoff–Brenner-type model potential originally designed for semiconductors [10]. The virtue of this approach, which allows a systematic and transparent fitting of the adjustable parameters, is that one analytic form can describe metallic and covalent interaction. The fitting software PONTIFIX developed by Erhart and Albe [26] allows

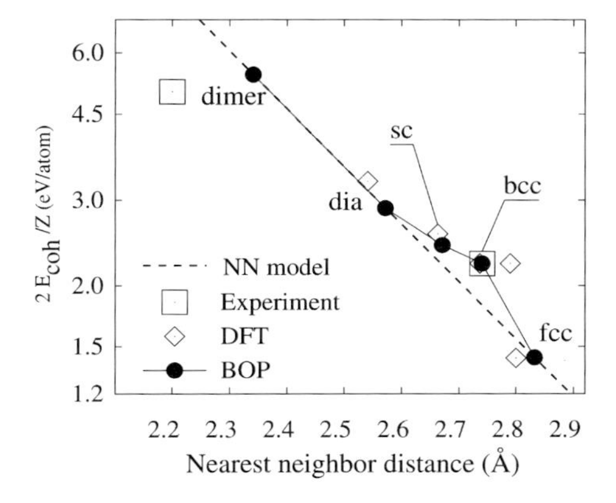

Fig. 16.1 Next neighbor distance and cohesive energy/next neighbor for several phases of solid tungsten. Compared are data obtained from calculations based on density-functional theory, experimental values and results of the tungsten potential [30]. The dashed line represents all possible minima for a next neighbor model. By considering second next neighbors, the bcc structure can be properly fitted.

one to efficiently adjust this scheme to the materials of interest. In the past, potentials for a variety of binary compounds have been derived based on this scheme, including Pt–C [10], Ga–As [27], Ga–N [28], and Si–C [29]. In doing so, not only the mixed phases but also the constituents were fitted, and it could be shown that low symmetry structures, like arsenic and gallium, can be well described. Recently, the method was also applied for modeling bcc tungsten and its carbides [30]. In this context, second next neighbor interactions were included. Figure 16.1 exemplifies one important feature of environment-dependent bond order potentials, namely the capability to properly describe crystal structure differences. In this semi-logarithmic plot, next neighbor distances and bond energies are compared. The BOP for tungsten properly reproduces the energetic differences and is capable of describing deviations from the equilibrium line for a next neighbor model (dashed line). Since volume changes in transitions between fcc and bcc structures are typically very small, while the bond angles change by more than 15%, the explicit inclusion of angular terms is obviously a necessary feature.

Another example, which is of importance for modeling mechanical properties, is the generalized stacking fault (GSF) energy curve. This is obtained by displacing one crystal half along the direction of a Shockley-partial from the position where the lattice is in perfect registry into the stable stacking-fault position. The maximum of this curve is a measure for the Peierls barrier of a moving dislocation and is a decisive quantity to determine the lattice resistance against dislocation nucleation. Both the energy of the stable and unstable stacking fault, however, are material parameters that are difficult to describe by simple pair potentials and pair functionals. This is especially true for materials with a high

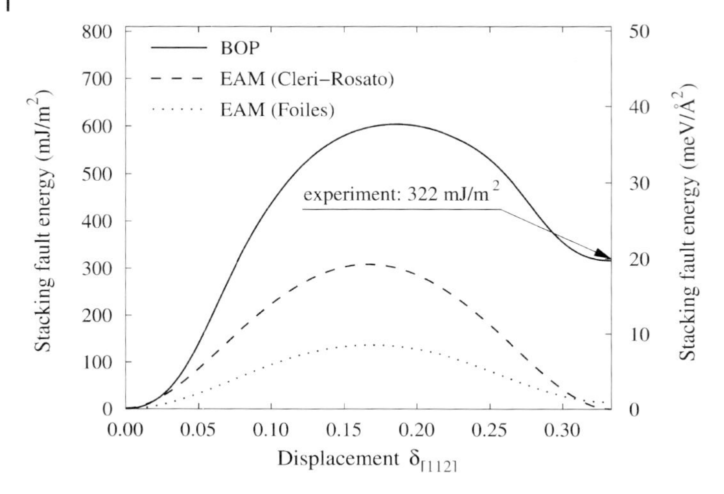

Fig. 16.2 Generalized stacking-fault energy curve for Pt. Shown are results for the EAM potentials of Foiles [32] and Cleri-Rosato [7] in comparison to a second next neighbor BOP for Pt [31].

stacking-fault energy, like platinum. Figure 16.2 shows a comparison of GSF curves calculated with two pair functionals by Foiles and Cleri-Rosato and our newly developed second next neighbor BOP for platinum [31]. The data show that both pair functionals largely underestimate the unstable and stable stacking-fault energies, whereas the BOP method allows one precisely to reproduce the experimental value for the stable stacking fault energy.

16.4 Concluding Remarks

Materials modeling by atomistic methods have matured to a level where a full set of materials properties that in return affect the phenomena observed in simulations can be considered by realistic analytic potentials. Although pair potentials and pair functionals remain a good choice in many circumstances, because they are physically transparent, efficient, and relatively simple to implement, there is a need for more refined methods that are able to reliably treat a wide variety of atomic configurations. In the realm of metals and metal compounds the modified embedded atom method as well as the analytic bond-order scheme provide sufficient flexibility to fulfill this requirement. Analytic interatomic potentials, which involve an approximate treatment of the electronic bonding energy, are certainly less accurate than fully quantum mechanical schemes. This is, however, counterbalanced by the enormous increase in computational efficiency, which allows one to study statistically relevant systems. Thus, atomic-scale simulations based on analytic potentials bridge the gap between the electronic structure calculations and continuum methods on the mesoscale. With increasing computational power there are bright prospects for this type of materials modeling. On the other hand it is evident that the increasing demand for quantitative predictions from computer simulations requires increasingly sophisticated potential descriptions. At the same time, however, we can expect that new physical effects will be observed which advance our current understanding of materials properties and processes. Although this chapter provides only a rough overview, which is far from being complete and by no means comprehensive, we hopefully attracted the reader's attention to an interesting field, which is exciting, but full of challenges and traps, and therefore deserves increasing research efforts.

Acknowledgment

The authors are grateful to the Deutsche Forschungsgemeinschaft for financial support of this work through contract Al 578/1 and SFB 595.

References

1 A.E. Mattsson, P.A. Schultz, M.P. Desjarlais, T.R. Mattsson, K. Leung, *Mod. Simul. Mater. Sci. Eng.* 13, 2005, R1.
2 C.M. Goringe, D.R. Bowler, E. Hernandez, *Rep. Prog. Phys.* 60, 1997, 1447.
3 A.E. Carlsson, in *Solid State Physics: Advances in Research and Applications*, ed. H. Ehrenreich, D. Turnbull, Academic Press, New York, 1990, p. 1.
4 J.M. Delaye, *Curr. Opin. Solid State Mater. Sci.* 5, 2001, 451.
5 F. Ducastelle, F. Cyrot-Lackmann, *J. Phys. Chem. Solids* 31, 1970, 1295.
6 M.W. Finnis, J.E. Sinclair, *Phil. Mag. A* 50, 1984, 45.
7 F. Cleri, V. Rosato, *Phys. Rev. B* 48, 1993, 22.
8 M.S. Daw, M.I. Baskes, *Phys. Rev. B* 29, 1984, 6443.
9 M.S. Daw, S.M. Foiles, M.I. Baskes, *Mater. Sci. Rep.* 9, 1993, 251.
10 K. Albe, K. Nordlund, R.S. Averback, *Phys. Rev. B* 65, 2002, 195124.
11 P.N. Keating, *Phys. Rev.* 149, 1966, 674.
12 F.H. Stillinger, T.A. Weber, *Phys. Rev. B* 31, 1985, 5262.
13 J. Tersoff, *Phys. Rev. B* 39, 1989, 5566.
14 A.E. Carlsson, *Phys. Rev. B* 44, 1991, 6590.
15 W. Xu, J.B. Adams, *Surf. Sci.* 301, 1994, 371.
16 J.A. Moriarty, *Phys. Rev. B* 42, 1990, 1609.
17 A.P. Horsfield, A.M. Bratkovsky, M. Fearn, D.G. Pettifor, M. Aoki, *Phys. Rev. B* 53, 1996, 12694.
18 M.I. Baskes, *Phys. Rev. B* 46, 1992, 2727.
19 M.I. Baskes, R.A. Johnson, *Mod. Simul. Mater. Sci. Eng.* 2, 1994, 147.
20 B.J. Lee, M.I. Baskes, *Phys. Rev. B* 62, 2000, 8564.
21 X.W. Zhou, H.N.G. Wadley, J.S. Filhol, M.N. Neurock, *Phys. Rev. B* 69, 2004, 035402.
22 G.V. Lewis, *Physica B & C* 131, 1985, 114.
23 T.E. Mitchell, M.I. Baskes, S.P. Chen, J.P. Hirth, R.G. Hoagland, *Phil. Mag. A* 81, 2001, 1079.
24 M.I. Baskes, J.E. Angelo, C.L. Bisson, *Mod. Sim. Mater. Sci. Eng.* 2, 1994, 505.
25 D.W. Brenner, *Phys. Rev. Lett.* 63, 1989, 1022.

26 P. Erhart, K. Albe, The software is freely distributed by the authors, 2006.
27 K. Albe, K. Nordlund, J. Nord, A. Kuronen, *Phys. Rev. B* 66, 2002, 035205.
28 J. Nord, K. Albe, P. Erhart, K. Nordlund, *J. Phys. Cond. Matter* 15, 2003, 5649.
29 P. Erhart, K. Albe, *Phys. Rev. B* 71, 2005, 035211.
30 N. Juslin, P. Erhart, P. Träskelin, J. Nord, K.O.E. Henriksson, K. Nordlund, E. Salonen, K. Albe, *J. Appl. Phys.* 98, 2005, 123520.
31 M. Müller, P. Erhart, K. Albe, unpublished.
32 S.M. Foiles, M.I. Baskes, M.S. Daw, *Phys. Rev. B* 33, 1986, 7983.

17
Selected Problems of Phase-Field Modeling in Materials Science

H. Emmerich and R. Siquieri

Abstract

We present a brief introduction to the field of phase-field modeling in computational materials science and discuss some of the latest challenges in the continuing development of this method. This will be illustrated by our own contributions to *quantitative* phase-field investigations of microstructure evolution in multicomponent alloys, as well as recent achievement in employing the phase-field method to investigate details of heterogeneous nucleation kinetics.

17.1
Introduction to Phase-Field Modeling in Materials Science

Microstructure evolution in materials science is usually characterized by two circumstances: first, the material sample as such is driven out of equilibrium, and, second, even more fundamental, at the microscale we can detect interfaces, which separate two in some sense physically unlike regions of the sample from each other.

Due to the nonequilibrium condition one of the two regions or phases will grow at the cost of the other. Examples are phase separation by spinodal decomposition or nucleation and subsequent growth of the nucleus in the nourishing phase [1]. Another example which has often been discussed as a paradigmatic problem is that of dendritic solidification [2–5]. The phenomenological description of these phenomena involves the definition of a precisely located interfacial surface on which boundary conditions are imposed. One of those boundary conditions typically yields a normal velocity at which the interface is moving. This is the so-called *sharp interface* approach, adopted both in analytical and numerical studies for a variety of contexts involving a moving boundary. The origin of such a description is often transparent, being obtained by symmetry arguments and common sense. Nevertheless the properties of sharp interface models can be quite subtle as in the case for dendritic growth. This is strongly coupled to the

Integral Materials Modeling: Towards Physics-Based Through-Process Models
Edited by Günter Gottstein
Copyright © 2007 WILEY-VCH Verlag GmbH & Co. KGaA, Weinheim
ISBN: 978-3-527-31711-0

question of how to view the interfacial surface. Already when introducing the notion of a surface quantity Gibbs implicitly entertained the idea of a diffuse interface [6]: any density of an extensive quantity (e.g. the mass density) between two coexisting phases varies smoothly from its value in one phase to its value in the other. The existence of a transition zone, though microscopically of atomic extent, underlies this definition of surface quantities as given by Gibbs. In phase transition phenomena, this notion has been employed in the spirit of Landau and Khalatnikov [7], who were the first to introduce an additional parameter to label the different phases in their theory on the absorption of liquid helium. Essentially diffuse interface or phase-field modeling, as it appeared subsequently in the literature in the context of phase transition phenomena [8, 9], is connected to such an additional order parameter. Clearly phase-field models have since advanced numerical treatment as well as understanding of microstructure evolution phenomena in materials science.

The most essential numerical gain of the phase-field approach compared to the sharp-interface approach is that it helps to overcome the necessity of solving for the precise location of the interfacial surface explicitly in each time step of a numerical simulation. This can be achieved by the introduction of one or several additional *phase-field* variables. They are the key difference of *phase-field* and *sharp-interface* models. These phase-field variables are continuous fields which are functions of space $\mathbf{r}$ and time t. They are introduced to describe the different physically unlike microstructure regions (phases) of the material sample. Typically these fields vary slowly in bulk regions and rapidly, on length scales of the order of the correlation length ξ, near interfaces; ξ is also a measure for the finite thickness of the interface. The free energy functional A determines the phase behavior. Together with the equations of motion this yields a complete description of the evolution of the system. In other contexts, such as critical dynamics [1, 10, 11], the fields are order parameters distinguishing the different phases. In a binary alloy, for example, the local concentration or sublattice concentration can be described by such fields. If one assumes a phase-field model to describe a physical situation, for which an established sharp-interface formulation exists as well, then, certainly, in the *sharp-interface limit* the phase-field model should correspond precisely to that sharp-interface formulation. However, keeping in mind that the interface can be understood to be of finite width in the sense of Gibbs denoted above, not only the *sharp-interface limit* of a phase-field model is a meaningful physical limit, but also the so-called *thin-interface limit* introduced by Karma and Rappel [12, 13]. To clarify the difference between the *sharp-interface limit* and this *thin-interface limit* here I will consider the growth of a dendrite with tip radius R into an undercooled melt [14]. Under more general circumstances, R might be representative of a typical macroscopic length scale such as the container size. For dendritic solidification at large undercoolings the growth is rapid and the radius of curvature of the dendritic tip is relatively small. As a consequence effects of capillary action and kinetics on the local interfacial temperature can be significant. In this regime, sharp-interface limits of the phase-field equations have been performed [15–20], which assume that the dimensionless interfacial temperature u is of the order of the small parameter ξ/R. Contributions from capillary effects and kinetics can be regarded to be of the same order. In this limit one also considers ξ to be small compared to the capillary length l_c, which presents a stringent resolution requirement for a numerical computation that aspires to describe this limiting case. At low undercoolings, on the other hand, dendrites grow more slowly and have a larger radius of curvature, so that it is reasonable to model capillary effects and kinetics as small corrections. Karma and Rappel refer to the corresponding analysis as the *thin-interface limit*. For this thin-interface limit one assumes $\xi \ll R$ but allows $\xi \sim l_c$. Almgren [21] has described this analysis as *isothermal asymptotics*, since to leading order in ξ/R the temperature is isothermal throughout the interfacial region with $u = O(\xi/R)$.

Momentarily diffuse interface modeling is a field in which numerical efforts as well as an intense focus on thermodynamic backgrounds and asymptotic behavior of the models drives the development to turn this approach into a more and more powerful technique with strong activities in the field we will briefly survey in Section 17.2.

17.2
Overview on Recent Issues in the Further Development of Phase-Field Modeling

The efforts described above have resulted into very elaborate model formulations [22] as well as very elaborate numerical implementations as for example described in [23] which by now allow one to simulate the growth of a single dendritic microstructure by at the same time taking into account long-range transport fields in reasonable time and high resolution. Thereby such approaches constitute an important contribution to carry out the relevant parameter studies for computational materials design, i.e. the parameter studies which allow one to understand the relation between processing parameters and microstructure evolution and – at best – also the materials properties at the macroscale of a material systems. This shows that there is an inherent challenge to computational materials design due to the nature of materials properties evolution in any processing step itself: The latter is essentially a multiscale dynamics, i.e. a dynamics where different evolution paths occurring at different length and time scales are strongly coupled to each other. On this background it is quite easy to understand that in the further development of the phase-field method in the context of computational materials design a lot of activities are concerned with this "scale-bridging" issue. Basically three ways have emerged in the community to do so. The first is to design innovative algorithms which couple different computational techniques originally designed for complementary scales as, for example, a DLA (Diffusion Limited Aggregation) or LBA (Lattice Boltzmann Automata) schemes to a phase-field model [24]. The second is to use advanced numerical techniques as multigrid, adaptivity, and parallelization to do fast computation for several scales based on a single-model approach. A third possibility arises from analytics, i.e. rigorous homogenization methods where one identifies the most relevant dynamical processes at each scale and develops a scale-bridging model based on these via expansion techniques [25]. Apart from these "scale-bridging" efforts, however, likewise noteworthy broader directions of further development of the phase-field method have emerged.

17.2.1
Evolution of Nanostructures in Condensed Matter Systems

What is remarkable about this point is that due to the continuum field nature of the phase-field approach one would claim that it should not be valid at the nanoscale. However, due to the successes of continuum approaches in microfluidics it appears to be justifiable to proceed with phase-field models for phase transition problems of similar physical nature at this scale, as well. Indeed quite successful studies could be carried out already [27]. Also the idea to employ phase-field models to investigate heterogeneous nucleation dynamics as described below in Section 17.3 is based upon this underlying physical picture.

17.2.2
Dynamics in Soft-Matter Systems at the Micro- and Nanoscale

Only recently colloids have started to establish themselves as model substances for material systems as e.g. metallic alloys. Colloids belong to the class of soft-matter materials, thus to a class which displays a structural length scale between several nanometers and one micrometer and appears macroscopically soft. From an experimental point of view the parameters which quantify the nucleation and growth kinetics of these material systems can be measured more easily than in, for example, metallic melts. This concerns among others the nucleation rate, the interfacial energies, as well as the kinetic and capillary anisotropies of a phase interface. For this reason the investigation of nucleation and initial growth dynamics in an interdisciplinary, colloidal, and traditional metallic material systems spanning comparative study effort, seems to be promising, of which phase-field modeling has to be an integral part, has been instituted [26].

Another soft-matter model system is that of a liquid capsule enclosed by a thin elastic shell. Looking at this model problem one realizes that modeling its small-scale dynamics shares again a lot of challenges with today's problems to obtain reliable micromechanical models for the dynamics in "traditional" material systems as metallic ones. For liquid capsules these problems have recently received considerable attention in cellular biology, bioengineering, and microencapsulation technology (e.g. [28] and references therein). Lately also in this field, phase-field models have been successfully applied, e.g. to the dynamics of microscopic vesicle membranes as red blood cell membranes: Du et al. [29] were able to compute the equilibrium configurations of a vesicle membrane under elastic bending energy, with prescribed volume and surface area based on the phase-field method. Biben et al. [30] extended this phase-field model for vesicle membranes to hydrodynamic flow.

Certainly the above survey cannot claim completeness; however, it reflects to a large extent the content of "phase-field-dominated" sessions at related conferences such as the EUROMAT, the annual meeting of the DPG and the APS spring meeting. In all of the above the further development of phase-field models benefits largely from strong collaborations with experimental partners to verify models but also to get a better feeling for the accuracy of model parameters and basic mechanisms essential to grasp in any model description. In the following we will explain in more detail challenges and achievements related to *quantitative* phase-field modeling, i.e. phase-field modeling in the thin interface limit, of multicomponent alloys for the example of peritectic solidification. We will then employ the same model approach to describe how phase-field modeling has only recently also been used by us to investigate heterogeneous nucleation kinetics.

17.3 "Quantitative Phase-Field Simulations" of Nucleation and Growth in Peritectic Material Systems

A great challenge doing "quantitative" simulation of microstructure evolution of material systems is to take into account the effect of hydrodynamics in the melt on the emerging structures. To simulate flow with free or moving surfaces quite enhanced schemes have been developed in the past decades (e.g. [31] and references therein). Including phase change in such problems makes the task more demanding. A further degree of difficulty arises in the context of peritectic solidification at the microscale, since in this context special attention has to be paid to the multiphase transformation dynamics under the influence of convection on the one hand, as well as to the heterogeneous nucleation kinetics of peritectic material systems on the other. Here we will demonstrate that the phase-field approach allows us to tackle both parts of the problem and thus qualifies very well as a comprehensive approach to simulate such systems quantitatively.

In modeling nucleation it is essential to realize that the solid–liquid interface is known to extend to several molecular layers. This has been indicated by experiments [32], computer simulations [33], and statistical mechanical treatments based on the density functional theory [34]. The need to pay particular attention to this diffuse interface results from the fact that for nucleation the typical size of critical fluctuations is comparable to the physical thickness of the interface. The success of such careful treatment can be seen in modern nucleation theories for homogeneous nucleation, which do consider the molecular scale diffuseness of the interface. These theories could remove the many orders of magnitude difference seen between nucleation rates from the classic sharp interface approach and experiment [35]. In heterogeneous nucleation we face a still more complex situation, since the principle degrees of freedom of the process are larger than in homogeneous nucleation: First of all each phase can nucleate separately. Moreover, several phases can nucleate jointly, i.e. approximately at the same space and time. Finally one phase can nucleate on top of the other. Here we are particularly interested in peritectic material systems. Even though many industrially important metallic alloy systems as well as ceramics such as the high-T_C superconductor YBCO are peritectics, much less is known about microstructural pattern formation in peritectic growth [36] than for example in eutectic growth. In such peritectic material systems it is particularly relevant to understand the nucleation of the peritectic phase on top of the properitectic phase in detail, since this is the nucleation process yielding the stationary growth morphology. For this specific nucleation process the precise configuration of the properitectic phase, i.e. its free energy on the one hand and its morphology on the other [37], should contribute to the precise nucleation rate. Nevertheless classic nucleation theory predicts the following nucleation rate for the nucleation of the peritectic phase on top of the properitectic one: if we assume the properitectic front to be planar, classically the nucleation rate reads $I = I_0 e^{-\Delta F^*/k_B T}$, where I_0 is a constant prefactor (with dimension equal to the number of nucleations per unit volume per unit time) and ΔF^* is the activation energy for heterogeneous nucleation. Assuming further that the critical nucleus is a spherical cap on a planar substrate (the spherical cap model), ΔF^* is given, respectively, in two and three dimensions by

$$\Delta F^* = \begin{cases} \dfrac{\gamma^2{}_{\beta L}}{\Delta F_B} \times \dfrac{\theta^2}{\theta - (1/2)\sin 2\theta} & \text{2D} \\ \dfrac{\gamma^3{}_{\beta L}}{\Delta F_B^2} \times \dfrac{16\pi(2+\cos\theta)(1-\cos\theta)^2}{12} & \text{3D} \end{cases} \tag{17.1}$$

where ΔF_B is the difference between the bulk free energy of the peritectic phase and of the liquid phase. The contact angle θ is determined by the balance of surface tensions parallel to the substrate, $\gamma_{\alpha L} = \gamma_{\alpha\beta} + \gamma_{\beta L}\cos\theta$. If the system is at local thermodynamic equilibrium, it can be shown that ΔF_B is proportional to $(T - T_P)$ [41], such that for a quasi-two-dimensional system the nucleation rate for the peritectic phase on the properitectic one can be written as

$$I = \begin{cases} I_{2D}\exp[-A/(T-T_P)^2] & \text{if } T < T_P \\ 0 & \text{if } T \geq T_P \end{cases} \tag{17.2}$$

where A is a constant and I_{2D} now has the dimension of number of nucleations per unit time and per unit length of the interface. The 3D form of ΔF^* is used in deriving Eq. (17.2) since, in practice, the size of a nucleus is still much smaller than the thickness of a thin sample. Equation (17.2) determines the classic local nucleation rate and hence the probability per unit time of a nucleus forming as a function of the local temperature at the solid–liquid interface. Thus morphological and energetical contributions to Eq. (17.2) resulting from the properitectic microstructure as discussed in [37] are neglected classically. In the following we derive for the first time a phase-field model approach for peritectic growth taking into account hydrodynamics in the molten phase, which is capable of treating this open issue. We will describe this new approach in detail in Section 17.3.1.

Also in Section 17.3.1 we investigate microstructure growth in a peritectic system under the influence of hydrodynamic convection in the melt. We will then report on first numerical investigations of the nucleation kinetics in such peritectic material systems, in particular on a morphological contribution from the properitectic phase to Eq. (17.2) in Section 17.3.2. Moreover, we will discuss the relation of our results to classic nucleation theory in Section 17.3.2. Finally we will conclude with a discussion of the general impact of our new approach for peritectic materials under the influence of convection and an outlook.

17.3.1 A Quantitative Phase-Field Model for Peritectic Growth Taking Into Account Hydrodynamic Convection in the Molten Phase

The starting point of our phase-field modeling approach for heterogeneous nucleation is the free energy functional of a representative volume of the investigated material system. This free energy functional is given by the volume integral

$$F = \int_V f\, dv \tag{17.3}$$

with the free energy density defined as

$$\begin{aligned} f = {} & \frac{W(\theta)^2}{2} \sum_i (\nabla p_i)^2 + \sum_i p_i^2 (1 - p_i)^2 \\ & + \tilde{\lambda} \left[\frac{1}{2} \left[c - \sum_i A_i(T) g_i(p) \right]^2 + \sum_i B_i(T) g_i(p) \right] \end{aligned} \tag{17.4}$$

where the interface width depends on the orientation of the interface and is given by $W(\theta) = W_0(1 + \varepsilon_4 \cos 4\theta)$, with ε_4 being the measure of the anisotropy, $\theta = \arctan \partial_y p_i / \partial_x p_i$, and $\tilde{\lambda}$ being a constant.

The coefficients are defined as

$$g_i = \frac{p_i^2}{4} \{15(1 - p_i)[1 + p_i - (p_k - p_j)^2] + p_i(9p_i^2 - 5)\} \tag{17.5}$$

$$A_i(T) = c_i \mp \frac{(k_i - 1)(T - T_p)}{m_i \Delta C}, \quad B_i(T) = \mp \frac{A_i(T - T_p)}{m_i \Delta C}$$

$$\text{and} \quad A_L = B_L = 0 \tag{17.6}$$

Here k_i are the partition coefficients and A_L and B_L are the liquid coefficients. We use three phase fields $p_i \in [0, 1]$, where $\sum_{i=1}^{3} p_i = 1$. The p_i label the properitectic, the peritectic, and the liquid phase, respectively, i.e. $i = \alpha$ (for the properitectic phase), $i = \beta$ (for the peritectic phase), and $i = L$ (for the liquid phase). Their dynamics are derived from the free energy functional F

$$\frac{\partial p_i}{\partial t} = \frac{1}{\tau}\frac{\delta F}{\delta p_i}$$

where τ is a relaxation time. The concentration field is given by

$$\frac{\partial c}{\partial t} + p_L v \cdot \vec{\nabla} c - \vec{\nabla} \cdot \left(M(p_i)\vec{\nabla}\frac{\delta F}{\delta c} + \hat{n}_L \frac{W}{2\sqrt{2}} \sum_{i=\alpha,\beta} A_i \frac{\partial p_i}{\partial t}(n_i \cdot n_L) \right) = 0 \tag{17.7}$$

where $M(p_i)$ is a mobility, $\hat{n}_i = \vec{\nabla} p_i / |\vec{\nabla} p_i|$ are unit vectors normal to i–L interface, and $\hat{n}_i \cdot \hat{n}_L$ prevents solute exchange between two solids. Moreover, the velocity of the hydrodynamic field is given by the equation

$$\frac{\partial p_L v}{\partial t} = -p_L v \cdot \vec{\nabla} v - p_L \vec{\nabla} p + \frac{1}{Re}\nabla^2 p_L v + M_1^2 \tag{17.8}$$

which is a modified Navier–Stokes equation, where $Re = \frac{\rho U}{\upsilon}$ with ρ being the liquid density, U the inflow velocity, and υ the liquid viscosity. M_1^2 is a dissipative interfacial force per unit volume and is modeled as in [45].

To our knowledge this is the first phase-field model convergent to the underlying sharp-interface problem in the thin-interface limit. The respective asymptotic analysis is summarized in [46]. This model allows us for the first time to investigate the peritectic transformation under the influence of convection quantitatively. Comparing peritectic growth with and without convection we find that hydrodynamic transport in the melt enhances the growth process considerably. This relation between melt flow and solidification dynamics is summarized in Fig. 17.1, where two pictures of growing microstructures are given at the same set of parameters except that the right microstructure opposed to flow whereas to the left growth proceeds purely diffusion limited. The dark circle indicates the properitectic phase, the light structure the peritectic phase, which is nucleating on top

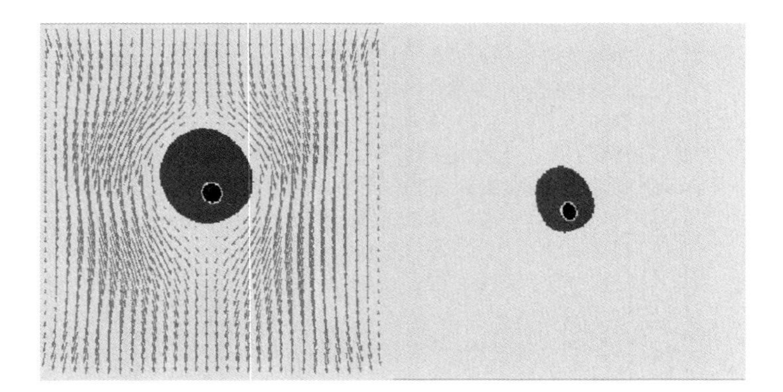

Fig. 17.1 Comparison of peritectic growth with and without convection.

of the properitectic one. The arrows on the left picture are indicating the velocity of the hydrodynamic field in the molten phase. These results are in qualitative agreement with experimental investigations of the peritectic material system Nd–Fe–B in [40].

17.3.2
Investigating Heterogeneous Nucleation in Peritectic Materials via the Phase-Field Method

In solidification experiments the final microstructure is determined by both the peritectic growth dynamics as well as the microstructure growth kinetics. Therefore, for a full quantitative comparison to experiments, it is essential to analyze the heterogeneous nucleation kinetics of the above peritectic material system as well. For such a system a nucleation event arises as a critical fluctuation, which is a nontrivial time-independent solution of the governing equations we can derive from the underlying free energy functional. Our derivation follows the standard variational procedure of phase-field theory (for a review see [42]). Solving the equations numerically under boundary conditions that prescribe bulk liquid properties far from the fluctuations ($p_i \rightarrow 1$, and $c \rightarrow c_\infty$ for $r \rightarrow \infty$) and zero-field gradients at the center, one obtains the free energy of the nucleation event as $\Delta F^* = F - F_0$. Here F is obtained by numerically evaluating the integration over F after having the time-independent solutions inserted, while F_0 is the free energy of the initial liquid. Based on ΔF^* the homogeneous nucleation rate is calculated as $I = I_0 \exp\{-\Delta F^*/kT\}$, where the nucleation prefactor I_0 of the classical kinetic approach is used, which proved consistent with experiments [43].

As introduced in Section 17.1, in a peritectic material sample it is particularly relevant to understand the nucleation of the peritectic phase on top of the properitectic phase in detail, since this is the nucleation process yielding the stationary growth morphology. As demonstrated previously via analytical predictions and Monte Carlo studies (e.g. [37, 39]), for this specific nucleation process the precise configuration of the properitectic phase, i.e. its free energy on the one hand and its morphology on the other, should contribute to the precise nucleation rate. This, as well as experimental evidence for deviations from classic nucleation theory in the system Nd–Fe–B [44], motivated us to study the effect of two morphological features of the properitectic phase on the nucleation rate of the peritectic one, namely (1) the effect of facettes and (2) the effect of its radius. In Fig. 17.2a and b we summarize our findings. As can be seen from Fig. 17.2a, the less facetted the properitectic phase, the larger is the nucleation probability for a peritectic nucleation on top of it. For the contribution resulting from the radius of the properitectic phase a similar relation is true: the larger the radius of the properitectic phase, the larger the probability of a peritectic nucleation on top of it. Both findings are in qualitative agreement with the following atomistic picture: unfacetted nuclei offer a great number of surface kinks for nucleation. This also holds for nuclei of small radii. However, small radius nuclei are also subject to large surface diffusion due to kink flow [38]. This overrides the first effect such

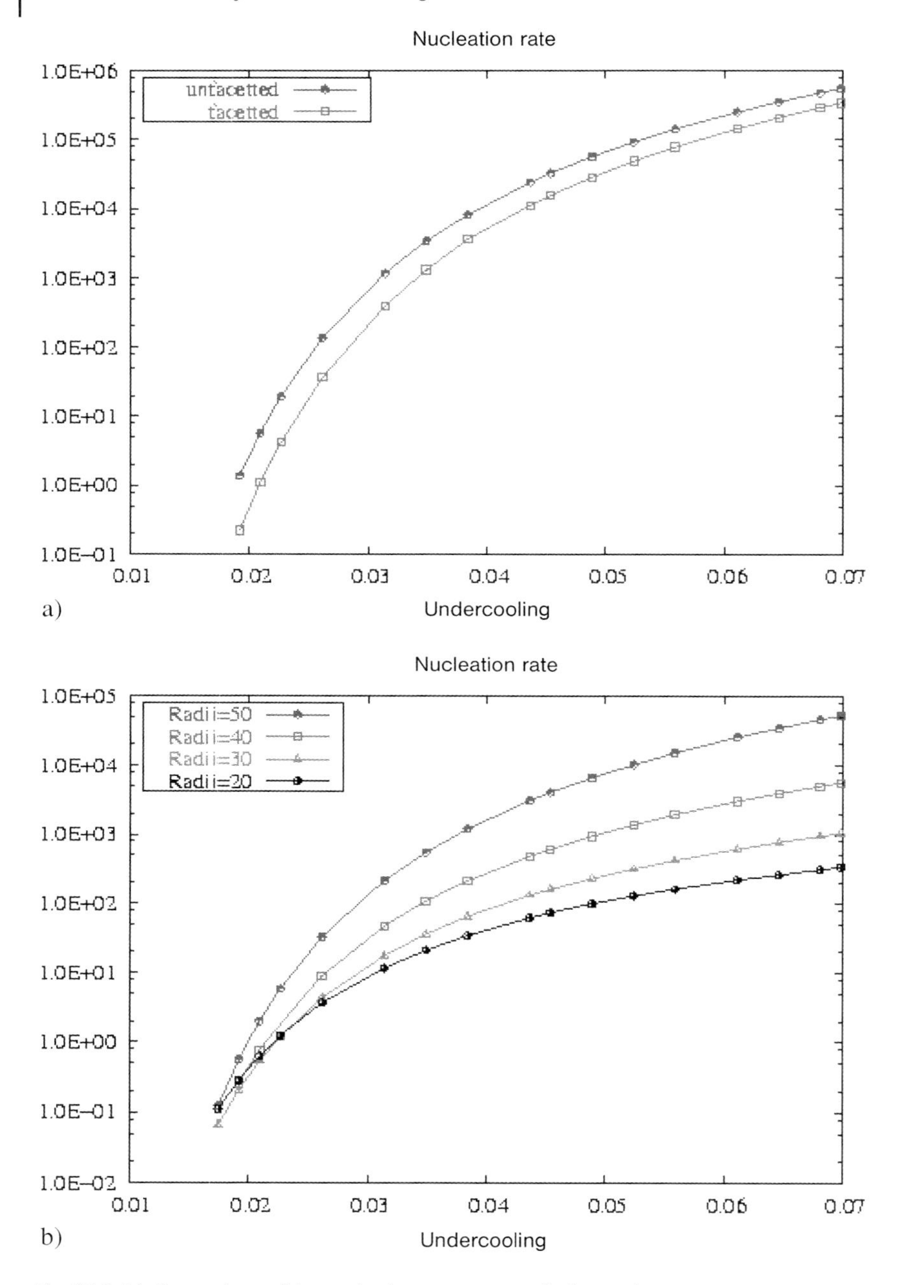

Fig. 17.2 (a) Comparison of the nucleation rate on top of a facetted nucleus to the one on top of an unfacetted nucleus. Obviously for the first case the rate is much larger than for the second. This is in qualitative agreement to analytical predictions in [41]. (b) Comparison of the nucleation rate on unfacetted nuclei of different radii. As can be seen, the larger the radius the larger the nucleation rate.

that the overall nucleation rate turns out to be smaller for smaller radii. Moreover, these findings are in qualitative agreement with [37] and thus provide a first qualitative validation for our new approach towards heterogeneous nucleation. However, it should be noted that the atomistic picture is just given for a common sense estimation of what our model should do. In the continuum picture underlying our investigations, the differences of the different curves arise due to the fact, that the total surface energy tied to the diffuse surface area of the properitectic nucleus depends on its morphology. Thus, the latter naturally has an impact on the nucleation rate just as indicated experimentally. This can be analyzed in more detail making use of the phase-field profiles at the stationary point [46].

17.4 Conclusions and Outlook

To summarize, in this chapter we have introduced basic issues where phase-field modeling can be expected to contribute to further understanding in the future and where at the same time the method itself will be developed further to tackle still more complex problems of computational materials design.

References

1 J.D. Gunton, M. San Miguel, P. Sahni, in *Phase Transitions and Critical Phenomena*, Vol. 8, ed. C. Domb, J.L. Lebowitz, Academic Press, London, 1983, p. 267.
2 E. Ben-Jacob, N. Goldenfeld, J.S. Langer, G. Schön, *Phys. Rev. A* 29, 1984, 330.
3 B. Caroli, C. Caroli, B. Roulet, in *Solids Far From Equilibrium*, ed. G. Godrèche, Cambridge University Press, Cambridge, 1992, p. 155.
4 J.B. Collins, H. Levine, *Phys. Rev. B* 31, 1985, 6119.
5 J.S. Langer, *Rev. Mod. Phys.* 52, 1980, 1.
6 J.W. Gibbs, *T. Conn. Acad.* 2, 1873, 382.
7 D. ter Haar, *Collected Papers of Landau*, Gordon and Breach, Washington, DC, 1967.
8 J.W. Cahn, J.E. Hilliard, *J. Chem. Phys.* 28, 1958, 258.
9 B.I. Halpering, P.C. Hohenberg, S. Ma, *Phys. Rev. B* 10, 1974, 139.
10 A.J. Bray, *Adv. Phys.* 32, 1994, 357.
11 P.C. Hohenberg, B.I. Halperin, *Rev. Mod. Phys.* 49, 1977, 435.
12 A. Karma, W.-J. Rappel, *Phys. Rev. E* 53, 1996, R3017.
13 A. Karma, W.-J. Rappel, *Phys. Rev. E* 57, 1998, 4323.
14 M.E. Glicksman, S.P. Marsh, in *Handbook of Crystal Growth*, Vol. 1b, ed. D.T.J. Hurle, Elsevier, Amsterdam, 1993, p. 1075.
15 G. Caginalp, *Arch. Rat. Mech. Anal.* 92, 1986, 205.
16 J.W. Cahn, P. Fife, O. Penrose, *Acta Mater.* 45, 1997, 4397.
17 P.C. Fife, in *CBMS-NSF Regional Conference Series in Applied Mathematics*, Vol. 53, SIAM, Philadelphia, PA, 1988, p. 11.
18 P.C. Fife, O. Penrose, *Electron. J. Diff. Equations* 1, 1995, 1.
19 E. Fried, M.E. Gurtin, *Physica D* 72, 1994, 287.
20 G.B. McFadden, A.A. Wheeler, R.J. Braun, S.R. Coriell, R.F. Sekerka, *Phys. Rev. E* 48, 1993, 2016.
21 R. Almgren, SIAM, *J. Appl. Math.* 59, 1999, 2086.
22 A. Karma, M. Plapp, *Phys. Rev. Lett.* 84, 2000, 1740.

23 J.-H. Jeong, N. Goldenfeld, J. Dantzig, *Phys. Rev. E* 64, 2001, 041602.
24 W. Miller, S. Succi, D. Mansutti, *Phys. Rev. Lett.* 86, 2001, 3578.
25 H. Emmerich, Ch. Eck, *Cont. Mech. Thermodynam.* 17, 2006, 373.
26 DFG Focus Program 1296 Nucleation and Growth Kinetics in Colloids and Metals: Steps Towards a Scale- and System-Bridging Understanding, instituted by the German Research Foundation, April 2006.
27 D. Kim et al., *Nanotechnology* 15, 2004, 667.
28 C. Pozrikidis, *J. Fluid Mech.* 440, 2001, 269.
29 Q. Du, C. Liu, X. Wang, *J. Comput. Phys.* 198, 2004, 450.
30 T. Biben, K. Kassner, C. Misbah, *Phys. Rev. E* 72, 2005, 041921.
31 M.F. Tomé and S. McKee, *J. Comp. Phys.* 110, 1994, 171.
32 W.J. Huisman, J.F. Peters, M.J. Zwanenburg, S.A. de Vries, T.E. Derry, D. Albernathy, J.F. van der Veen, *Nature* 390, 1997, 379.
33 R.L. Davidchack, B.B. Laird, *J. Chem. Phys.* 108, 1998, 9452.
34 R. Ohnesorge, H. Löwen, H. Wagner, *Phys. Rev. E* 50, 1994, 4801.
35 L. Gránásy, F. Iglói, *J. Chem. Phys.* 107, 1997, 3634.
36 W.J. Boettinger, S.R. Coriell, A.L. Greer, A. Karma, W. Kurz, M. Rappaz, R. Trivedi, *Acta Mater.* 48, 2000, 43.
37 K.A. Jackson, in *Growth and Perfection of Crystals*, ed. R.H. Doremus, B.W. Roberts, D. Turnbull, Wiley, New York, 1958, p. 319.
38 O. Pierre-Louis, M.R. D'Orsogna, T.L. Einstein, *Phys. Rev. Lett.* 82, 1999, 3661.
39 V.A. Shneidman, K.A. Jackson, K.M. Beatty, *Phys. Rev. B* 59, 1999, 3579.
40 O. Filip, R. Hermann, L. Schultz, *J. Mater. Sci. Eng. A* 375–377, 2004, 1044.
41 J.W. Christian, *The Theory of Transformations in Metals and Alloys*, 1st edn, Pergamon Press, Oxford, 1965, p. 537.
42 H. Emmerich, *The Diffuse Interface Approach in Material Science: Thermodynamic Concepts and Applications of Phase-Field Models*, Springer monograph, Lecture Notes in Physics LNPm 73, 2003.
43 K.F. Kelton, *Solid State Phys.* 45, 1991, 75.
44 J. Strohmenger, T. Volkmann, J. Gao, D.M. Herlach, *J. Mater. Sci. Eng. A* 375–377, 2004, 561.
45 C. Beckermann, H.-J. Diepers, I. Steinbach, A. Karma, X. Tong, *J. Comput. Phys.* 154, 1999, 468.
46 H. Emmerich, R. Siquieri, in preparation.

18
Prediction of Microstructure and Microporosity Development in Aluminum Gravity Casting Processes

M. Schneider, W. Schaefer, G. Mazourkevitch, M. Wessén, I.L. Svensson, S. Seifeddine, J. Olsson, C. Beckermann, and K. Carlson

Abstract

A practical tool for the modeling of microstructural and microporosity development during the solidification of aluminum gravity castings is presented. The microstructure model combines consideration of the macroscopic energy transport in the casting with factors governing microstructural growth such as the local thermodynamic equilibrium based on the alloy composition, growth kinetics for those phases that are stable and can grow at a given temperature, and microsegregation of the alloying elements. An approach based on microsegregation of gas dissolved in the melt is used to model pore formation and growth during solidification. Gas species transport in the melt is coupled with the simulation of feeding flow and calculation of the pressure field in the melt. The rate of pore growth is calculated based on the local level of gas super saturation, taking into consideration the local microstructure. Preliminary comparisons with measurements of both predicted microstructure and porosity distributions are used to illustrate the use of the simulation tool.

18.1
Introduction

The increasing demands being placed on cast components, for example in the automotive industry, mean that it is increasingly important to take advantage of the full potential of cast alloys. In the last decade, casting process simulation has established itself as an accepted tool in the foundry industry. By analyzing the layout of gating and feeding systems, casting process simulation is used as an important tool for quality assurance and process optimization within the foundry. However, current developments in casting simulation go well beyond filling and feeding problems to address the quality of the cast component, considering all of the relevant parameters (melt composition, inoculation, local solidification condi-

Integral Materials Modeling: Towards Physics-Based Through-Process Models
Edited by Günter Gottstein
Copyright © 2007 WILEY-VCH Verlag GmbH & Co. KGaA, Weinheim
ISBN: 978-3-527-31711-0

tions, etc.). Information about the local properties in a casting is of interest to both the foundry operative and the casting designer and allows the generation of casting designs that meet service requirements and can be produced cost effectively.

In this chapter, a practical tool for the modeling of microstructural development during the solidification of aluminum gravity castings is presented. The model combines consideration of the macroscopic energy transport in the casting with factors governing microstructural growth such as the local thermodynamic equilibrium based on the alloy composition, growth kinetics for those phases that are stable and can grow at a given temperature, and microsegregation of the alloying elements. The result is a prediction of the three-dimensional distribution of the primary and eutectic phase fractions and dendrite arm spacings. Using information about the distribution of phases and their structures throughout the casting, the local mechanical properties can also be determined.

The presence of defects such as porosity leads to a reduction in both the static and dynamic properties of a casting. For this reason, the prediction of the local porosity level in an aluminum casting is critical for quantifying the local casting quality. To reach this goal, an approach based on microsegregation of gas dissolved in the melt is used to model pore formation and growth during solidification. Gas species transport in the melt is coupled with the simulation of feeding flow and calculation of the pressure field in the melt. The rate of pore growth is calculated based on the local level of gas supersaturation, taking into consideration the local microstructure. With the model, the porosity distribution in the solidified casting can be predicted. Preliminary comparisons with measurements of both predicted microstructure and porosity distributions are used to illustrate the use of the simulation tool.

18.2
Microstructure Modeling

Depending on the local cooling conditions in the casting and the alloying element content, a diverse range of phases and structures may form in Al–Si-based alloy castings. In the work discussed here, a deterministic modeling approach is used where the phase transformations are described using kinetic models on the microscopic level. As illustrated schematically in Fig. 18.1, these models are coupled with phase diagram routines, consideration of local microsegregation, and the macroscopic heat transfer in the casting system (including the local latent heat evolution) [1]. At present, the phases considered in the simulation are the liquid, α-Al, Si, β-AlFeSi, Al_2Cu, and Mg_2Si.

The macroscopic transport of heat during solidification is described by the energy equation in the form

$$\rho C_p \frac{\partial T}{\partial t} = \nabla \cdot \{k \nabla T\} + \rho \Delta H \frac{\partial f_s}{\partial t} \tag{18.1}$$

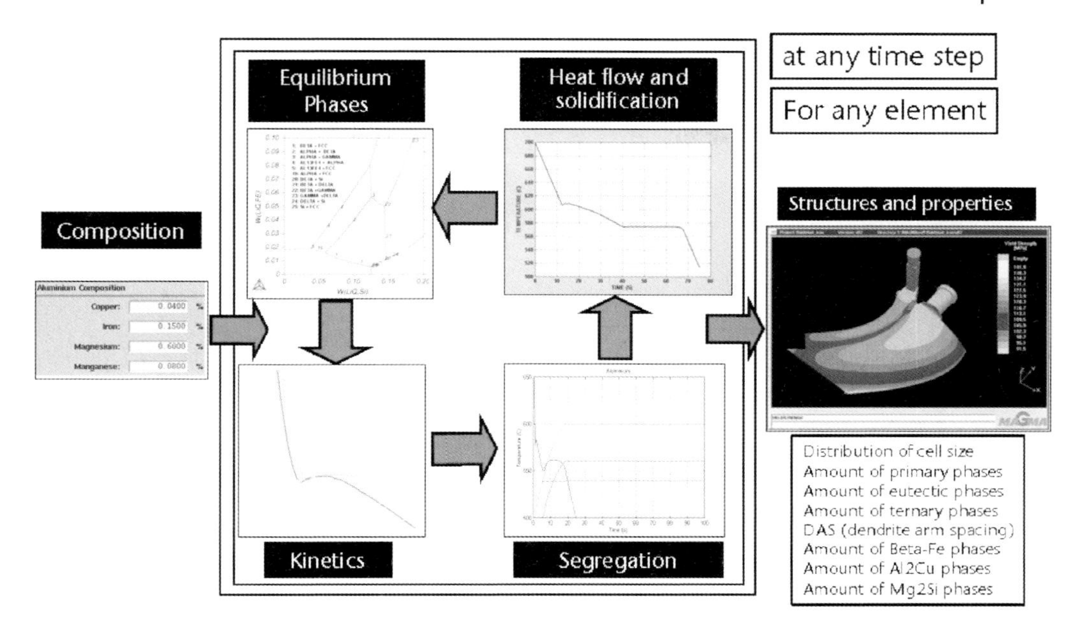

Fig. 18.1 Schematic illustration of the phenomena that need to be accounted for when considering the modeling of microstructure development for Al–Si-based cast alloys.

where k is the thermal conductivity, ρ is the mixture density, C_p is the mixture specific heat capacity, ΔH is the latent heat, and f_s is the local fraction of solid. The evolution of the solid fraction can be obtained from the grain growth velocity, $\mathrm{d}R/\mathrm{d}t$, and the number of grains N_v by

$$\frac{\mathrm{d}f_s}{\mathrm{d}t} = N_v 4\pi R^2 (1 - f_s) \frac{\mathrm{d}R}{\mathrm{d}t} \tag{18.2}$$

where the term $(1 - f_s)$ accounts for grain impingement. Based on comparisons between measurements and simulation, it has been determined that the nucleation of grains can best be described using a time dependent fading law as described by Svensson and Millberg [2]

$$N_\mathrm{v} = At^m \tag{18.3}$$

where the constants A and m need to be determined experimentally and t is the time from mold filling until the equilibrium temperature is reached.

The kinetics of the grain growth are described as a function of the undercooling below the equilibrium temperature T_eq by

$$\frac{dR}{dt} = B(T - T_{eq})^b \tag{18.4}$$

where the constants B and b again need to be determined for each phase from experiments as a function of alloy content.

Changes in the melt composition during solidification are modeled by assuming Scheil-like microsegregation

$$dC_l^i = C_l^i \frac{(1-\kappa^i)}{1-f_s} df_s \tag{18.5}$$

where C is the weight fraction of the alloying element i in the liquid phase and κ^i is the corresponding partition coefficient.

Since a significant number of phases, and therefore complex eutectics, can form in the alloy systems under consideration, it is not feasible to approximate the phase diagram in a "quasi-binary" manner. Instead, nonlinear equations have been used to describe the liquidus surface for each phase p as a function of liquid composition

$$T_p^{eq} = T_o + \sum_{i=1}^{j} m_{1i} C_l^i + m_{2i} (C_l^i)^2 \tag{18.6}$$

Depending on the local solidification path, a new eutectic is allowed to nucleate and grow when the equilibrium temperature of the previous eutectic or primary precipitated phase falls below any other phase considered. In this way, it is possible to simulation the fraction of the different phases formed.

Extensive experiments have been performed (cylindrical castings with different mold materials and DSC measurements) in order to determine the parameters describing nucleation and grain growth in Eqs. (18.3) and (18.4). Samples that were virtually free from defects were also produced using a gradient furnace and then used for tensile testing. The details of the model and the corresponding experiments are documented elsewhere [1, 3, 4].

Using the gradient furnace samples, the dependence of the mechanical properties on the microstructure *alone* (with a very low content of defects such as oxide films or shrinkage and gas porosity) could be determined. In this way, the local mechanical properties were correlated to the alloy content, phase fractions and local mechanical properties [1].

18.3
Microporosity Modeling

The model for porosity formation builds on earlier work by Carlson et al. [6], using a multiphase approach where it is assumed that the porosity and the solidified metal phases are stationary. The mixture continuity equation and a liquid phase momentum equation are solved to determine the pressure and feeding flow velocity fields in the solidifying casting. The mixture continuity equation, considering the liquid (l), solidified metal (s), and porosity (p) phases, can be written as [6]

$$\nabla \cdot \mathbf{v} = -\frac{1}{\rho_l}\left\{\frac{\partial}{\partial t}[f_s(\rho_s - \rho_l) + \rho_l - f_p(\rho_l - \rho_p)] + \mathbf{v} \cdot \nabla \rho_l\right\} \tag{18.7}$$

where f designates the volume fraction of the respective phase, ρ the density of the phases and $\mathbf{v}$ denotes the superficial liquid velocity, $\mathbf{v} = f_l \mathbf{v}_l$. The terms on the right-hand side of Eq. (18.7) show that the feeding flow (divergence in velocity) is balanced by solidification contraction, liquid density change due to for example superheat, the formation and growth of porosity, and gradients in the liquid density (which are typically small).

If the inertial terms are neglected, the liquid momentum equation is given as [6]

$$\nabla^2 \mathbf{v} = \frac{f_l}{K}\mathbf{v} + \frac{f_l}{\mu_l}\nabla P - \frac{f_l}{\mu_l}\rho_{\text{ref}}\mathbf{g} \tag{18.8}$$

where P is the melt pressure, $\mathbf{g}$ is the gravity vector, ρ_{ref} is a reference liquid density, and μ_l is the dynamic viscosity of the liquid metal. The last two quantities are assumed to take the values of the corresponding properties at the liquidus temperature. The mushy zone permeability is modeled as $K = (\lambda_2)^2(f_l)^3/[180(1-f_l)^2]$, where λ_2 is the secondary dendrite arm spacing (SDAS). Note that in purely liquid regions, the permeability becomes very large and Eq. (18.8) reduces to Stokes' equation. In the mushy zone, the left-hand side of Eq. (18.8) becomes very small when compared to the permeability term, and the equation reduces to Darcy's law.

As for the other elements present in the melt, the concentration of hydrogen changes during solidification due to solubility differences in the solid and liquid. Similar to Eq. (18.5) the average concentration of hydrogen dissolved in the melt is obtained from the mixtures species conservation equation [6]

$$\frac{\partial}{\partial t}(f_s\rho_s C_s^H + f_l\rho_l C_l^H + f_p\rho_p C_p^H) + \nabla \cdot (\rho_l C_l^H \mathbf{v}) = 0 \tag{18.9}$$

where for fast diffusion in the solid $C_s^H = \kappa^H C_l^H$, and if the porosity is composed of hydrogen only, the hydrogen concentration in the pore is $C_p^H = 1$.

Previous models of porosity have typically assumed the local diffusion of hydrogen in the melt to be infinitely fast (see Lee et al. [7] for a review or [8, 9] for recent examples). However, Lee and coworkers [10–14] have shown, both experimentally and theoretically, that diffusion of hydrogen to the pore can be the rate limiting factor in porosity growth for aluminum alloys.

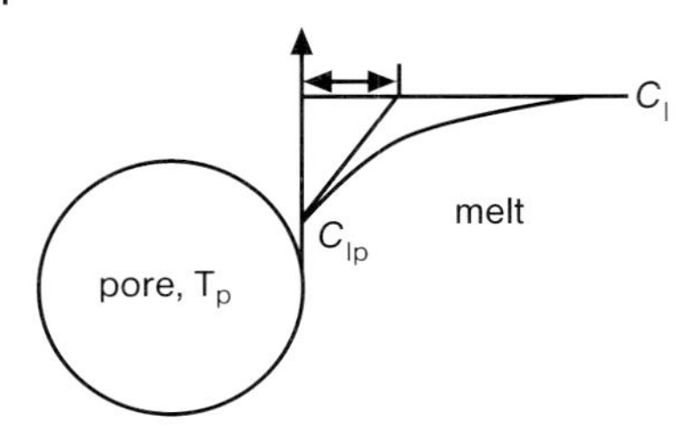

Fig. 18.2 Hydrogen diffusion boundary layer around a growing pore.

In the present model, these effects are accounted for using an equation describing hydrogen microsegregation derived by combing the gas species conservation equation for the pore phase with the gas species balance at the interface between the pore and liquid. Using the fact that the gas concentration in the pore is unity (see above), the resulting equation is

$$\frac{\partial}{\partial t}(f_p \rho_p) = \frac{S_{lp} \rho_l D_l}{l_{lp}} \frac{(C_l^H - C_{lp}^H)}{1 - C_{lp}^H} = \gamma_{lp} \Omega_l \tag{18.10}$$

where C_{lp}^H is the equilibrium concentration of hydrogen in the liquid at the pore/liquid interface, S_{lp} is the interfacial area concentration of the pores, D_l is the mass diffusivity of hydrogen in the melt, and l_{lp} is a liquid diffusion length at the pore/liquid interface. The second equality expresses the equation in terms of a pore growth factor $\gamma_{lp} = S_{lp} \rho_l D_l / l_{lp}$ and a normalized liquid supersaturation $\Omega_l = (C_l^H - C_{lp}^H)/(1 - C_{lp}^H)$. Figure 18.2 illustrates the boundary layer in concentration around a growing pore, including the interfacial concentration and the diffusion length.

The interfacial area as well as the diffusion length can both be expressed in terms of the pore radius [15]. In the model, the effective pore radius is determined using a model from Pequet et al. [9]

$$r_p = MAX[r_o; MIN(r_{\text{sphere}}; r_{\text{dend}})] \tag{18.11}$$

where r_o is a negligibly small initial pore radius, $r_{\text{sphere}} = [3\varepsilon_p/(4\pi n)]^{1/3}$ (n is the pore number density), and $r_{\text{dend}} = \lambda_2(11 - \varepsilon_s)/(2\varepsilon_s)$. Using Eq. (18.11), all regimes of pore growth can be covered [15]: initial spherical growth, ellipsoidal growth between dendrite arms, and pore merging (creating so-called connected porosity).

From Eq. (18.10) it can be seen that when the average hydrogen concentration in the melt is greater than the concentration at the pore/liquid interface (i.e. a finite supersaturation exists), hydrogen diffuses to the pore and it grows (i.e. f_p increases). If the pore growth factor is large, which occurs for large pore interfacial areas (e.g. a high pore number density) and/or for short diffusion lengths, the two concentrations become equal since the pore growth rate must remain finite. This means that the melt becomes well mixed. Note that the pore volume fraction also changes with the hydrogen density, which is calculated using the ideal gas law.

Equation (18.10) is not solved directly for the pore fraction but is actually solved for the equilibrium hydrogen concentration at the pore/liquid interface. Rather, as described in [2], the pore fraction is determined using continuity, Eq. (18.7). In this way, the effects of feeding flow and solidification shrinkage on the pore fraction are taken into account.

The pressure in the pore is calculated using Sievert's law

$$P_p = \left[100C_{lp}\left(\frac{f}{K_e}\right)\right]^2 \times 101{,}325 \quad (18.12)$$

where f and K_e are the activity and equilibrium coefficients of the hydrogen gas. When no pores are present, the hydrogen partial pressure in the melt is calculated using Eq. (18.12) using C_l instead of C_{lp}. The activity coefficient can be calculated as a function of the silicon and copper concentrations in the melt and the equilibrium coefficient as a function of temperature [15]. The capillary pressure of the pore is modeled as

$$P_\sigma = \frac{2\sigma}{MAX(r_p; r_{nuc})} \quad (18.13)$$

where σ is the surface tension. At present, the effective pore radius at nucleation, r_{nuc}, is used as a parameter to control the maximum capillary pressure at which pores nucleate or start to grow. Porosity is assumed to nucleate if $P \leq P_p - P_\sigma$ and once porosity forms, the melt pressure at that location is forced to $P = P_p - P_\sigma$.

18.4 Results and Discussion

The models described above have been implemented in the commercial MAGMASOFT casting process simulation code for a full coupling to the macroscopic heat flux, mold filling, etc.

18.4.1 Microstructure and Properties

For the prediction of microstructure, depending on the initial composition, one of the phases α-Al, Si, β-AlFeSi will be predicted to form as the primary precipitate (primary growth of Al_2Cu and Mg_2Si has not been modeled due to lack of com-

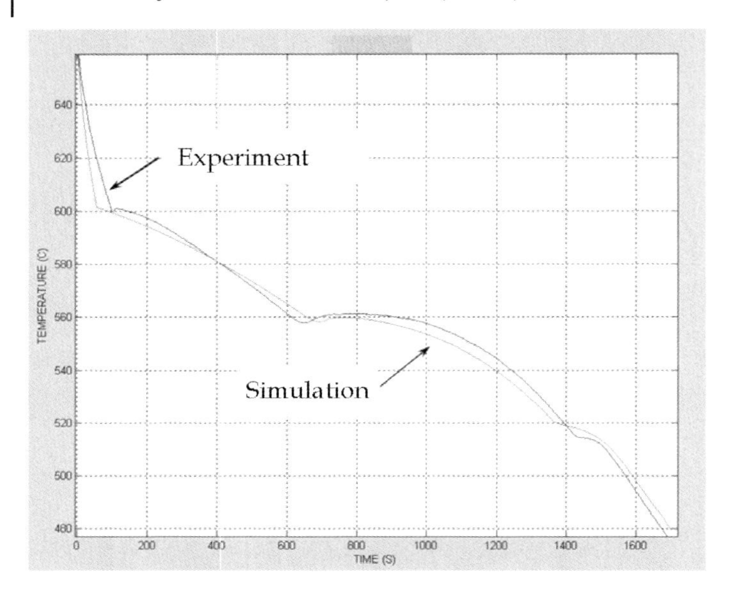

Fig. 18.3 Comparison of measured and simulated cooling curves for an Al–7.16% Si–3.51% Cu alloy.

mercial interest in such alloys). After formation of the primary phase, any of the other phases may form depending on the solidification path, forming up to quinternary eutectics.

As an example, Fig. 18.3 shows a comparison of simulated and measured cooling curves for a 30 mm diameter cylindrical casting in an insulated mold. The alloy is an Al–7.16% Si–3.51% Cu with traces of other elements. In the cooling curves, it can be observed that the recalescence and growth of the α-Al/Si eutectic at approx. 560 °C as well as the formation of Al_2Cu together with α-Al/Si as a ternary eutectic at approx. 520 °C are both well captured by the model. This provides confidence that phase diagram description as well as the growth kinetics in the model is correctly described.

A second example is the simulation of the solidification of a step casting, shown in Fig. 18.4, with thicknesses of 5, 10, 20, 40, and 80 mm cast with the alloy A354. Figure 18.4 also shows the variation of the solid fraction with temperature in the middle of the step cross-section for the various thicknesses. As can be seen, the solidification path varies quite dramatically between the thinnest and the thickest wall sections. This figure illustrates an important aspect regarding microstructure simulation – the fact that the local solid fraction varies based on the local solidification conditions. This is in contrast with a purely macroscopic approach, where kinetic effects are ignored and the solid fraction is specified to be a function of temperature. This is of importance as the solid fraction distribu-

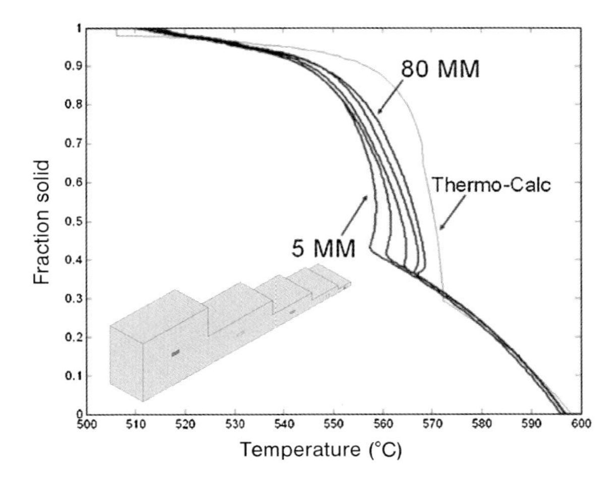

Fig. 18.4 Simulated variation in solid fraction with temperature at the center of the different wall thicknesses in a step-shaped A354 casting.

tion in the casting plays a critical role in the prediction of feeding flows and porosity formation.

Clearly, it is the distribution of the mechanical properties, and not necessarily the microstructure itself, that is of primary interest to the foundry operative or casting designer. Figure 18.5 shows the predicted tensile strength distribution for a sand cast fan wing made of an A354 alloy. The experimentally measured strengths for several probe locations are also shown in Fig. 18.5. It should be

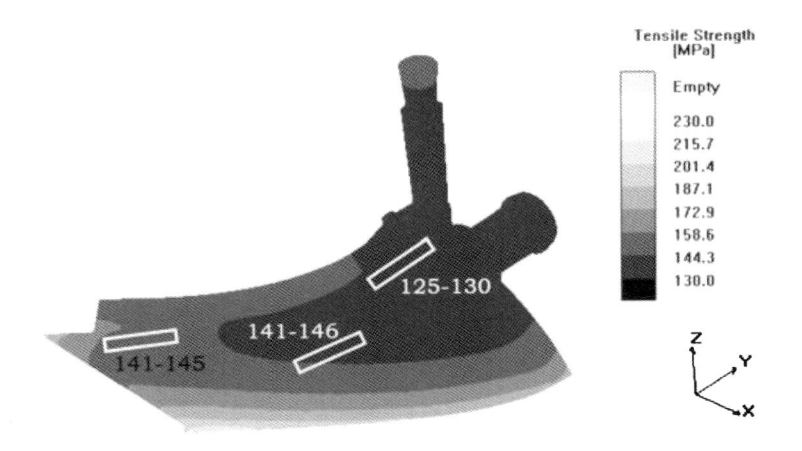

Fig. 18.5 Predicted tensile strength distribution in a sand wing casting made from an A354 alloy.

noted that at present an adjustable quality parameter is used to describe the "defect content" of the melt and the influence of these defects on the mechanical properties [1]. It is anticipated that this quality parameter will, in the future, be coupled to predicted defect distributions, e.g. porosity predictions as described below.

18.4.2 Microporosity

In order to perform parameter studies using the microporosity model, the directional solidification of an A356 alloy has been simulated. It is assumed that there is a linear temperature gradient G that moves through the domain with a constant speed, which results in a constant cooling rate $\dot{T}$. Atmospheric pressure is applied at the right boundary where inflow of feed metal is allowed, and a no flow condition is applied at the left boundary. It should be noted that the porosity results shown here are based on a prescribed solid fraction versus temperature curve and do not currently use the results from the micromodel described above.

Figure 18.6 illustrates the results of the model with profiles of some of the calculated quantities based on a pore density of 10^9 m^{-3}, and an initial pore size of approximately 10 μm (a pore nucleation pressure of 1.6 bar). As can be seen from the liquid fraction curve on the left side of Fig. 18.6, the temperature decreases from right to left, and at the time shown the mushy zone occupies the space from approximately 0.025 to 0.15 m, with a completely solidified layer near the left wall. The figure also shows how porosity starts to form near the beginning of the eutectic reaction where the liquid fraction is around 35% (at approximately 0.065 m). The pore fraction continues to increase until solidification is complete.

The melt pressure drops very little up to the eutectic point, but then drops very steeply due to the rapid decrease in the permeability as the liquid fraction goes to zero. The pore pressure also reflects this change as porosity grows. Up to the point where porosity forms, the pore pressure profile shows how the partial pres-

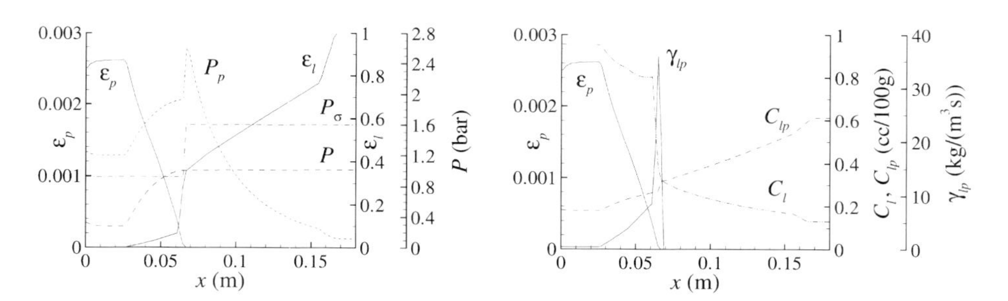

Fig. 18.6 Profiles of phase fractions, pressures and concentrations during the directional solidification of an A356 alloy with a temperature gradient of 500 °C m^{-1}, a cooling rate of 0.5 °C s^{-1}, and an initial hydrogen content of 0.13 cm^3 H_2/100g Al.

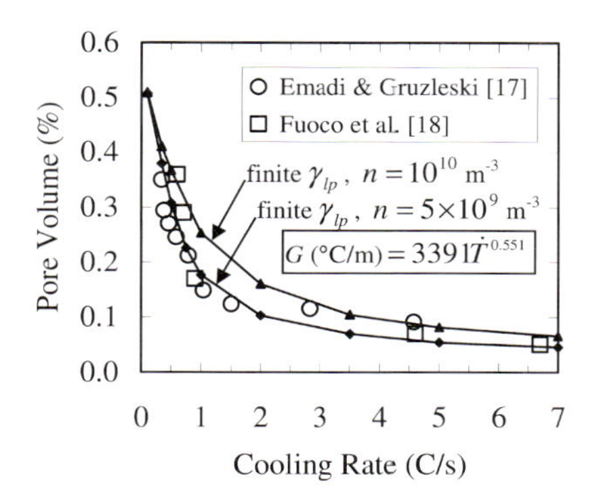

Fig. 18.7 Variation of the pore fraction with cooling rate with a variable temperature gradient, compare with experimental data [16, 17].

sure of hydrogen in the melt increases with decreasing temperature according Sievert's law.

The right side of Fig. 18.6 shows how the hydrogen content of the melt increases up to the eutectic point due to partitioning of hydrogen from the solid to the melt. At the same time, the equilibrium hydrogen content of the melt decreases primarily due to the decreasing temperature. At the point where the two concentrations are equal, porosity starts to form. After this point, the difference between the two concentrations corresponds to the hydrogen supersaturation during pore growth. More extensive details of parameter studies using the model can be found elsewhere [15].

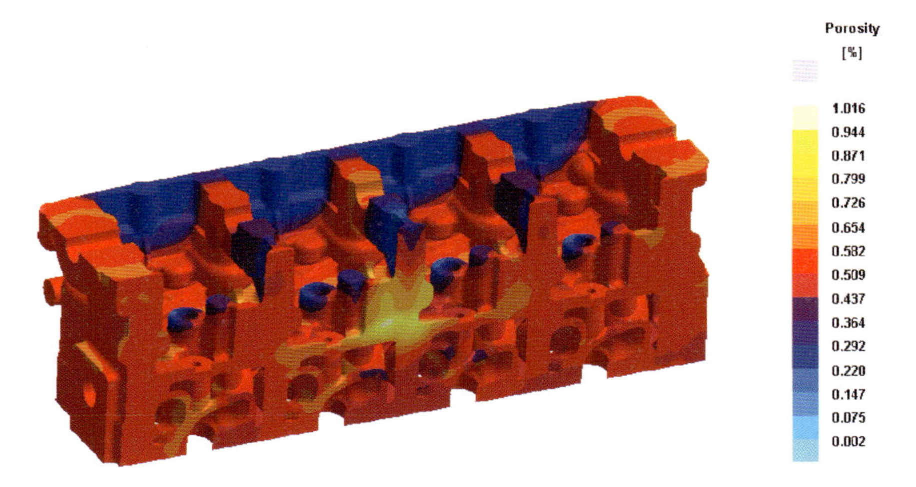

Fig. 18.8 Preliminary simulation results for the predicted porosity in a crankcase casting.

Figure 18.7 shows a comparison of model predictions with measured pore volume fractions [16, 17]. In order to match the conditions in the experiments, which were performed using solidification upwards against a strong chill with well-insulated side walls, the experiments were first simulated without porosity predictions to determine the change in temperature gradient with distance from the chill. The corresponding relationship between temperature gradient and cooling rate is shown in the figure. Using this information, two sets of simulations were performed with initial pore densities of 10^{10} and 5×10^9 m^{-3}. These densities were chosen to approximately match the image analysis data for the cast samples [16]. The predictions for the two simulations fit the measured data quite well. It is important to realize that the steep decrease in pore fractions for cooling rates up to 1 °C s^{-1} could not be predicted if the finite-rate diffusion of hydrogen in the liquid was not taken into account. Also here, more details can be found in [15].

The microporosity model has also been used to simulate the pore fraction and pore size distributions in complex three-dimension castings. A preliminary result for a crankcase casting is shown in Fig. 18.8. The validation of the model for real shape castings in ongoing.

18.5 Conclusions

Models for microstructure and microporosity development in aluminum castings have been presented and shown to reproduce observed distributions. Clearly, further validation of the models using well-documented experiments is required. Work is ongoing to realize a much closer coupling of the models (i.e. consideration of the solid fraction and density distributions from the microstructure model to determine the feeding flow and pressure distributions in the microporosity model or using information from the microstructure predictions in the pore size model). Further, the consideration of local defect predictions in the determination of mechanical property distributions will be addressed in future work. It is anticipated that the tools described here will support the design of castings to meet service requirements while simultaneously supporting a cost-effective production.

Acknowledgment

Portions of this work (microstructure modeling) have been performed within the IDEAL project supported by the European Commission (5th Framework Program), which is gratefully acknowledged.

References

1 M. Wessén, I.L. Svensson, S. Seifeddine, J. Olsson, W. Schaefer, *Modelling of Casting, Welding and Advanced Solidification Process X*, in press.
2 I.L. Svensson, A. Millberg, A. Diószegi, *Int. J. Cast Met. Res.* 16, 2003, 1.
3 H. Cao, M. Wessén, *Metall. Mater. Trans. A* 35A, 2004, 304.
4 S. Seifeddine, J. Olsson, to be published.
5 T. Sjögren, Master thesis, Jönköping University, Sweden, 2001.
6 K.D. Carlson, Z. Lin, R.A. Hardin, C. Beckermann, G. Mazourkevitch, M.C. Schneider, in *Modelling of Casting, Welding and Advanced Solidification Processes X*, ed. D.M. Stefanescu et al., TMS, Warrendale, PA, 2003, p. 295.
7 P.D. Lee, A. Chirazi, D. See, *J. Light Metals* 1, 2001, 15.
8 A.S. Sabau, S. Viswanathan, *Metall. Mater. Trans. B* 33B, 2002, 243.
9 Ch. Pequet, M. Gremaud, M. Rappaz, *Metall. Mater. Trans. A* 33A, 2002, 2095.
10 P.D. Lee, J.D. Hunt, in *Modeling of Casting, Welding and Advanced Solidification Processes VII*, ed. M. Cross, J. Campbell, TMS, Warrendale, PA, 1995, p. 585.
11 P.D. Lee, J.D. Hunt, *Acta Mater.* 45, 1997, 4155.
12 R.C. Atwood, S. Sridhar, W. Zhang, P.D. Lee, *Acta Mater.* 48, 2000, 405.
13 R.C. Atwood, P.D. Lee, *Metall. Mater. Trans. B* 33B, 2002, 209.
14 R.W. Hamilton, D. See, S. Butler, P.D. Lee, *Mater. Sci. Eng.* A343, 2003, 290.
15 K.D. Carlson, Z. Lin, C. Beckermann, G. Mazurkevich, M. Schneider, *Modelling of Casting, Welding and Advanced Solidification Processes X*, in press.
16 D. Emadi, J.E. Gruzelski, *AFS Trans.* 100, 1992, 307.
17 R. Fuoco, H. Goldenstein, J.E. Gruzelski, *AFS Trans.* 100, 1992, 307.

19
Enhanced 3D Injection Molding Simulation by Implementing Applied Crystallization Models

M. Thornagel

Abstract

Shrinkage and warpage of semicrystalling thermoplastic polymers are dominated by four different mechanisms: differential temperature, differential pressure, differential fiber orientation, and differential crystallinity. In addition to the generally considered first three mechanisms 3D-SIGMA has implemented two crystallization models to accurately simulate the crystallization process. The result is an improved shrinkage and warpage simulation and an extensive insight in the crystallization process. Especially the archivation of the melt particle history gives the plastics expert new opportunities to really understand plastics phenomena like surface defects.

19.1
Introduction

Most injection-molded plastics articles for advanced applications are designed for and made of semicrystalline thermoplastic materials (e.g. polyamides used for engine covers, for pump housings or for mobile phone housings). Those polymers are characterized by a relatively high stiffness, high impact strength, and good processability among others. Today all technical plastics applications are subject to strict requirements concerning geometrical tolerances and dimensional stability. To control the article dimensions reproducibly is one of the main challenges to successfully apply semicrystalline polymers to advanced technical applications.

Plastics articles show warpage due to infolded stresses. Typical factors influencing warpage are the packing pressure profile, nonuniform cooling processes, and fiber orientations for short-fiber-reinforced polymers. Additionally, semicrystalline thermoplastics show a considerable decrease of the specific volume due to the development of a partly crystalline morphology while cooling down from melt temperature to ejection temperature and further down to ambient. The significant re-

Integral Materials Modeling: Towards Physics-Based Through-Process Models
Edited by Günter Gottstein
Copyright © 2007 WILEY-VCH Verlag GmbH & Co. KGaA, Weinheim
ISBN: 978-3-527-31711-0

duction in the specific volume causes over-proportional volumetric shrinkage and induces additional internal stresses.

The implementation of two applied crystallization models in 3D-SIGMA, the original Nakamura–Hieber model and a modified Weibull-based model, makes it now possible to really get a look into and understand crystallization effects. 3D-SIGMA archives the complete history of a unit-mass particle starting with entering the runner until it has traveled to its end-position within the cavity. This available particle history, including all experienced phenomena like shear heating in the gate, high cooling rates near the cavity wall or near metal inserts, etc., is the basis for a detailed simulation of the crystallization process. This presentation illustrates the differences between both implemented crystallization models and discusses the simulated crystallization results.

19.2
Current Situation and Motivation

Figure 19.1 shows a typical pvT diagram for semicrystalline polymer materials. Important for thermoplastics is that the specific volume (reciprocal of density) of the polymer strongly depends on temperature and especially on the pressure. The presented two-dimensional form of the pvT diagram is often used to illustrate the injection molding process. The injection molding process starts from the 1 bar isobar with a steadily increasing pressure until the cavity is filled completely. The process continues with a packing phase where the pressure is kept on a certain level to compensate the thermally induced material shrinkage. After sealing of the gate the cooling phase begins with an isochors pressure release until the 1 bar isobar is reached again. The difference between the level of the specific

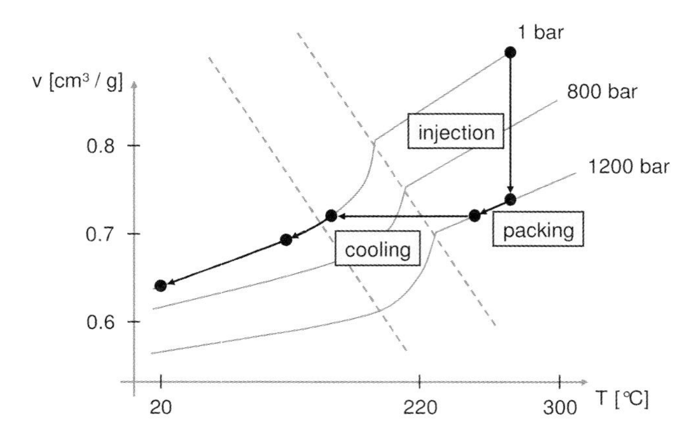

Fig. 19.1 pvT diagram for semicrystalline polymers: injection process.

volume when the 1 bar isobar is reached again to the level at ambient temperature gives a good impression of the remaining shrinkage potential of the article.

Due to the thermal behavior of the mould locally different temperatures appear inside the article (transient temperature distribution), and additionally a transient pressure distribution has to be considered. Both mechanisms lead to differential shrinkage of the part resulting in internal stresses.

Furthermore the reduction of the specific volume makes that the part shrink on mould areas (cores). Due to this geometrical restriction of the shrinkage additional internal stresses inside the article are built up. After ejection of the article from the mould the geometrical restrictions are suddenly removed, and the article strives for a new mechanical equilibrium. This leads to an initial article deflection directly after ejection. Subsequently, the article continues cooling outside the mould at ambient temperature which results in an ongoing shrinkage and warpage process. However when the article is cooled down to ambient completely still a change of the deformation can be measured for some materials even weeks later.

The mechanism behind these short- and long-term warpage phenomena is the crystallization process of semicrystalline thermoplastic materials. The pvT diagram of Fig. 19.1 does not show a bandwidth of the specific volume when the 1 bar isobar at ambient temperature is reached and pressure and temperature are kept constant. In fact for real applications a specific volume distribution can be found inside the article depending on the reached degree of crystallinity. Therefore, the conventional pvT diagram is not sufficient to describe the shrinkage and warpage behavior of semicrystalline thermoplastics completely.

The pvT data are measured with a certain cooling rate, normally at 2 °C min^{-1} (Fig. 19.2). This is at least two orders of magnitude slower compared to the cooling situation inside an injection mold (Fig. 19.3). If the pvT data are measured at higher cooling rates (only few laboratories are able to realize this at the moment)

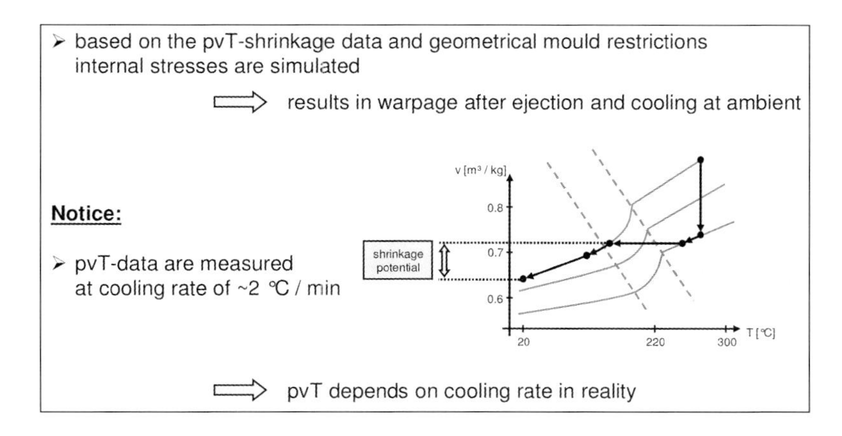

Fig. 19.2 Application of pvT diagram in injection-molding software today.

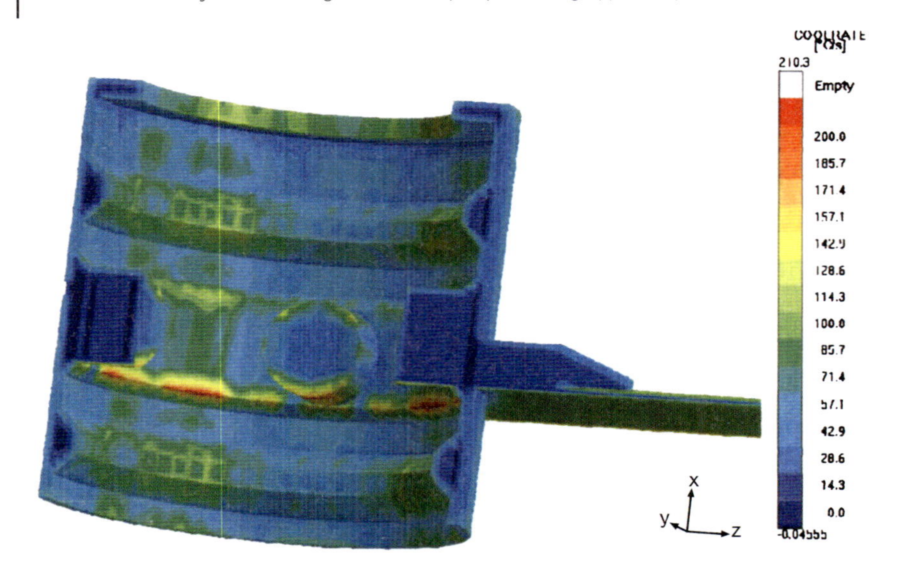

Fig. 19.3 Cooling rates at beginning of cooling phase: cut view.

one can observe a significant change in the pvT behavior of the semicrystalline materials. The transition regions of the isobars are shifting to lower temperatures, and additionally one can get different levels of specific volume at 1 bar/ambient temperature depending on the experienced cooling rates.

19.3
Implementation of Crystallization Models Into 3D-SIGMA

3D-SIGMA takes two actions understanding now the dependencies of the pvT-behavior of semicrystalline polymers on the cooling rates and the interaction with the shrinkage and warpage phenomena. Firstly, the thermal history of any melt particle is simulated and stored now. Secondly, two crystallization models are implemented into the software.

The thermal history of melt particles is needed because a plastics article does not experience a constant cooling rate as often assumed, but every melt particle traveling through the cavity is passing through dynamic cooling rate fields strongly varying with time and location. Particles experience specific pvT situations for every location in the cavity in every time step depending on both the actual thermal/rheological situation and their own history! From this the consequence is that the common global simulation view is no longer sufficient to increase the accuracy of the shrinkage and warpage simulation for semicrystalline materials.

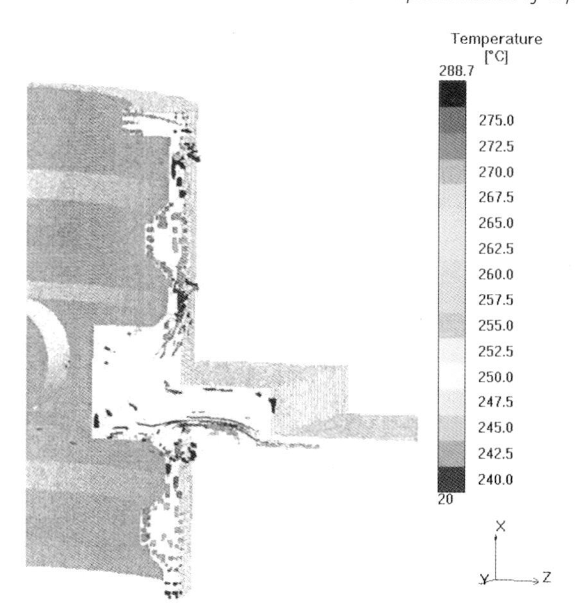

Fig. 19.4 Thermal history of particles of significant importance.

Just to take the specific volume from the conventional pvT diagram (Fig. 19.1) depending on the current level of pressure and temperature in a simulated cell is not enough. This external observation point does not represent the dynamics of the process and the transport of properties (particle history). To take the next step the observation point has to be switched to an internal one, sit onto a melt particle and collect the information it experiences during its travel (Fig. 19.4). The final properties (here: specific volume/crystallinity) of a particle in an article are, therefore, a summation of the experienced states of this particle within a plenty of different pvT-situations. The two in 3D-SIGMA implemented crystallization models both incorporate the complete particle history.

A couple of applied crystallization models were evaluated concerning accuracy of results and effort to feed the models with appropriate material data [1–4]. At the end, the original Nakamura–Hieber crystallization model was chosen. This model is based on an integral kernel representing a symmetric Gaussian distribution which can be interpreted as trigger for the crystallization speed of the polymer material (Fig. 19.5).

This symmetric crystallization speed curve (crystallization speed at start and end are equal) fits to many polymer materials. However, there are also polymers which show completely different crystallization behavior. This was the reason to implement a second 3D-SIGMA crystallization model which is based on a shape-flexible Weibull distribution (Fig. 19.6).

By adjusting the model parameters the shape of the Weibull integral kernel can be adjusted to the real crystallization behavior of the polymer. Figure 19.6 shows a

crystallisation degree: $\xi(t) = 1 - e^{-\left(\int_0^t K(T(t'),p(t')) - K_d(T(t'),\xi(t))dt'\right)^n}$

Integral kernel:
$$\begin{cases} K(T,p) = k_0 * e^{-k_1\left(T-k_2-T_{u1}p-T_{u2}p^2\right)^2} \\ K_d(T,\xi) = 0.01\xi \max(0, T - T_{u0} - T_z * 0.1) \end{cases}$$

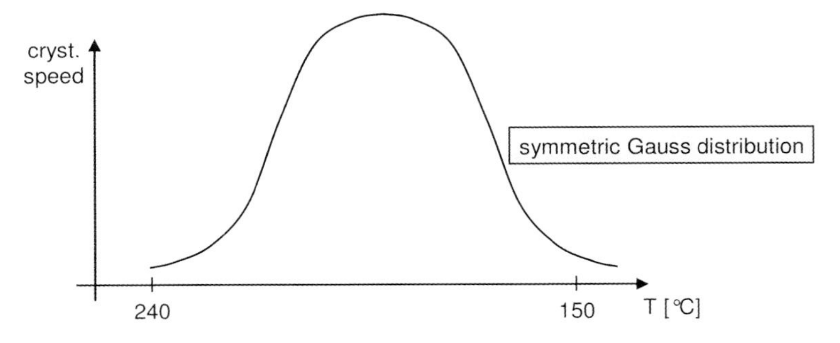

Fig. 19.5 Nakamura–Hieber crystallization model.

cryst. deg: $\xi(t) = 1 - e^{-\left(\int_0^t K(T(t'),p(t'))dt'\right)^n}$

kernel:
$$\begin{cases} K(T,p) = \begin{cases} k(T_{u0} + T_{u1}p + T_{u2}p^2 - T) & T < T_{u0} + T_{u1}p + T_{u2}p^2 \\ 0 : \text{else} \end{cases} \\ k(x) = k_2\,(k_3 x)^{m-1}\, e^{(k_3 x)^{3+m}} \end{cases}$$

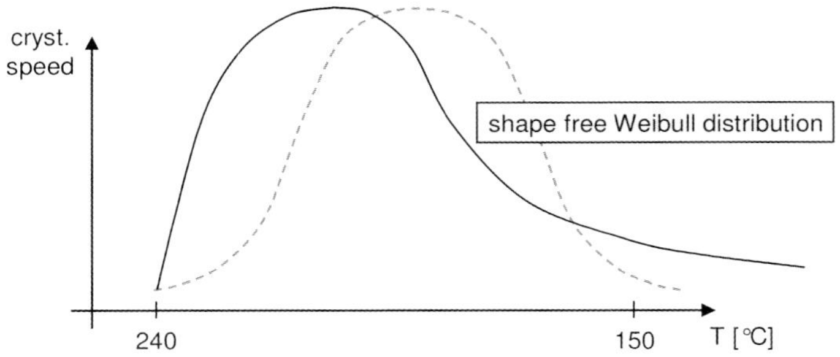

cryst. speed

shape free Weibull distribution

240 150 T [°C]

Fig. 19.6 3D-SIGMA Weibull crystallization model.

Weibull distribution for a polymer which rapidly starts to crystallize at high temperatures but it continues crystallizing even at low temperatures; therefore, the crystallization speed slows down extremely at the end.

19.4 Crystallization Results: Parameter Study

Figure 19.7 shows an example geometry which is used to study effects of the crystallization models. The melt is injected at the backside of an insert; it flows around the insert and subsequently into the plate shaped area with an additional

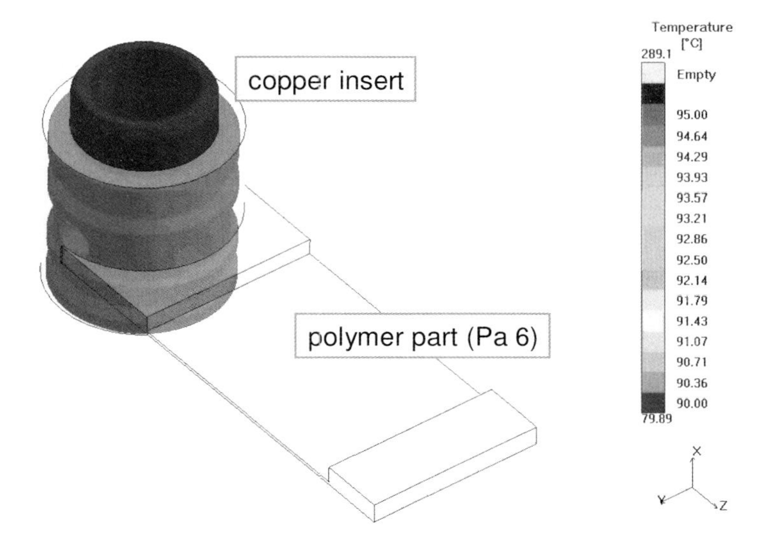

Fig. 19.7 Example geometry: insert molding.

thickness variation. For the insert material copper was chosen and for the plastics material a standard polyamide 6 grade. The initial melt temperature was 285 °C, the initial mold temperature 80 °C, and the initial temperature of the insert was varied in two steps. The cavity was filled by a constant flow rate of 20 $cm^3\ s^{-1}$, and subsequently a packing pressure of 500 bar was applied for 2 s. The mold was opened after 10 s additional cooling time and the part ejected.

The differences between the two implemented crystallization models are shown in Fig. 19.8. Presented are the crystallization degrees at ejection time in a sliced view, i.e. the article is cut into half. Both results refer to the same legend on the left-hand side. The presented crystallization degree is a relative number corresponding to the really achievable amount of crystallinity in the considered polymer material (polyamide 6 grade). For example if a polymer reaches a maximum crystallinity of 63% (37% are still amorphous), the simulated crystallization number of 1 for 100% has to be interpreted in such a way that the maximum possible 63% crystallinity is reached now.

Both results presented in Fig. 19.8 show qualitatively the same crystallinity distribution. The crystallization starts in the thin-walled areas of the part and additionally near the copper insert. The rest of the article is still on such a high temperature level that crystallization is not possible yet. In fact this is a strong indicator that the cooling time is too short, and two consequences can be predicted. Firstly, the article will show a considerable out of the mould shrinkage and, therefore, a high warpage potential. Secondly, the ejection of the part becomes a challenge due to the still low stiffness.

The comparison of both result plots show that the article simulated with the Weibull-based model shows a significantly higher crystallinity at ejection time. Having the underlaying crystallization speed curves of both models (Gauss versus

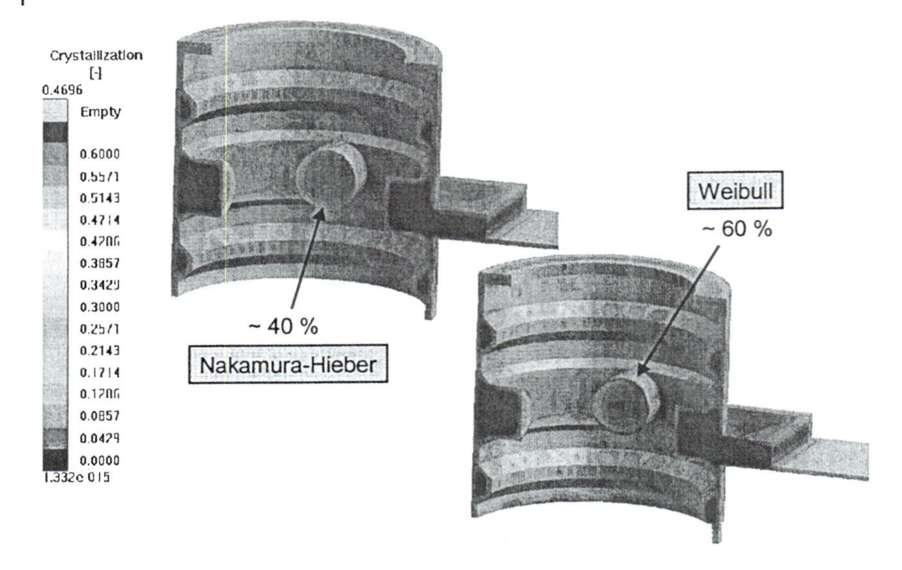

Fig. 19.8 Degree of crystallization at ejection: comparison of both models.

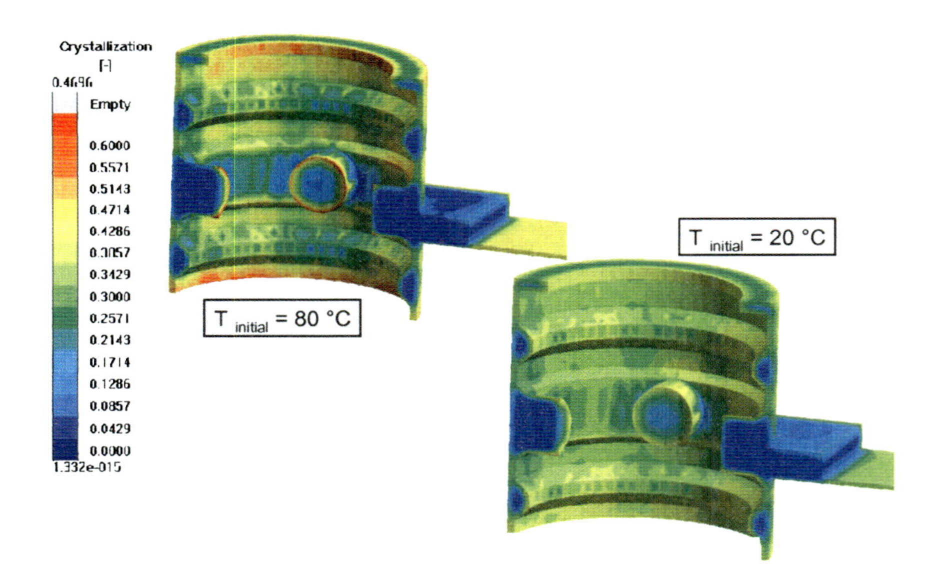

Fig. 19.9 Variation of initial temperature of copper insert at ejection.

Weibull) in mind these differences are not surprising. Figure 19.8 clearly shows the effect of the rapidly starting crystallization speed of the Weibull model.

Figure 19.9 presents the influence of a preheated metal insert on the crystallinity distribution. The article on the left was simulated with the copper insert preheated to 80 °C; the insert already has the same temperature as the mould at beginning of the filling phase. The article on the right was simulated with the copper insert starting at 20 °C. Both simulations were performed with the 3D-SIGMA Weibull crystallization model. Again both result plots refer to the legend on the left-hand side.

The qualitative crystallinity distribution is quite comparable between both results. However the absolute values vary considerably. The left article reaches a maximum relative crystallization number of 60% whereas the right article has not exceeded 40%. These differences can have a significant influence on the part quality and especially on the geometrical dimensions (tolerances).

The preheating of the insert does not only affect the article properties nearby the insert but also downstream the flow path. The thin-walled area of the plate-shaped geometry is a good example to show the effect of the particle history. The article with the preheated insert (80 °C) shows a yellow to orange color map which indicates a crystallinity of approximately 45% in the thin-walled area. The right article with the insert at ambient temperature shows a green to yellow color map representing a crystallinity of approximately 35%. That means particles passing the hot insert experienced more pvT situations where crystallization was possible than those particles which had passed the cold insert.

After 70 s in the cooling phase outside the mold the situation has changed (Fig. 19.10; preheated insert on the upper right side; insert starting from 20 °C on the lower left side). Both articles had time to cool down at ambient temperature

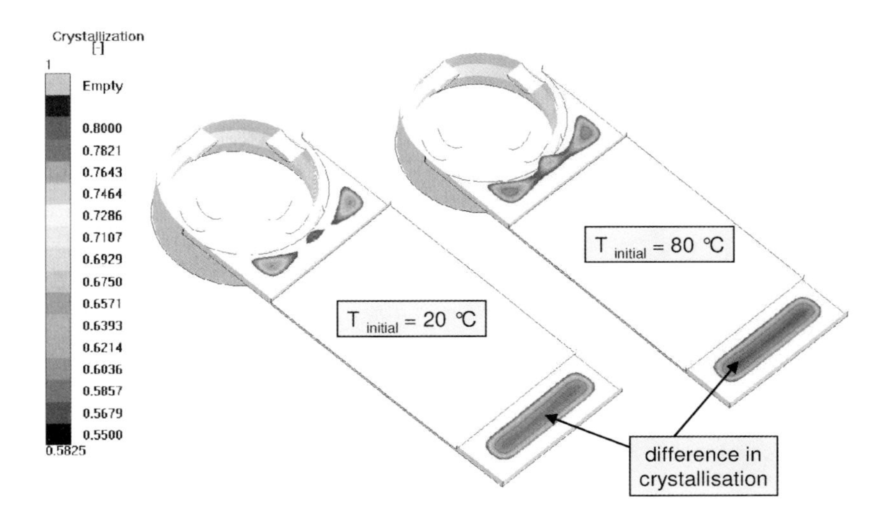

Fig. 19.10 70 s after ejection: regions with crystallinity above 80% not visible.

outside the mold and to continue crystallizing. All regions with crystallinity above 80% are defined transparently in Fig. 19.10. For both simulations regions of crystallinity below 80% are detected in the thick-walled areas of the plate. The article with the cold insert has crystallized further than the article with the hot insert due to a different temperature distribution near the insert. The article with the hot insert has currently started crystallizing in this area while the other article is already in the dropping part of the Weibull distribution (Fig. 19.6). Additionally there is a slight difference in crystallinity at the thick-walled area at the end of the plate between both simulations. These differences are again a result of the different particle histories.

19.5 Conclusions

An accurate shrinkage and warpage prediction for semicrystalline polymers is still an ongoing challenge in plastics technology due to the complex crystallization process of these materials. 3D-SIGMA has taken the step and implemented two applied crystallization models successfully; an original Nakamura–Hieber and a self-developed Weibull model.

To simulate the crystallization process properly the particle history of every melt particle has to be considered. Therefore, 3D-SIGMA has developed an efficient algorithm to track melt particles and collect their experienced pvT situations. Also new is that the pvT data are formulated depending on both pressure/temperature and on the actual cooling rate for every time step and location.

The impact of the now available crystallization results can be split into two effects: on one hand an accurate prediction of the crystallization process during filling, packing, and cooling inside and outside the mold will significantly increase the accuracy of the shrinkage and warpage simulation. On the other hand the crystallization results itself are really interesting. Different shrinkage potentials inside a part become obvious, and the actual causes can be studied. The influence on the part quality of preheating an insert can be investigated, and describing the particle history can be the key to understand and control surface defects.

References

1 M. Moneke, Doctoral dissertation, DKI, TU Darmstadt, Germany, 2001.
2 K. Nakamura, T. Watanabe, K. Katayama, T. Amano, *J. Appl. Polym. Sci.* 10, 1972, 1077.
3 K. Nakamura, K. Katayama, T. Amano, *J. Appl. Polym. Sci.* 17, 1973, 1031.
4 S. Hoffmann, Doctoral dissertation, IKV Aachen, RWTH Aachen, Germany, 2002.

20
Modeling Shearing of γ' in Ni-Based Superalloys

C. Shen, J. Li, M.J. Mills, and Y. Wang

Abstract

The activation energies associated with the formation of a superlattice intrinsic stacking fault (SISF) from an antiphase domain boundary (APB) and the formation of a superlattice extrinsic stacking fault (SESF) from a complex stacking fault (CSF) in γ' phase of Ni-based superalloys are calculated using a microscopic phase field dislocation model in combination with the nudged elastic band (NEB) method. The model incorporates the generalized stacking fault (GSF) energy and hence allows for dislocation dissociations. The results suggest that formation of SESF from CSF rather than formation of SISF from APB is most likely the operating mechanism for the isolated faulting of γ' precipitates observed in the experiments.

20.1
Introduction

In precipitation-hardened alloys, dislocation–precipitate interactions include Orowan looping, precipitate shearing, and dislocation climbing over the precipitates. In Ni-based superalloys all these mechanisms have been observed under different deformation or microstructural conditions [1–3]. The transition in deformation mode depends on precipitate size, applied stress, deformation temperature, and time at temperature (due to microstructural coarsening). The classic strengthening mechanism for Ni-based superalloys has been the shearing of γ' precipitates by pairs of $a/2\langle 110 \rangle$ matrix dislocations coupled by an antiphase domain boundary (APB). Recent experimental observations [4], however, have shown that stacking faults rather than APBs are created in the γ' precipitates during deformation. Some of the possible mechanisms producing either a superlattice intrinsic stacking fault (SISF) or a superlattice extrinsic stacking fault (SESF) rather than an APB in the γ' phase have been proposed, including Kear [5] and Condat [6]. However, no quantitative calculations of the kinetics of these mechanisms exist.

Integral Materials Modeling: Towards Physics-Based Through-Process Models
Edited by Günter Gottstein
Copyright © 2007 WILEY-VCH Verlag GmbH & Co. KGaA, Weinheim
ISBN: 978-3-527-31711-0

Recently the energetics for the formation of SISF and SESF in γ' particles have been investigated using a quantitative microscopic-level phase field dislocation model [7]. In the present contribution the major features of the model and the main results on the activation energies for the formation of these two types of stacking fault are summarized. The results may shed light on the possible mechanisms of γ' shearing and microstructure sensitivity observed in disk alloys during their creep deformation.

20.2 Method

A dislocation can be considered as, instead of a singular defect line, a boundary that divides a glide plane into two areas that differ in inelastic shear displacement. The phase field models for dislocation glide [8, 9] utilize a field description of the inelastic shear displacement for dislocations. Although the inelastic displacement field is generally defined in a three-dimensional (3D) crystal, it is mostly concentrated in glide planes for dislocations and becomes nearly zero in magnitude in the remaining "good" crystal. In the glide plane, the inelastic displacement field varies rapidly only near the dislocation, by which the dislocation core is identified. The use of inelastic displacement or, more generally, the localized transformation strain field simplifies the description of dislocation in several aspects. Since dislocations are merely identified at the positions where the shear displacement changes rapidly on glide planes, the use of inelastic displacement field can naturally treat dislocations of arbitrary number and geometry in a 3D crystal. Introduction of the generalized stacking fault (GSF) energy to the phase field dislocation model [9, 10] allows the system to sample not only the "good" crystal and the dislocations, but also various stacking faults associated with dislocation activities. For example the creation and annihilation of APBs in γ' due to $a/2\langle 110\rangle$ γ-matrix dislocations has been simulated with no *ad hoc* treatment for the stacking faults [11].

20.2.1 Dislocation and Stacking Fault Modeling

Recently a microscopic-level phase field dislocation model was developed to study quantitatively dislocation core structures and various stacking faults [7]. The model uses the inelastic shear displacement field, $\mathbf{u}(\mathbf{r})$, to describe dislocations and stacking faults. The equilibrium state, as well as the dynamic evolution, of dislocations is governed by minimization of a total energy functional

$$E^{total}[\mathbf{u}(\mathbf{r})] = E^{cryst}[\mathbf{u}(\mathbf{r})] + E^{elast}[\mathbf{u}(\mathbf{r})] + W[\mathbf{u}(\mathbf{r})] \tag{20.1}$$

that consists of E^{cryst}, the crystalline energy, E^{elast}, the elastic energy, and W, the mechanical work by external loading. The crystalline energy (known as misfit energy in Peierls models [12, 13]) is a spatial integral of a nonconvex local density function that characterizes the interplanar misfit potential between the crystal halves at the two sides of the glide plane

$$E^{cryst} = \int d\mathbf{r} \frac{\gamma(\mathbf{u}(\mathbf{r}))}{d} \tag{20.2}$$

where γ is the GSF energy and d the interplanar distance. The elastic energy, E^{elast}, based on 3D linear elasticity Green's function solution, is given in a close form in Fourier space as [14]

$$E^{elast} = \int \frac{d\mathbf{g}}{(2\pi)^3} [C_{ijkl}\tilde{\varepsilon}_{ij}(\mathbf{g})\tilde{\varepsilon}_{kl}(\mathbf{g}) - g_i\tilde{\sigma}_{ij}(\mathbf{g})\Omega_{jk}\tilde{\sigma}_{kl}(\mathbf{g})g_l/g^2] \tag{20.3}$$

where $\mathbf{g}$ is the reciprocal space vector, $\tilde{\varepsilon}_{ij}(\mathbf{g})$ is a Fourier transform of the inelastic (transformation) strain tensor $\varepsilon_{ij}(\mathbf{u}(\mathbf{r})) \equiv \mathbf{u}(\mathbf{r}) \otimes \mathbf{n}/d$, with $\mathbf{n}$ being the glide plane normal. $\tilde{\sigma}_{ij}(\mathbf{g}) \equiv C_{ijkl}\tilde{\varepsilon}_{kl}(\mathbf{g})$, $[\mathbf{\Omega}^{-1}]_{jk} = g_i g_l C_{ijkl}/g^2$, and C_{ijkl} is the elastic modulus tensor. The integral, converted to a discrete sum, is taken in the entire Brillouin zone except for a volume element of $(2\pi)^3/V$ at $\mathbf{g} = 0$, with V being the volume of the crystal. The mechanical work under applied stress σ_{ij}^{appl} is given by

$$W = \int d\mathbf{r}\sigma_{ij}^{appl}(\mathbf{r})\varepsilon_{ij}(\mathbf{u}(\mathbf{r})) \tag{20.4}$$

Whereas the traditional gradient energy is absent in the model, the balance between the elastic energy and the crystalline energy naturally gives rise to a diffuse dislocation core at the length scale of Burgers vector. Additionally, relaxation of the inelastic shear displacement field is allowed only within the glide plane, which is the same as in the Peierls dislocation model [12, 13]. However, the phase field model offers the ability of treating arbitrary dislocation configurations. It enables an efficient computation of planar dislocation structures and stacking faults over a large length scale while maintaining a quantitative treatment of dislocation cores [7].

20.2.2
Calculation of Activation Energy

The APB and complex stacking fault (CSF) energies are much higher than that of the SISF and SESF in the γ' phase. However, the formation of SISF or SESF involves the creation of a new $a/6\langle 112\rangle$ dislocation loop. The change in the total energy may also include the interaction energy of the stacking fault with neighboring dislocations and the applied mechanical work, which are all included in Eq. (20.1). The minimum energy pathway and the corresponding activation energy can then be determined by locating the corresponding saddle point on the total energy surface.

An intuitive way is to sample the total energy surface by a series of trial stacking faults with predefined size and geometry. If the "planned" configurations are sufficiently close to that along the actual minimum energy path (MEP) of the fault formation, the total energy mapped out will initially increase with the area of the stacking fault and decrease afterwards, where the saddle point gives the activation energy. Furthermore, relaxation of each configuration via Ginzburg–Landau equation

$$\frac{\partial \mathbf{u}(\mathbf{r})}{\partial \tau} = -\frac{\delta E^{total}}{\delta \mathbf{u}(\mathbf{r})} \tag{20.5}$$

allows producing a more realistic relaxed dislocation core structure, and thus predicts more accurately the activation energy. This approach is used to study the formation of SISF. In Eq. (20.5) τ is the reduced time.

A more rigorous approach is to use the nudged elastic band (NEB) method [15–18], which is now widely applied in theoretical chemistry and condensed matter physics in finding MEP and saddle point on a high dimensional potential energy surface. With the phase field total energy functional (Eq. 20.1) the NEB method can be applied to identify the critical configuration and activation energy of the stacking fault without pre-assumption of its geometry. An example of using this approach to study the formation of SESF is also included.

20.3 Results

20.3.1 Nucleation of SISF from APB

Figure 20.1 shows the GSF energy for γ' phase on a (111) plane and various stacking faults with their associated Burgers vectors. The GSF energy is expressed in a 2D Fourier series on a (111) crystallographic plane with the base vectors being on the reciprocal lattice of the crystal [19], which naturally preserves the crystal symmetry. The Fourier series is fitted to the fault energies: $\gamma_{CSF} = 221$ mJ/m^2, $\gamma_{SISF} = 12.0$ mJ/m^2, and $\gamma_{APB} = 172$ mJ/m^2. The elastic moduli used are $C_{11} = 224.3$ GPa, $C_{12} = 148.6$ GPa, $C_{44} = 125.8$ GPa [20].

The calculation of the activation energy of homogeneous nucleation of an SISF is performed in a cubic computational cell of edge length 200 nm, with periodic boundary conditions applied in all three dimensions. A circular $a/6\langle 112\rangle$ dislocation loop is placed on a prerelaxed APB (of which the shear displacement is automatically determined by the GSF energy) that extends throughout the entire simulation cell on a (111) glide plane. The system size is found sufficiently large that the effect from image dislocations is negligible. Each calculation uses a different radius for the loop, and the initial core profile of the dislocation is single-step sharp. The dislocation core is also relaxed with Eq. (20.5) for a certain number of

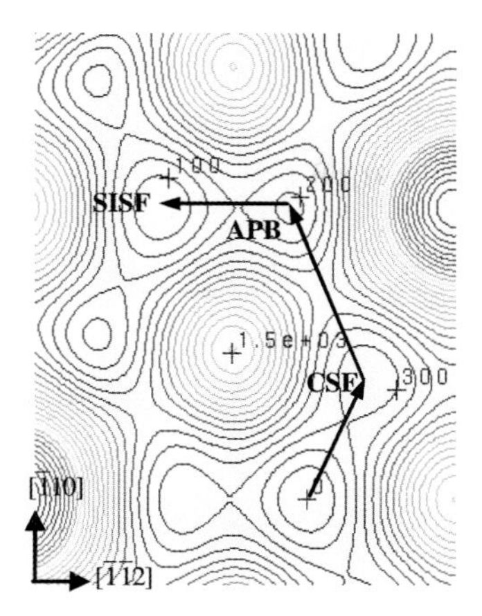

Fig. 20.1 GSF energy (in mJ m^{-2}) of Ni3Al for (111) plane.

iterations that have been found to be sufficient by a separate test on straight dislocations. The total energy of the system is then calculated using Eq. (20.1) for both unrelaxed and relaxed cases, and the results are plotted with respect to the loop radius in Fig. 20.2. In the plot the total energy is shifted by a reference energy of a configuration with pure APB. The activation energies are found to be 27.7 and 17.0 eV with and without core relaxation, respectively.

A more realistic configuration is illustrated in Fig. 20.3, where nucleation of an SISF is considered in a finite APB region with an additional interaction with the

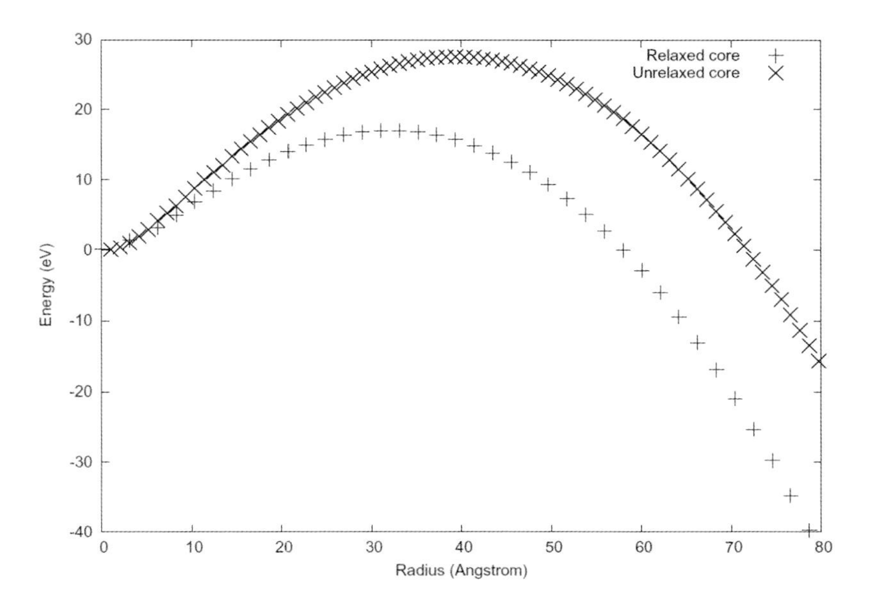

Fig. 20.2 Activation energy of homogeneous nucleation of SISF from APB.

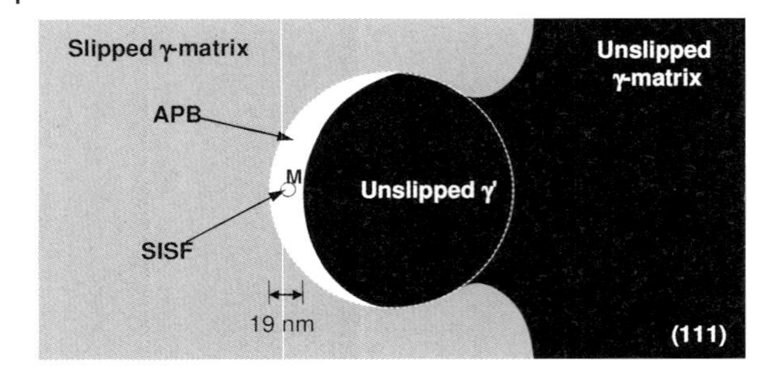

Fig. 20.3 Nucleation of SISF in a finite APB (at the marker M). Slip regions are shown in different shade. The γ' particle is outlined by the dashed circle.

leading $a/2\langle 110\rangle$ matrix dislocation. The simulation size is $424 \times 212 \times 212$ nm^3. The radius of the γ' particle and the interparticle spacing (due to the periodic boundary condition) are 140 and 72 nm, as provided from experimental observation. The initial configuration is produced by applying a resolved shear stress of 693 MPa on the matrix dislocation. The dislocation is forced through the γ matrix between two γ' particles, while in the meantime it partially cuts into the γ' particle. The configuration is found nearly in equilibrium under the applied stress. Afterwards a circular SISF is introduced in the same way as before. The activation energies are found to be respectively 22.7 and 17.7 eV with and without core relaxation (Fig. 20.4).

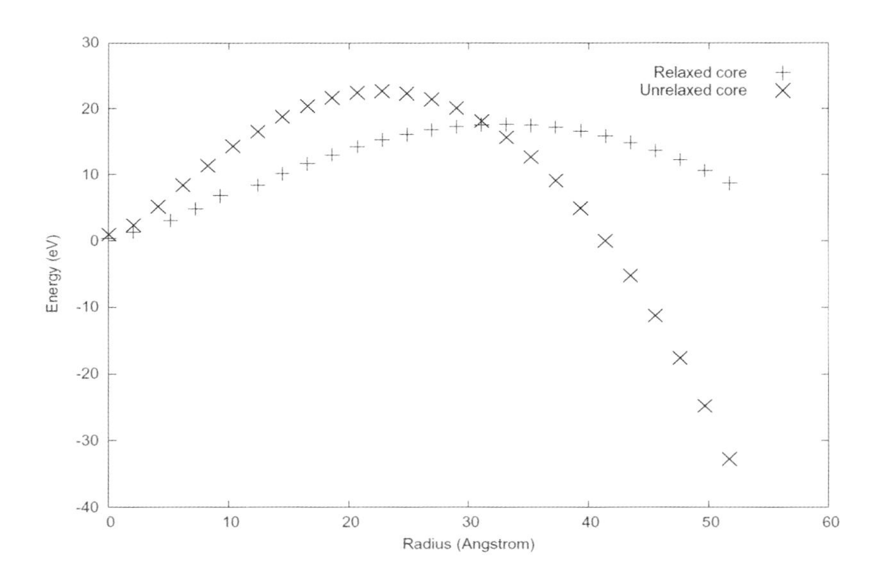

Fig. 20.4 Activation energy of nucleation of SISF in finite APB.

20.3.2
Nucleation of SESF from CSF

The calculated activation energies for the formation of SISF on APB seem too high for the process to be a plausible mechanism. Therefore, we consider the formation of SESF by successive $a/6\langle 112\rangle$ slip in two adjacent (111) glide planes and a subsequent short-range diffusive reordering between the two atomic planes [21]. Furthermore, we assume that the slip in one plane occurs much later than the slip in the other plane, so that the formation of SESF is equivalent to a homogeneous nucleation in the plane one atomic step above an existing infinite CSF. The second assumption may be considered reasonable since the repulsive force between the two identical $a/6\langle 112\rangle$ dislocations, if the leading one is still in the neighborhood, will considerably raise the barrier for SESF formation. The GSF energy in this case involves a two-layer configuration. Prior to a full planar potential being available, we use a GSF energy that is constrained along the $\langle 112\rangle$ direction, as shown in Fig. 20.5. The portion of the energy from CSF to pseudo twin corresponds to a slip on the second plane. At this moment the energy is simply rescaled to match the pseudo twin energy to the known SESF energy that accounts for the short-range reordering. A more accurate calculation is underway.

The system size considered for the calculation is 200 nm × 200 nm × 200 nm. Fifty images are used with the NEB calculation for the MEP and the activation energy. The two ending images correspond to an infinite CSF and an infinite SESF, respectively. All the remaining images are initially created by linear interpolation. We also applied an external stress along the $\langle 112\rangle$ direction with various magnitude to assist nucleation. Figure 20.6 shows the energies under an applied stress of 500 MPa. Since the geometry of the critical configuration is no

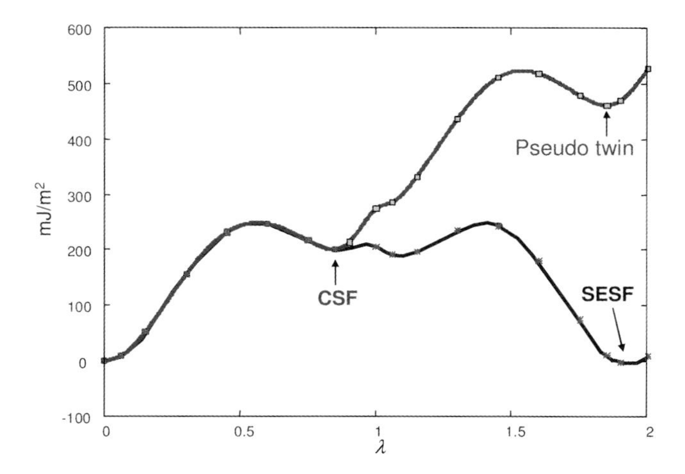

Fig. 20.5 Ni_3Al pseudo twinning multiplane generalized stacking fault energy (along $1/6\langle 112\rangle$) and the rescaled portion (approximately from $\lambda = 1$ to 2) for SESF after short-range reordering. The energy is calculated in VASP with 108 atoms.

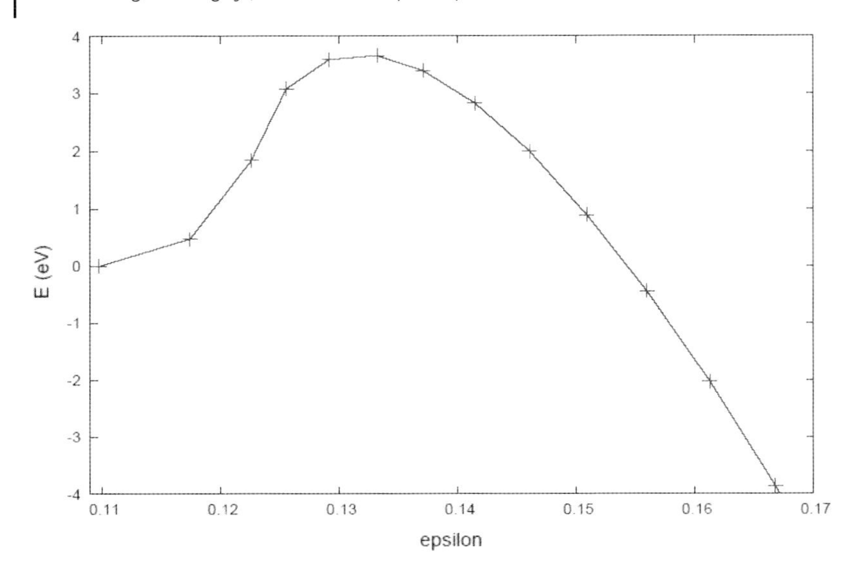

Fig. 20.6 Activation energy of formation of SESF from an infinite CSF.

longer an assigned parameter, we use the total inelastic strain in the glide plane as the indication of the size of the SESF nucleus. The activation energy is found to be 3.7 eV (Fig. 20.6). Moreover, as the applied stress increases to 600 MPa the activation energy reduces to 0.5 eV (Fig. 20.7).

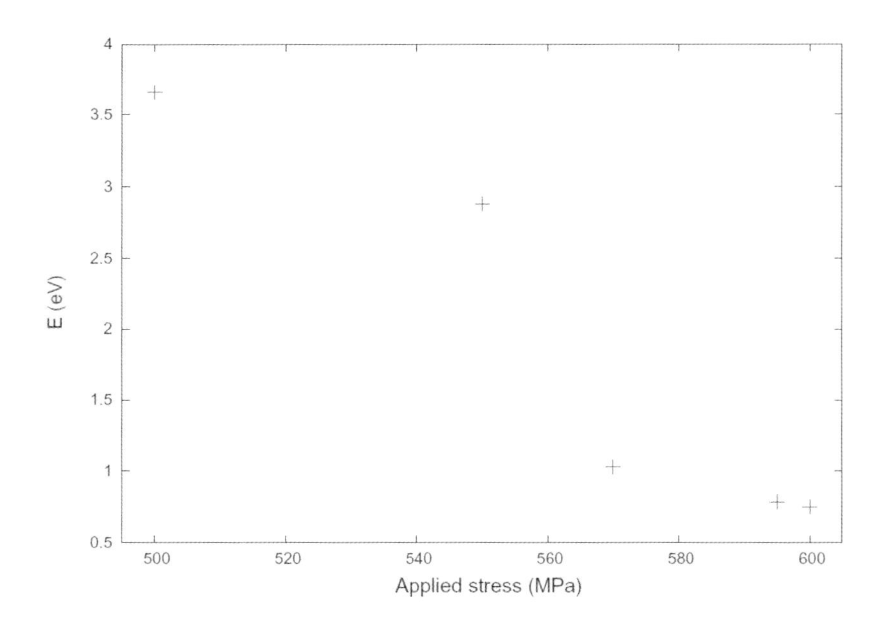

Fig. 20.7 Variation of the activation energy of SESF with applied stress.

20.4 Summary

Activation energies for nucleation of SESF in CSF and nucleation of SISF in APB are calculated using a microscopic phase field dislocation model combined with the NEB method. The results show that the activation energy for formation of an SESF is considerably lower than that for the formation of an SISF. Presence of dislocations in the neighborhood may alter the nucleation barrier. In the case of an SESF, since the leading $a/6\langle 112\rangle$ partial dislocation that creates the first layer CSF has the same Burgers vector as the dislocation that subsequently produces the SESF, their mutual repulsive elastic interaction is expected to increase the activation energy. In the case of SISF, however, the Burgers vector of the leading $a/2\langle 110\rangle$ matrix dislocation is nearly perpendicular to the one that connects APB to SISF. Their elastic interaction has little effect on the activation energy, as indicated by the calculation. Under an applied stress the activation energy for SESF can be reduced considerably, e.g., from 3.7 eV at 500 MPa to 0.5 eV at 600 MPa. Since the two partial dislocations in the two layers of CSF have the same Burgers vector, the optimal stress direction is parallel to the Burgers vector, because it can facilitate the formation of both CSF and SESF. For SISF, since the two Burgers vectors are normal to each other, the effect of the applied stress is reduced. With a rough estimate, the activation energy with an applied stress in the direction 45° to both Burgers vectors (with no relaxation of dislocation core) can be reduced by 30%.

Acknowledgments

This work is supported by US Air Force Office of Scientific Research through the MEANS2 program (Grant No. FA9550-05-1-0135) and by the Office of Naval Research through the D3D program (Grant No. N00014-05-1-0504).

References

1 D. Mukherji, F. Jiao, W. Chen, R. P. Wahi, *Acta Metall. Mater.* 39, 1991, 1515.
2 Y. H. Zhang, Q. Z. Chen, D. M. Knowles, *Mater. Sci. Technol.* 17, 2001, 1551.
3 B. Decamps, S. Raujol, A. Coujou, F. Pettinari-Sturmel, N. Clement, D. Locq, P. Caron, *Phil. Mag.* 84, 2004, 91.
4 G. B. Viswanathan, P. M. Sarosi, M. F. Henry, D. D. Whitis, W. W. Milligan, M. J. Mills. *Acta Mater.* 53, 2005, 3041.
5 B. H. Kear, J. M. Oblak, A. F. Giamei, *Metall. Trans.* 1, 1970, 2477.
6 M. Condat, B. Décamps, *Scripta Metall.* 21, 1987, 607.
7 C. Shen, J. Li, Y. Wang, manuscript in preparation.
8 Y. U. Wang, Y. M. Jin, A. M. Cuitino, A. G. Khachaturyan, *Acta Mater.* 49, 2001, 1847.
9 C. Shen, Y. Wang, *Acta Mater.* 51, 2003, 2595.
10 C. Shen, Y. Wang, *Acta Mater.* 52, 2004, 683.

11 C. Shen, M. J. Mills, Y. Wang. *Mater. Res. Soc. Symp. Proc.*, 753, 2002, BB5.21.21.
12 R. E. Peierls, *Proc. Phys. Soc.* 52, 1940, 23.
13 F. R. N. Nabarro, *Proc. Phys. Soc.* 59, 1947, 256.
14 A. G. Khachaturyan, *Theory of Structural Transformations in Solids*, John Wiley, New York, 1983.
15 G. Mills, H. Jonsson, *Phys. Rev. Lett.* 72, 1994, 1124.
16 G. Mills, H. Jonsson, G. K. Schenter, *Surf. Sci.* 324, 1995, 305.
17 H. Jonsson, G. Mills, K. W. Jacobsen, in *Classical and Quantum Dynamics in Condensed Phase Simulations*, ed. B. J. Berne, G. Ciccotti, D. F. Coker, World Scientific 1998, p. 385.
18 G. Henkelman, H. Jonsson, *J. Chem. Phys.* 113, 2000, 9978.
19 G. Schoeck (private communication).
20 F. Kayser, C. Stassis. *Phys. Stat.* 64, 1981, 335.
21 M. Kolbe, *Mater. Sci. Eng. A* 319, 2001, 383.

21
Minimal Free Energy Density of Annealed Polycrystals

M. E. Glicksman and P. R. Rios

Abstract

Polycrystalline structures are of paramount importance to materials science and engineering. Polycrystals are examples of space-filling irregular network structures that also occur in foams as well as in certain biological tissues. Therefore, accurate description of the characteristics of polycrystals is of fundamental importance. Recently, one of the authors (M.E.G.) has published a paper in which a method of representation of irregular networks by regular polyhedra with curved faces was devised. In Glicksman's method a whole class of irregular polyhedra with a given number of faces, N, is represented by a single symmetrical polyhedron with N curved faces. In this paper these special polyhedra are used to determine the minimum free energy density of annealed polycrystals. Moreover, it is shown that the critical average N-hedra (ANH), that is, the ANH which has $\sim$13.4 flat faces, provides the theoretical minimum partition area cost in three dimensions.

21.1
Introduction

A distinct feature of metallurgy and materials science is that it integrates many areas of knowledge, including chemistry, physics, and mathematics, to solve materials-related technological problems. Polycrystalline networks represent an interesting subject that attracts attention from both the basic and applied ends of metallurgy and materials science. For example, modern steel production is increasingly making use of ever finer grain sizes to improve mechanical properties, and one's knowledge of the geometry and behavior of polycrystalline networks assumes yet greater practical value. The subject of efficient partitioning of space in connected networks also covers basic scientific areas that for many years have attracted the attention of scientists who work on the fundamentals of polycrystals or foams. Indeed, the efficient partitioning of 3D space ($\mathbb{R}^3$) is an old problem,

Integral Materials Modeling: Towards Physics-Based Through-Process Models
Edited by Günter Gottstein
Copyright © 2007 WILEY-VCH Verlag GmbH & Co. KGaA, Weinheim
ISBN: 978-3-527-31711-0

initially discussed by Lord Kelvin in 1885 [1]. Kelvin proposed that the most efficient partition of $\mathbb{R}^3$ – the one yielding the lowest area or free energy "cost" per unit of volume – would be achieved by a body-centered cubic (BCC) stacking of 14-sided tetrakaidecahedra. Kelvin's conjecture prevailed for over a century, until, in 1994, Weaire and Phelan [2] demonstrated another more efficient solution to the partition problem (see Fig. 21.5). Another important result bearing on this problem was published by Kusner [3] in 1992. Kusner proved that the lower bound to the average number of faces in a minimally partitioned network in $\mathbb{R}^3$ space is equal to

$$\langle N \rangle_{\min} = \frac{6 \arcsin(3)}{3 \arcsin(3) - \pi} = 13.3973326\ldots \tag{21.1}$$

The importance of the Kusner mathematical proof for the present work will become clear in what follows.

As already mentioned, an accurate, quantitative description of the characteristics of polycrystals is of fundamental importance. A complete description of such a network includes knowledge of the geometric characteristics of the individual component crystals, their crystallographic orientation, their texture, and the nature of the interfaces between individual crystals. Not only is it important to know these network properties at a point in time, but one must be able to predict dynamic behavior when such networks evolve over time because of grain growth. Nonetheless, even a purely geometric characterization of an irregular network of grains is not easy. One needs to know, for example, the volume and shape of the individual grains. This requirement alone requires tedious and time-consuming experimental techniques such as serial sectioning, or more demanding methods such as "disintegrating" a polycrystal, for example, by adding gallium to an aluminum polycrystal and categorizing all the individual grains. In practice, of course, one seldom resorts to such "heroic" endeavors, and just settles for a single measurement, such as the "grain size", using either the ASTM number or the mean intercept length. Even if detailed measurements of all the grains were available, making sense of a vast variety of irregular polyhedra would still be a daunting task.

Recently, one of the authors (M.E.G.) published a paper in which a method was devised of representing irregular networks by regular polyhedra with curved faces [4]. In Glicksman's method an entire class of irregular polyhedra with a given number of faces, N, is represented by a single symmetrical polyhedron with N faces. These polyhedra are "regular polyhedra" with curved faces, constructed in such a way that they satisfy the topological constraints imposed by a space-filling network. Glicksman called these regular polyhedra with curved faces "average N-hedra" or ANHs. In this work they will also be called ANHs, for brevity. This approach simplifies the mathematical treatment of irregular 3D networks. The geometric properties of the ANHs can be calculated exactly. In essence, ANHs act as "proxies" for analyzing irregular network grains, allowing rigorous treatment of problems pertaining to these networks [5–7].

In this chapter the ANH concept is applied to the problem of minimal free energy density of annealed polycrystals, although the results presented here equally apply to foams.

21.2 Construction and Properties of ANHs

A polycrystalline structure is not just a contiguous aggregate of polyhedral grains. Grain boundary energies, or surface tensions, are imposed throughout the network, which must adhere to certain topological constraints. These "network" conditions must also be satisfied by the ANHs, which we use as proxies or representatives of the actual space-filling irregular polyhedral grains.

Any polyhedron, with trivalent vertices, must satisfy Euler's formula relating the number of edges, E, vertices, V, and faces, N,

$$N - E + V = 2 \tag{21.2}$$

In a polycrystalline network, three grain faces meet along a common triple line, and four faces meet at common quadruple point. For an isolated polyhedron this is equivalent to requiring that two polyhedral faces meet along a common edge, and that three edges meet at a vertex. Smith found an expression for the average number of edges per face, p, for an N-faced polyhedron having trihedral vertices [6]

$$p = 6 - \frac{12}{N} \tag{21.3}$$

Moreover, equilibrium conditions require that the three faces meet with interior dihedral angles all equal to $2\pi/3 = 120°$, and four triple lines meet at their common quadruple point, or node, with the "tetrahedral angle" equal to $\arccos(-1/3) \cong 109.47°$. These equilibrium conditions are known as "Plateau's rules", and are overly restrictive, inasmuch as they apply to a polycrystal only if it is in *full* equilibrium at every face, edge, and vertex. A more general and far less restrictive condition is to require instead that three faces meet with the average dihedral angle equal to $2\pi/3$, and four triple lines meet with the average tetrahedral angle equal to $\arccos(-1/3)$. These *average* network topological conditions are obeyed by ANHs. The construction of an ANH may be exemplified by constructing an average 4-hedron, as shown in Fig. 21.1. The starting point for the construction of a symmetrical curved-face 4-hedon is the underlying "primitive" polyhedron, with four flat faces: i.e., the regular tetrahedron. In order to make the tetrahedron satisfy the average network constraints, each of the flat faces must be curved uniformly. When this is done the edges formed by intersection of the curved faces must also be curved. The angle, α, between the normals, $\mathbf{N}$, located at the center of adjacent faces remains the same after they have been curved, but

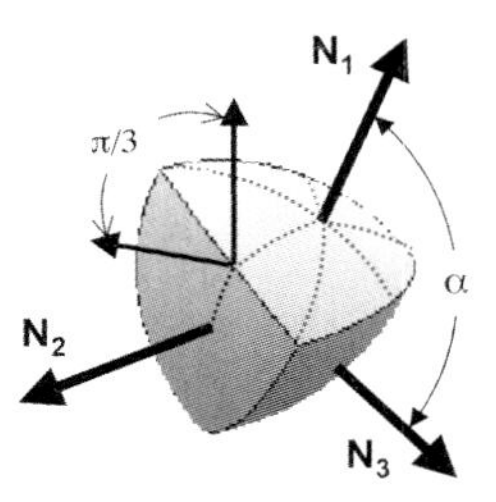

Fig. 21.1 Geometry of the average 4-hedron. Depiction of the interfacial angle, $\alpha \approx 109.5^\circ$, is shown between adjacent face-centered normals, N1 and N2. Owing to the curved faces, the face normals rotate and meet at the edge with a dihedral angle equal to 60° prescribed by network averages.

the angle between normals adjacent to an edge are now smaller than α. If the faces are sufficiently curved, this angle can be made to equal 60°, corresponding to an interior dihedral angle equal to 120°. It is also necessary to curve the edges so that they meet at their common vertex with the correct tetrahedral angle, $\sim 109.47^\circ$. This process may be repeated for every integer flat-faced regular polyhedron with $3 \leq N \leq \infty$. Unfortunately, not every regular polyhedron can be constructed. The reason for this restriction can be understood using Smith's formula, Eq. (21.2). The number of sides per face, p, is determined by the number of faces. For a polyhedron to be constructible, p evidently must be an integer. This condition is satisfied when $N = 3, 4, 6$, and 12, giving in turn $p = 2, 3, 4$, and 5. All of these "constructible" ANHs are illustrated in Fig. 21.2. In Fig. 21.2 the volumes of these ANHs are scaled so that the distances between nearest vertices, λ, remain equal in each case.

Even though ANHs cannot be constructed for other values of N, it is nevertheless possible to find both the topological and metric properties, including curvature, volume, and area edge length. Formulas for these can be found elsewhere

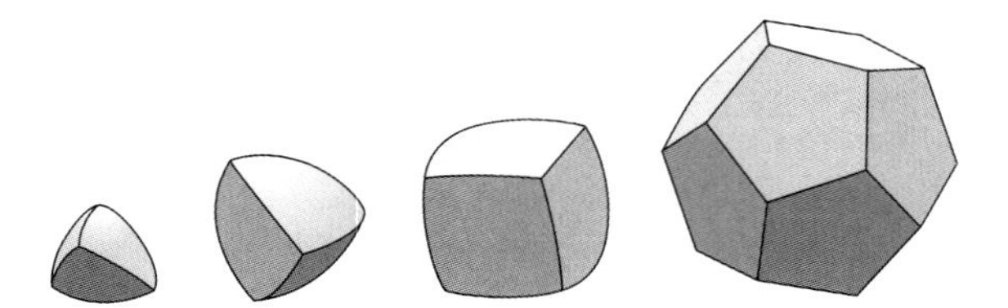

Fig. 21.2 The four constructible ANHs of unit vertex to vertex distance. From left to right: $N = 3, 4, 6$, and 12, with their volumes approximately proportional to 0.350, 0.764, 2.11, and 9.25, respectively. Each unitary ANH in the interval $3 \leq N \leq 13$ consists of a fixed volume enclosed by N identical convex curved faces intersecting at identical curved edges of unit chordal length between vertices. Edges meet three at a time at identical trihedral vertices. Vertices are symmetrically disposed about the volume centroid. All ANHs for $N \geq 14$ have concave faces, and none are constructible.

[4, 5, 7]. Here we present only a few formulae that are needed for the present purposes. An ANH with N identical curved faces has $3(N-2)$ identical edges and $2(N-2)$ identical vertices. The mean curvature of the faces of an ANH, H, is

$$H = \frac{2}{\lambda} \sin\left(\frac{\pi}{6} - \frac{\alpha}{2}\right) \left[\left(\cot\frac{\omega}{2} \sin\frac{\alpha}{2} - \csc\frac{\omega}{2} \sin\frac{\alpha}{2} - \cot\frac{\pi}{p}\right)\right]^{-1} \tag{21.4}$$

where λ is the linear gauge, or scale factor, chosen here as the distance between neighboring vertices on the regular polyhedron. The parameters appearing in Eq. (21.4) are α, the angle between normals located at the center of the ANH's adjacent faces; p, the number of sides per face; and ω, the angle through which an edge turns between vertices, in order that three edges meet at a common vertex with the "tetrahedral angle" of $\sim 109.47°$.

These parameters (α, p, and ω) are analytic functions of N, which may be found through Eq. (21.3) and the following auxiliary functions:

$$\alpha = 4 \arctan\sqrt{1 - 2\sec\left(\frac{\pi}{2(N-2)}\right)\cos\left(\frac{\pi(2N-3)}{6(N-2)}\right)} \tag{21.5a}$$

$$\omega = \pi + 2\arctan\left(\sin\frac{\alpha}{2}\tan\frac{\pi}{p}\right) - 2\arccos\left(-\frac{1}{3}\right) \tag{21.5b}$$

The mean curvature is positive for $N < 13$ and negative for $N > 14$. The exact point at which H changes sign occurs where $N = N_C$. The value N_C is noninteger, and is given by the exact expression

$$N_C = 2 + \frac{\pi}{2\arctan\sqrt{\tan^3\frac{\pi}{12}}} \cong 13.397332571438\ldots \tag{21.6}$$

One notes that Eqs. (21.1) and (21.6) yield identical results; therefore, the critical ANH exhibits exactly the minimum number of faces predicted from Kusners' theorem. The ANH possessing N_C faces is termed the "critical" ANH or N_C-hedron. All faces of the N_C-hedron possess zero mean and Gaussian curvature; in other words, it is an abstract polyhedron, the faces of which are both minimals, and flat. If it were possible to construct a network just using N_C-hedra as the "unit cell" there would not be any pressure difference across any of the flat interfaces and, equivalently, the chemical potential would also remain uniform throughout the polycrystal. Spontaneous interface motions would not occur. Such a network would thereby exhibit the minimum free energy, and it would remain a in state of metastable equilibrium so long as topological transitions were precluded.

Although the N_C-hedron is not constructible, one can calculate its key properties exactly. For example, the area and volume of an N_C-hedron with a gauge λ are given by the expressions

$$A_C = \lambda^2 \frac{3\pi}{2} \left(\frac{\tan\left(\frac{1}{2} \arccos \frac{1}{3}\right)}{\arctan\sqrt{\tan^3 \frac{\pi}{12}}} \right) \cong \lambda^2 \times 24.1773934\ldots \tag{21.7}$$

$$V_C = \lambda A_C \frac{\cot\left(\frac{1}{2} \arccos \frac{1}{3}\right)}{6 \tan \frac{\pi}{6}} \cong \lambda^3 \times 9.8703795\ldots \tag{21.8}$$

21.3
Assessing Network Partitioning: Isoperimetric Quotient and Dimensionless Energy Cost

Two dimensionless parameters have been used to assess the efficiency or free energy cost of certain polyhedra or combinations of polyhedra: (1) the isoperimetric quotient, Q, and (2) the dimensionless energy cost, G [8]. These quantities are in some sense equivalent; we will mainly use G, but Q will also be included for completeness. The isoperimetric quotient, Q, for any polyhedron, is a dimensionless reciprocal measure of the partition area, A, per unit of enclosed volume, V. The isoperimetric quotient may be calculated using its standard definition

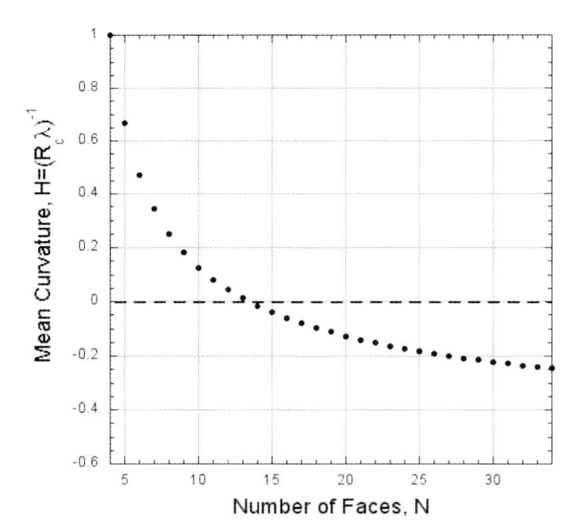

Fig. 21.3 Mean curvature H versus number of faces on ANHs. The values of N cover the practical range encountered in network structures such as polycrystals and foams. The mean curvature divides the population of ANHs: for $N \leq 13$ the mean curvatures are positive, implying shrinkage, whereas for $N \geq 14$ the mean curvatures are negative, implying growth. It is interesting to note that none of the ANHs have exactly zero mean curvature, so that in three dimensions network cells are, on average, either shrinking or growing; i.e. none are conditionally stable polyhedra.

$$Q = 36\pi \frac{V^2}{A^3} \tag{21.9}$$

The particular normalization chosen for Eq. (21.9) sets the largest possible isoperimetric quotient for spheres, for which the partition efficiency achieves its maximum value of $Q_{\text{sphere}} = 1$. A rigorous discussion based on geometric measure theory of the isoperimetric inequality and the existence of area-minimizing surfaces can be found in [9]. All other enclosing surfaces exhibit lower, i.e. less efficient, Q-values than that for spheres [10]. Substituting the expressions derived above for the area and volume of the critical ANH, Eqs. (21.7) and (21.8), respectively, into the formula for the isoperimetric quotient, Eq. (21.9), yields a theoretical measure for the most efficient polyhedron that is statistically capable of partitioning a space-filling network in $\mathbb{R}^3$, namely

$$Q^* = \frac{2}{3}\left(\frac{\cot^3\left(\dfrac{\arccos\frac{1}{3}}{2}\right)}{\tan^2\frac{\pi}{6}} \arctan\sqrt{\tan^3\frac{\pi}{12}}\right) \approx 0.7796356\ldots \tag{21.10}$$

Following the work of Cox and Fortes [8], we define the dimensionless energy cost, G, to construct a polyhedron with N faces as

$$G(N) = \frac{1}{2}\frac{A(N)}{(V(N))^{2/3}} \tag{21.11}$$

The sum of $G(N)$ for all grains in a polycrystal, divided by the number of grains, yields the average dimensionless energy per grain, $\langle G \rangle$, to assemble the polycrystalline network, which is given as

$$\langle G \rangle = \frac{1}{n}\sum_{i=1}^{i=n} G_i(N) \tag{21.12}$$

Figure 21.4 displays a composite plot of dimensionless energy costs, G, versus the number of faces, N, calculated for a variety of constructible polyhedra. Areas and volumes for these polyhedra were obtained by high-accuracy simulations and measurements performed by Cox [11], using Brakke's Evolver program [12], and then converted to dimensionless energy costs via Eq. (21.11). The values calculated for all integer unitary ANHs in the range $4 \leq N \leq 34$ are also shown in Fig. 21.4. The G-values for the ANHs were calculated on the basis of their theoretically derived areas and volumes as explained in Section 21.3. Note the nearly perfect agreement achieved between the simulated and theoretical G-values for the three constructible ANHs that appear on this plot, where $N = 4, 6$, and 12.

Two important observations may be made from Fig. 21.4. The first is that the range of values assumed by the dimensionless energy costs is narrow –

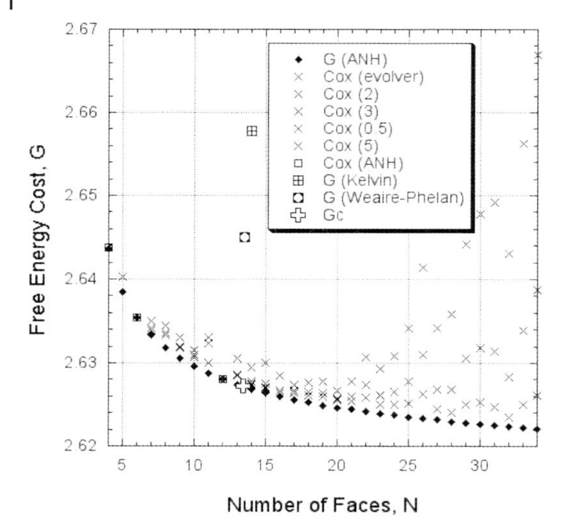

Fig. 21.4 Dimensionless energy cost, G, versus number of faces, N, on a polyhedral cell. Data are shown for a narrow ($\approx 2.6\%$) range of G-values for ANHs (black diamonds); for a selection of constructible irregular polyhedra and ANHs (crosses and open squares, respectively), as measured by Cox [11] using computer simulation; for Kelvin's orthic tetrakaidecahedron at $N = 14$; for the Weaire–Phelan duplex tiling at $N = 13.5$; and for the critical ANH at $N = 13.397$. The G-values for highly symmetric ANHs form a lower bound for every class of polyhedra tested in various topological face combinations. The range of G-values exhibited by various polyhedra with a fixed number of faces increases with N, as many more topological combinations become possible.

amounting to only 2.6% of the average value. The second observation is that the G-values for the set of ANHs shown form a sharp bound to the lowest values calculated from Cox's Surface Evolver data for the selection of constructible polyhedra simulated in each topological class. The most efficient constructible integer polyhedra fall to within about 0.1% of the dimensionless energy cost of their corresponding ANHs. In not a single instance, however, do any constructible polyhedra have a lower dimensionless energy cost than its ANH counterpart. Where the ANHs happen to be constructible, the simulation data and the analytical value agree within the numerical tolerances of the Evolver program.

As a result, one may conclude that ANHs provide both a lower bound as well as an excellent approximation to constructible irregular polyhedra. In other words, these data support the concept that ANHs are an accurate representation or "proxy" for irregular polyhedra encountered in polycrystals. Of special significance is the dimensionless energy cost for the critical ANH

$$G(N_c) = \frac{1}{2}\frac{A(N_c)}{(V(N_c))^{2/3}} = 2.62718\ldots \tag{21.13}$$

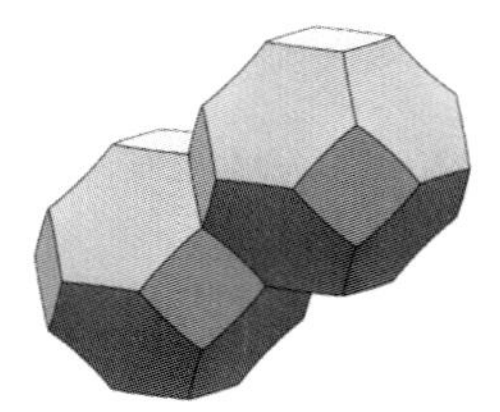
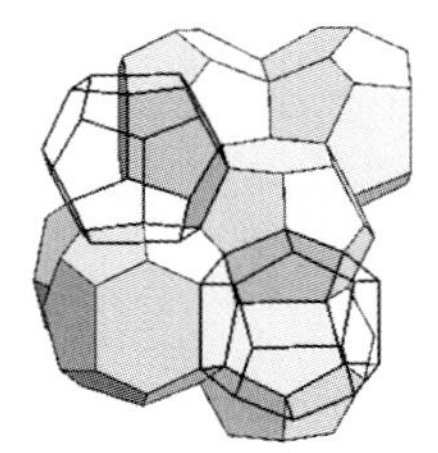

Fig. 21.5 Left: Kelvin's "relaxed" tetrakaidecahedra. Illustration suggests how the flat faced version of these polyhedra would stack by translation to tessellate 3D space and form a body-centered cubic (BCC) lattice. BCC lattice points would be located at the centroids of each Kelvin cell. Right: Weaire–Phelan duplex tiling. This tiling consists of eight stacked polyhedra: two irregular pentagonal dodecahedra (shown as open frames) each isolated by six medial tetrakai-decahedra (shown as solid polyhedra). Cell renderings were adapted from the following URLs: http://www.susqu.edu/brakke/kelvin/kelvin.html and http://www.queenhill.demon.co.uk/polyhedra/wp/wp.htm.

If one imagines a network constructed entirely of critical ANHs its dimensionless energy cost would be $\langle G \rangle = G_c$. The critical ANH value, $G_c = 2.62718$, is lower than the dimensionless energy cost of tiling space with Kelvin's tetrakaidecahedron [1], 2.65737, and with Weaire–Phelan tiling [2], 2.64417. The Weaire–Phelan tiling consists of two, irregular pentagonal dodecahedra and six tetrakaidecahedra. When combined into a unit cell, these eight polyhedra pack to form a duplex unit with the average value of $\langle N \rangle = 13.5$ faces. Kelvin's and Weaire–Phelan's tiling are illustrated in Fig. 21.5. The dimensionless energy cost of the critical ANH is significantly smaller than the others.

Table 21.1 provides a listing of all the unit cells discussed here, along with their associated isoperimetric quotients, Q, and partition cost, G, for the network produced by translation of the unit cell. Other TCP structures have been identified that are not included in Table 21.1. These include the C-15 structure ($MgCu_2$) and the binary Z-phase (Zr_4A_{l3}) [13], which exhibit coordination numbers, or lattice $\langle N \rangle$-values, that equal 13.3333 and 13.4286, respectively. These TCP struc-

Table 21.1 Comparison of partition cost and isoperimetric quotients of unit cells and the partition costs of their respective networks or lattices.

Cell	Q	Network	$\langle N \rangle$	$\langle G \rangle$	$\langle G \rangle / G_c$
KO	0.753367	BCC	14	2.65737	1.01150
KR	0.756977	Const. V	14	2.65314	1.00988
WP	0.764704	A15	13.5	2.64417	1.00648
ANH_c	0.779636	Random	13.397	2.62718	1

KO, Kelvin (orthic); KR, Kelvin (relaxed); WP, Weaire and Phelan; ANH_c, critical ANH.

tures tightly bracket Kusner's bound, $\langle N \rangle_{\min}$ (Eq. 21.1). The relative network partition cost, listed in the last column, is the ratio of the average cost, $\langle G \rangle$, to the minimal cost, $G_c = 2.62718$. As noted in Table 21.1, the constructible relaxed Weaire–Phelan tiling is a scant 0.6% above the theoretical minimal partition for a random network with the critical ANH as its unit cell. Remarkably, even Kelvin's original suggestion in 1887 of the 14-sided Voronoi partition of the BCC lattice as nature's most efficient is only slightly more than 1% above the limiting minimal partition of space found in this work.

21.4 Conclusions

1. The analysis of network structures in $\mathbb{R}^3$ led to the conclusion that average N-hedra, which comprise an infinite set of symmetrical polyhedra, always exhibit the most efficient area partitioning in every topological class. A theorem by Kusner proves that the average number of faces for the minimally partitioned network is approximately 13.397... Kusner's result is identical to the number of faces that we find for the "critical" ANH, which has zero face curvature (both mean and Gaussian). The implication of this finding is that the critical ANH – which is not constructible – provides the statistical unit cell for forming a minimal network with the least partition area (grain boundary area) and the lowest free energy "cost" per unit volume.

2. The implications of this finding to real polycrystals is interesting, insofar as the minimum free energy cost determined here represents an irreducible energy density below which a polycrystal cannot achieve. Grains in polycrystals may undergo complex topological changes that allow them to evolve in time, fill space contiguously, and reduce their boundary areas and energy. Polycrystals undergoing grain growth might approach, or even hover near, the limiting minimum energy density we have calculated. No combination of physical polyhedral (space-filling) grain shapes, however, can exceed the geometric or energetic efficiency of the critical ANH.

Acknowledgments

The authors are deeply indebted to Dr. Simon Cox, Physics Department, Trinity College, Dublin, Ireland, for sharing his Evolver simulation data with us. The authors express their appreciation for the financial support of this study derived from the John Tod Horton Distinguished Professorship in Materials Science and Engineering, at Rensselaer Polytechnic Institute. One of the authors (P.R.R.) is grateful to the Conselho Nacional de Desenvolvimento Científico e Tecnológico, CNPq, and to the Fundação de Amparo à Pesquisa do Estado do Rio de Janeiro, FAPERJ for his financial support.

References

1 W. Thomson (Lord Kelvin), *Phil. Mag.* 24, 1887, 503.
2 D. Weaire, R. A. Phelan, *Phil. Mag. Lett.* 69, 1994, 107.
3 R. Kusner, *Proc. R. Soc. Lond.* 439, 1992, 683.
4 M. E. Glicksman, *Phil. Mag.* 85, 2005, 3.
5 M. E. Glicksman, P. R. Rios, *Z. Metallkd.* 96, 2005, 1099.
6 P. R. Rios, M. E. Glicksman, *Acta Mater.* 54, 2006, 1041.
7 P. R. Rios, M. E. Glicksman, *Mater. Res.* in press.
8 S. J. Cox, M. A. Fortes, *Phil. Mag. Lett.* 83, 2003, 281.
9 F. Morgan, *Geometric Measure Theory*, 3rd edn, Academic Press, 2000.
10 R. Kusner, J. M. Sullivan, *AMS Math. Rev.* 11, 1996, 233.
11 Simon Cox, Trinity College, Dublin, Ireland, personal communication, 2004.
12 K. Brakke, *The Motion of a Surface by its Mean Curvature*, Princeton University Press, Princeton, NJ, 1977.
13 J. M. Sullivan, in *Proc. Eurofoam 2000*, Delft, 2000, p. 111.

22
Modeling Dynamic Grain Growth and its Consequences

P. S. Bate

Abstract

Superplastic forming makes use of a high rate sensitivity of flow stress found in certain fine-grained alloys at elevated temperature. The main microstructural feature of superplastic alloys is a fine grain size, and that fine grain size is invariably stabilized at the high temperatures involved by Zener pinning. Grain growth does occur during superplastic deformation, however, and a mechanism has been proposed for this phenomenon which is based on the deformation perturbing the geometry of the Zener-pinned state. The practical consequence of this growth is in it causing strain hardening, and so it must be included in constitutive laws in some form. Modeling of dynamic grain growth and superplastic deformation is discussed in this chapter.

22.1 Introduction

The phenomenon of superplasticity has been known for many years, and its main features and history have been dealt with in detail elsewhere [1]. It is characterized by very high ductilities in some metal alloys (and some ceramics) with fine grain sizes at relatively high temperatures and slow strain rates. The principal reason for the high ductility is a high degree of sensitivity of flow stress, σ, to plastic strain rate, $\dot{\varepsilon}^{\mathrm{p}}$. A power law linking those quantities is often assumed, but is limited in applicability. Such a relationship does, however, inspire the usual measure of rate sensitivity, m, defined as:

$$m = \left(\frac{\partial \ln \sigma}{\partial \ln \dot{\varepsilon}^{\mathrm{p}}}\right)_{T,\{\Lambda\}} \qquad (22.1)$$

The partial derivative is defined at constant temperature, T, and microstructure, the latter being quantified by a set of state variables, $\{\Lambda\}$. Values of m for superplastic behavior are typically greater than about 0.4, and can be as high as 0.7 or even more.

Integral Materials Modeling: Towards Physics-Based Through-Process Models
Edited by Günter Gottstein
Copyright © 2007 WILEY-VCH Verlag GmbH & Co. KGaA, Weinheim
ISBN: 978-3-527-31711-0

It is notable that superplasticity was observed many years ago, and that the large elongations were considerably in excess of those characteristic of ductile metals at room temperature. This has been largely responsible for the notion of superplasticity as a distinct phenomenon, separate from "normal" deformation and with its own special mechanisms. In fact, there is no clear differentiation and certainly no accepted definition of superplasticity. Materials are reported as superplastic with m values of 0.3 and tensile elongations as low as 200%, indeed, in one case room-temperature superplasticity of fine-grained copper was reported on the basis of its ability to be cold rolled to large reductions [2]. The idea of a distinct microstructural mechanism, or mechanisms, is also less clear than is often assumed. Grain boundary sliding is a feature of high-temperature deformation, and many models of superplasticity assume that this becomes dominant to the point where grains move relative to one another as, essentially, rigid bodies to achieve the deformation. Such relative grain translation is not, however, consistent with certain observations such as the preservation of marker alignments, either artificially introduced [3] or resulting from orientation banding in the microstructure [4, 5]. It does provide a superficially satisfactory explanation of the fact that the intensity of crystallographic texture reduces during superplastic deformation and that grains do not become as elongated as they would in "conventional" plastic deformation. It should be noted that the view that grains remain equiaxed during superplastic deformation is simply incorrect. While the evolution of texture is an interesting subject, it will not be considered here, but the observations of grain shape change are of both academic and practical significance and are related to a phenomenon commonly associated with superplastic deformation, dynamic grain growth.

Nearly all superplastic alloys have a fine grain size, of the order of 1 to 10 μm, and this needs to be stabilized at the relatively high forming temperatures involved. In aluminum alloys, the forming temperature is typically 450–550 °C, and static grain growth would usually be rapid. In all superplastic alloys, the grain size is stabilized by the Zener mechanism [6], using either a small volume fraction of fine particles – as in the commercial aluminum alloys – or a large volume fraction of second phase, which is typical of superplastic titanium alloys. Despite having fine metastable grain sizes, grain growth invariably occurs during superplastic deformation. This is dynamic grain growth, and has been suggested as one common feature of microstructural development in superplasticity [7]. This grain growth is of interest for two reasons: first because it has an important consequence on the mechanical behavior of the material and, second, because of its potential significance in the microstructural mechanisms of superplasticity.

22.2 Mechanical Behavior in Superplasticity

Grain size is the dominant microstructural parameter in superplasticity, and has a large effect on both the flow stress and the strain rate sensitivity. In general

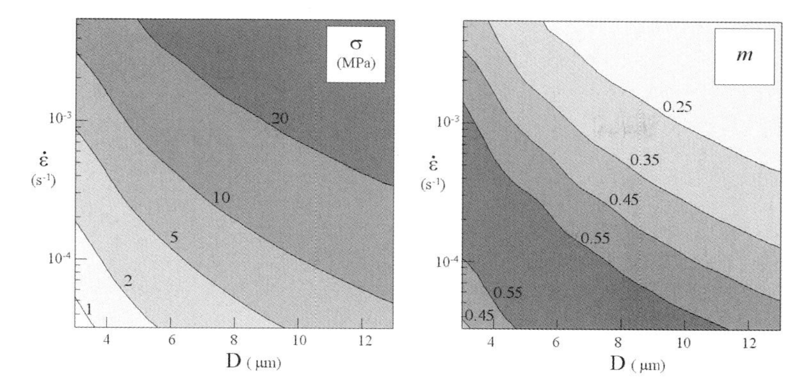

Fig. 22.1 Surfaces of flow stress (left) and m value (right) as functions of grain size, D, and strain rate, $\dot{\varepsilon}$, for superplastic AA5083 (Al–4.5Mg) in uniaxial tension at 530 °C.

terms, increasing the grain size increases the flow stress in the superplastic regime, and reduces the strain rate at which the high values of m occur. The overall relationship can best be demonstrated using surface plots. D'Oliveira et al. [8] used a surface plot to show the m value as a function of stress and grain size, but it is perhaps easier to use a pair of such diagrams giving the flow stress and m value as functions of grain size and strain rate, as shown in Fig. 22.1.

For any given grain size, there is a regime of best (optimum) m value, and as grain size increases due to the dynamic grain growth, this regime shifts to lower values of strain rate. This behavior not only provides the impetus for producing fine-grained superplastic alloys, but also has consequences on the superplastic forming cycle. Because of the low flow stresses involved, quite modest gas pressures can be used in forming sheet, and the pressure–time sequence used naturally influences the strain rates in the component being formed. In general terms, it is sensible to start the forming sequence with relatively fast strain rates – especially as there is little shape complexity at that stage – and reduce strain rates when details of the component are being formed and high m values are most desirable.

In practice, optimization of the forming cycle (forming an adequate product in the shortest time) requires finite element modeling, and that in turn requires adequate constitutive laws. A constitutive law that gives adequate performance – in terms of mimicking mechanical behavior in tensile tests under a variety of strain rate histories – has been given by Ridley et al. [9]. In this, the flow stress at a constant temperature is given by:

$$\sigma = \sigma_0 + C \sinh^{-1}(B\dot{\varepsilon}^{\beta}) \qquad (22.2)$$

where σ_0 is a "threshold" stress, C and β are material-dependant constants, and B depends on the microstructure. Specifically, B is given by a simple power-law relationship involving the grain size, D. This relationship was used to give the data for Fig. 22.1. It follows the rule that total strain should not be included in a constitutive law, but that the effect of prior deformation should be included via the evolution of microstructural state. In this case, as noted above, the microstructural state can be adequately represented by the grain size, and so the evolution is principally determined by the dynamic grain growth.

22.3 Dynamic Grain Growth

The dependence of flow stress on grain size and the evolution of grain size with strain lead to strain hardening. This is of a very different cause than strain hardening in conventional deformation, which is associated with the accumulation of slip dislocations, but its practical consequence in resisting strain localization and increasing overall ductility is the same.

Dynamic grain growth has been reported and discussed by many workers in several different alloys [7, 10–14]. The significance of the dynamic grain growth on strain hardening in superplastic deformation was recognized by Hamilton [15], who, following Wilkinson and Cáceres [14], used a growth rate that was linear with strain. This unfortunately leads to a constitutive law with a dependence on total strain, and a form where the growth rate depends only on the current grain size and strain rate is preferred. In [9], the following form was used:

$$\dot{D} = \lambda D \dot{\varepsilon}^{q} \tag{22.3}$$

The growth parameter, λ, defines the rate of strain-controlled growth, and in cases where there is no static grain growth at the temperature involved, the index q would be unity. However, in many cases the pinning particles coarsen with time. In such cases, $q < 1$. There is a strong possibility that this coarsening is accelerated by deformation, but the reasons for that effect are not clear.

As pointed out by Hamilton [15], the contribution to ductility of the grain growth hardening can be very significant. This can be seen in an experimental result shown in Fig. 22.2, where the exponential strain hardening resulting from dynamic grain growth is indicated.

In fact, if the constitutive laws outlined above are extrapolated to finer initial grain size, then the grain growth hardening becomes even more significant. If grain growth is not affected by the generation of new boundaries during deformation – which eventually occurs – then the Considère quantity determining the onset of diffuse necking,

$$\frac{1}{\sigma}\frac{\partial \sigma}{\partial \varepsilon} \tag{22.4}$$

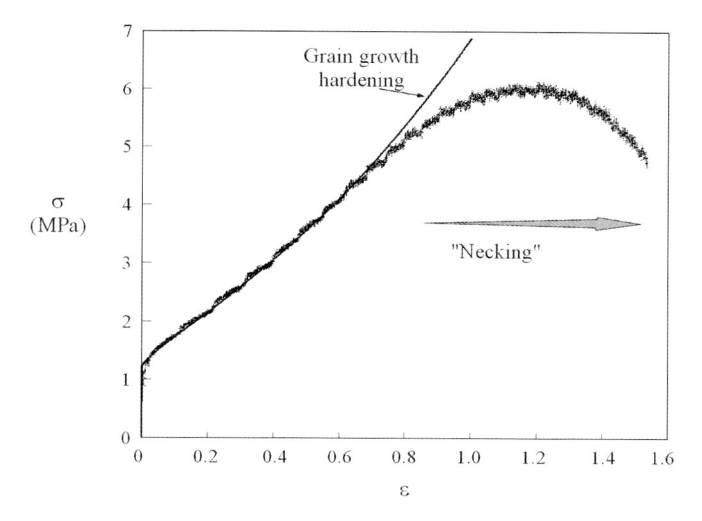

Fig. 22.2 Experimental stress–strain curve for superplastic AA5083 deformed at a constant rate of 10^{-4} s^{-1} at 530 °C. The stress and strain were calculated on the assumptions of constant gauge volume and uniform section: deviations from that behavior, essentially "necking", lead to the downward curve. Prior to that, however, the exponential grain growth hardening is observable.

can be greater than unity for strains in excess of 1.3. Irrespective of the m value, and in the absence of cavitation, that in itself would indicate tensile ductilities greater than 250%. This is potentially as great a factor in the development of fine-grained materials as the effect on the m value mentioned above.

In developing the model of mechanical behavior encapsulated in Eqs. (22.2) and (22.3), no microstructural measurements were made. Of course, the phenomenology of dynamic grain growth was used in Eq. (22.3), but the grain size itself was not measured until after the mechanical model was developed and fitted to experimental mechanical data. It turns out, in fact, that the grain growth parameters agree quite well with those determined by metallography. This method of using the fact that microstructure must determine the internal state, but developing constitutive laws only from mechanical data with *a postiori* correlation with microstructure has been used before, typically in cases of industrial expedience [16]. A fundamental issue is that simple metallographic parameters can differ from the parameters which actually have major influences on mechanical behavior: the "projections" of the microstructure into the metallographic and the mechanical categories are frequently different.

There is a common feeling that it is necessary to justify work on microstructurally based constitutive laws in terms of practical consequence, and it is often thought that detailed microstructural mechanisms can be justified on the basis of allowing generality and application to cases beyond the domain of experimental mechanical test information. This is rather dubious. For example, models of simple work hardening based on dislocation accumulation have been developed for many years now, but none of them would have predicted the orthogonal strain path change effect (e.g. [17]). What should not be ignored is the fact that scientific investigation is a perfectly valid and respectable human pursuit: any practical benefits – and there usually are some – can be considered as a bonus. These comments are valid for dynamic grain growth, and the modeling of the phenomenon at the microstructural scale is the next topic.

22.4 Microstructural Modeling of Dynamic Grain Growth

Two factors have influenced the development of the model for dynamic grain growth described here. First, the phenomenon can be observed in nonsuperplastic alloys as well as superplastic ones, and second, it is observed where the grain size is stabilized by Zener pinning. This effectively rules out an association with any special superplastic mechanism, and indicates that the effect of deformation on the Zener-pinned state needs to be considered. The simplest effect of deformation is to change the geometry of the microstructure, and this was initially investigated using a two-dimensional "vertex" model [18] with a uniform, i.e. homogenized, drag field. This showed that geometric perturbation could lead to dynamic grain growth of the form and magnitude observed in practice. Inhomogeneity of deformation – predicted using a simplified crystal plasticity finite element method – and a reduction of the Zener drag with time – representing particle coarsening – were both shown to increase the growth rate.

Although some results were obtained using a localized Zener drag field to represent the effect of discrete particles [19], simulation in three dimensions with a more realistic representation of particles necessitated using the Monte Carlo Potts (MCP) type of model. Deformation was restricted to simple shear, achieved by unit slips along a sequence of planes in the model cell array. This again showed similar behavior to that observed in practice, including that in a nonsuperplastic Al–6Ni alloy [20]. An example of the prediction of microstructures of dynamic grain growth using MCP modeling is shown in Fig. 22.3, and an example of the predicted growth in mean grain diameter is given in Fig. 22.4.

As noted above, the reason for the growth in these models is simply that the geometry is being changed. The Zener-pinned state is metastable, such that the overall curvature is zero at all locations, at least where all boundaries have the same surface energies. Changing geometry – the configuration of particles and boundaries – will naturally change curvatures and allow growth. Ultimately, the local pinning events need to be considered, but the fact that the vertex model can predict the effect indicates that a coarser-scale view can be taken.

Grains which are initially equiaxed will approximate to spheres, and so to ellipsoids when deformed. At a coarse scale, the curvatures of such ellipsoids should correlate to some degree with the driving pressure for grain growth and so the limiting grain size when Zener drag is present. One interesting aspect of this is

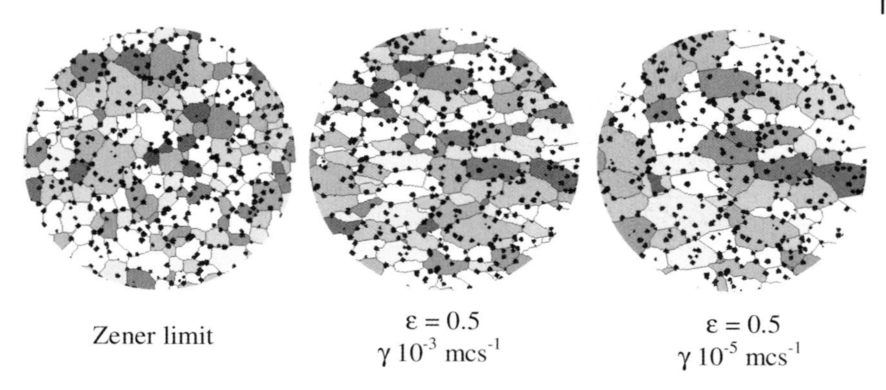

Fig. 22.3 MCP "microstructures" from a three-dimensional simulation of grain growth with a volume fraction of particles $f_v = 0.1$. The gray scales represent different grain orientations; particles are shown as black. The structures are shown at the static Zener limit, and following subsequent deformation at a fast and a slow strain rate. The strain rates are relative to time scales of Monte Carlo steps (mcs); one step is when, on average, all of the domain cells have been checked for transition.

that the development of curvatures with straining depends on the straining state. This is shown in Fig. 22.5 for plane strain and axisymmetric tension.

A given elongation in plane strain generates a wider difference in curvature than in axisymmetric deformation. An effect of straining state on dynamic grain growth was observed in Al–6Ni [20], with plane strain giving growth parameters

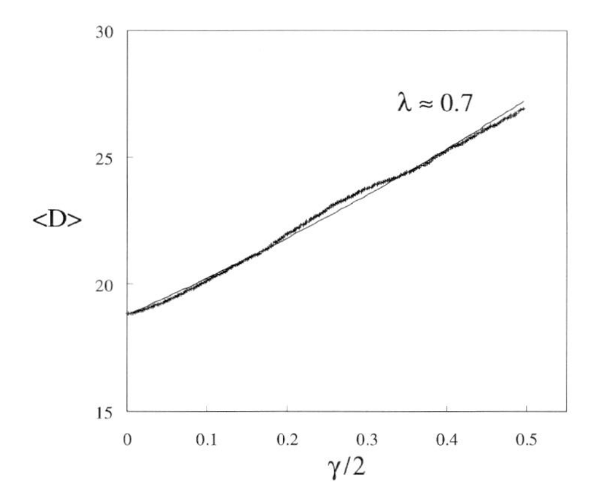

Fig. 22.4 An example of predicted growth in mean diameter, $\langle D \rangle$, in units of Monte Carlo Potts cell size, with half the simple shear strain, γ, for a slow strain rate. The growth parameter, γ, is similar to those observed in experiments.

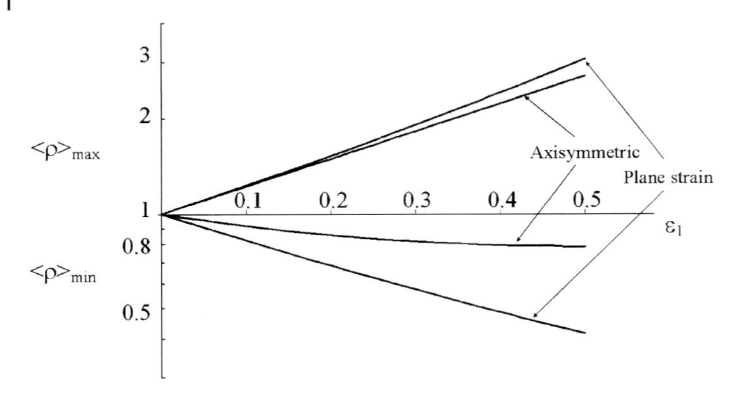

Fig. 22.5 Maximum and minimum mean curvatures, $\langle\rho\rangle$, of ellipsoids resulting from the deformation of a unit sphere in axisymmetric and plane strain tension, as functions of tensile strain, ε_1.

of $\lambda \approx 0.8$, compared to axisymmetric values of $\lambda \approx 0.4$. This is in line with the overall development of curvature. If this difference is a general phenomenon, it will translate into a straining-state dependence of strain hardening, and this will have an effect on practical forming.

This practical consequence can be modeled. The straining state can be quantified by using the second and third invariants of the plastic strain rate, I'_2 and I'_3, to give a parameter ξ:

$$\xi = 1 - \frac{(1/2I'_3)^2}{(-1/3I'_2)^3} \tag{22.5}$$

which varies from 0 in axisymmetric deformation to 1 in plane strain. This can then be used in a modified form of the constitutive law described above by replacing the constant grain growth parameter in Eq. (22.3) by one that depends on the straining state. A simple linear form is:

$$\lambda = \lambda_u(k\xi) \tag{22.6}$$

where λ_u is the axisymmetric growth parameter and k determines the magnitude of the straining state effect. The result of this on predicted thickness strains and grain sizes in a simple free-formed dome of super plastic AA5083 is shown in Fig. 22.6.

The effect on strain distribution in this case is relatively small, but is likely to be more significant in more complex shapes. The effect on grain size distribution is more marked, and can be tested by comparison with experimental results. Preliminary investigation shows that there appears to be an effect of straining state on dynamic grain growth in formed parts.

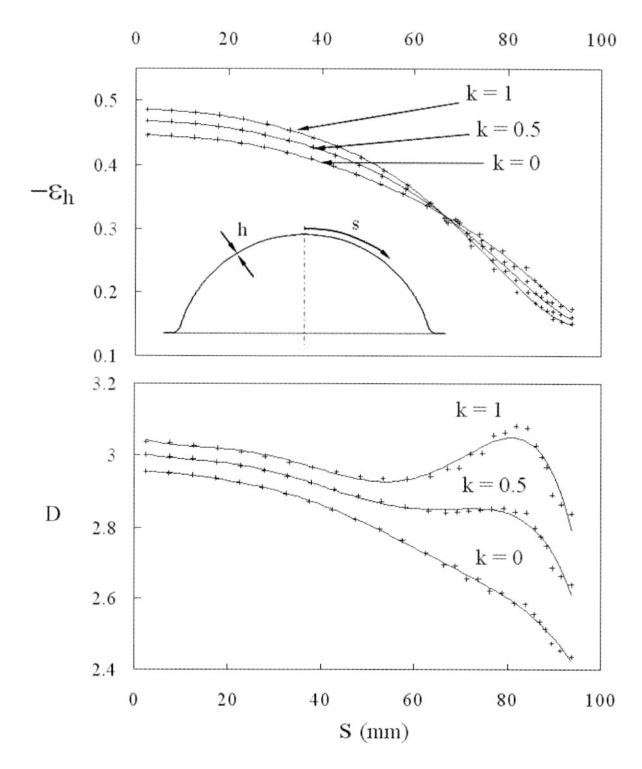

Fig. 22.6 The thickness strain (top) and mean grain size (bottom) predicted using finite element modeling of the free super plastic bulging of a dome with different degrees of strain-state-dependent dynamic grain growth, quantified by *k*. The distribu-tions are shown as functions of perimetric length from the pole of the bulge, as indicated on the inset schematic.

22.5 Conclusion

Dynamic grain growth is an important phenomenon, because it has significant consequences on mechanical behavior in super plastic deformation. As well as shifting the regime of maximum strain rate sensitivity to lower strain rates, it causes strain hardening which makes a very important contribution to ductility and strain uniformity in practical forming. The microstructural phenomenon can be modeled on a simple basis, where the changes in geometry associated with deformation perturb the metastabile Zener-pinned state. Whilst it is not necessary to use explicit metallographic information in constitutive models, a knowledge of microstructural development is beneficial for developing such models and appreciating more subtle effects, such as the possibility of straining-state-dependent hardening.

References

1 J. Pilling, N. Ridley, *Superplasticity in Crystalline Solids*, Institute of Metals, London, 1989.

2 L. Lu, M. L. Sui, K. Lu, *Science* 287, 2000, 1463.

3 P. L. Blackwell, P. S. Bate, *Metall. Mater. Trans.* 27A, 1996, 3747.

4 P. L. Blackwell, P. S. Bate, *Metall. Trans.* 24A, 1993, 1085.

5 P. S. Bate, F. J. Humphreys, N. Ridley, B. Zhang, *Acta Mater.* 53, 2005, 3059.

6 C. Zener; quoted in C. S. Smith, *Trans. Met. Soc. AIME* 175, 1948, 15.

7 J. R. Seidensticker, M. J. Mayo, *Scripta Mater.* 38, 1998, 1091.

8 A. S. D'Oliveira, W. T. Roberts, P. S. Bate, *Mater. Sci. Technol.* 12, 1996, 735.

9 N. Ridley, P. S. Bate, B. Zhang, *Mater. Sci. Eng. A* 410–411, 2005, 100.

10 D. Lee, W. A. Backofen, *Trans. Met. Soc. AIME* 293, 1967, 1034.

11 B. M. Watts, M. J. Stowell, D. M. Cottingham, *J. Mater. Sci.* 6, 1971, 228.

12 M. A. Clark, T. H. Alden, *Acta Metall.* 21, 1973, 1195.

13 M. Suery, B. Baudelet, *J. Mater. Sci.* 8, 1973, 363.

14 D. S. Wilkinson, C. H. Cáceres, *Acta Metall.* 32, 1984, 1335.

15 C. H. Hamilton, *Metall. Trans. A* 20, 1989, 2783.

16 P. L. Blackwell, J. W. Brooks, P. S. Bate, *Mater. Sci. Technol.* 14, 1998, 1181.

17 W. B. Hutchinson, R. Arthey, P. Malmström, *Scripta Metall.* 10, 1976, 673.

18 P. Bate, *Acta Mater.* 49, 2001, 1453.

19 F. J. Humphreys, P. S. Bate, *Mater. Sci. Forum* 357–359, 2001, 477.

20 K. B. Hyde, P. S. Bate, *Acta Mater.* 53, 2005, 4313.

23
Modeling of Severe Plastic Deformation: Evolution of Microstructure, Texture, and Strength

Y. Estrin

Abstract

A model designed to describe deformation of metallic materials up to very large strains is presented. It is based on the concept of a "phase mixture", in which a dislocation cell-forming metallic material is considered to consist of two phases: the dislocation cell walls and the cell interiors. A simple phase mixture rule for stresses – under the assumption that the total strain in the two phases is the same – is then used. The evolution equations for the dislocation densities in the cell walls and the cell interiors, which are treated as the principal internal variables, are at the core of the model, along with certain assumptions regarding scaling between the average dislocation cell size and the variation of the volume fraction of cell walls. A provision for describing texture evolution is also made.

The model was applied to simulate the evolution of microstructure, texture, and strength of several metallic materials under severe plastic deformation by equal channel angular pressing. The results of the simulations confirming a good predictive capability of the model are presented. A most recent development of the model with regard to the evolution of the misorientation angle between the neighboring dislocation cells is also discussed.

23.1
Introduction

While the theory of early stages of strain hardening (stages II and III) has long been established and is generally accepted, there is still some controversy with regard to late strain hardening stages (see [1] for a review). The present author, together with colleagues, has suggested a dislocation-based model that embraces all strain hardening stages in a unified way [2]. In the model, the dislocation cell walls and the cell interiors are considered as separate phases and a simple rule of mixtures for the two phases is used. The model, which was initially two-dimensional, was later extended to three dimensions [3]. It was tested for various

Integral Materials Modeling: Towards Physics-Based Through-Process Models
Edited by Günter Gottstein
Copyright © 2007 WILEY-VCH Verlag GmbH & Co. KGaA, Weinheim
ISBN: 978-3-527-31711-0

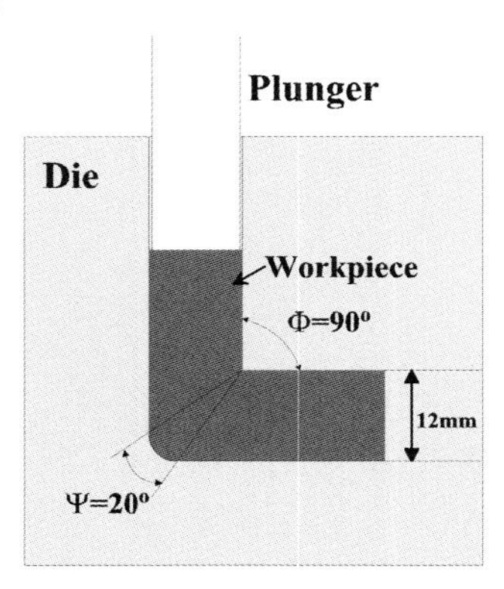

Fig. 23.1 Principal layout of one of the ECAP setups used in the author's laboratory.

large strain situations, such as torsion deformation [4], and was found to represent the large strain material behavior adequately.

Severe plastic deformation by equal channel angular pressing (ECAP) offers itself as a test case to verify the predictive capabilities of the model. Indeed, due to the possibility of repeated pressing through an angular channel (Fig. 23.1) accumulation of strain up to hundreds or even thousands of percent is possible [5, 6]. Besides, ECAP has recently emerged as a very promising technique to produce bulk materials with extremely refined structures [7, 8], and there is a recognized need for ECAP process optimization by means of computer simulations. Below, we shall outline the aforementioned model and illustrate its potential with regard to ECAP simulations vis-à-vis experiment.

23.2 Fundamentals of the Model

The model will be presented in its most advanced 3D form [3]. The cells are assumed to have a cubic shape with cell size d and wall thickness w (Fig. 23.1). The volume fraction of the cell wall then reads:

$$f = \frac{d^3 - (d-w)^3}{d^3} \tag{23.1}$$

Denoting the dislocation densities in the cell walls and the cell interior as ρ_w and ρ_c, respectively, and considering various dislocation generation and annihilation processes in the two "phases", one obtains the evolution equations for these two internal variables:

$$\dot{\rho}_w = \frac{6\beta^* \dot{\gamma}(1-f)^{2/3}}{bdf} + \frac{\sqrt{3}\beta^* \dot{\gamma}(1-f)\sqrt{\rho_w}}{fb} - k_0 \left(\frac{\dot{\gamma}}{\dot{\gamma}_0}\right)^{-1/n} \dot{\gamma}\rho_w \tag{23.2}$$

$$\dot{\rho}_c = \alpha^* \frac{1}{\sqrt{3}} \frac{\sqrt{\rho_w}}{b} \dot{\gamma} - \beta^* \frac{6\dot{\gamma}}{bd(1-f)^{1/3}} - k_0 \left(\frac{\dot{\gamma}}{\dot{\gamma}_0}\right)^{-1/n} \dot{\gamma}\rho_c \tag{23.3}$$

In the above equations it was assumed that the shear rates in both "phases" are the same (denoted $\dot{\gamma}$). Interpretation of the various terms in Eqs. (23.2) and (23.3) is straightforward. Thus, the first term in Eq. (23.2) is associated with dislocations coming from cell interior and getting "entrapped" in the walls, the second term represents dislocation generation by wall sources, and the last term describes dynamic recovery of wall dislocations. Similarly, in Eq. (23.3), the first term stems from injection of dislocations generated at wall sources into cell interiors; the second term (which is a counterpart of the first term in Eq. (23.1)) represents the loss rate of cell interior dislocations running into the walls, and the third term represents dislocation losses due to dynamic recovery in cell interiors. The quantity $\dot{\gamma}_o$ denotes a reference shear rate and b the magnitude of the Burgers vector. The quantities α^* and β^* are numerical constants. The exponent n in the dynamic recovery term is inversely proportional to the absolute temperature in the temperature range below half the melting temperature, while it assumes a constant value in the high-temperature case, the temperature dependence then residing in the coefficient k_o. The average cell size d (which is a "precursor" of the developing grain size) was assumed [2, 3] to scale with the inverse square root of the total dislocation density ρ:

$$d = K/\sqrt{\rho} \tag{23.4}$$

Here K is a constant and the total dislocation density is defined as the weighted sum of the dislocation densities in the two separate "phases":

$$\rho = f\rho_w + (1-f)\rho_c \tag{23.5}$$

Strain hardening is considered by relating the equivalent resolved shear stresses τ_c^r and τ_w^r in the cell interiors and the cell walls, respectively, to the equivalent resolved plastic shear rates $\dot{\gamma}_c^r$ and $\dot{\gamma}_w^r$ (both assumed equal to $\dot{\gamma}$) and the dislocation densities:

$$\tau_c^r = \alpha G b \sqrt{\rho_c} \left(\frac{\dot{\gamma}_c^r}{\dot{\gamma}_0}\right)^{1/m} \tag{23.6}$$

$$\tau_w^r = \alpha G b \sqrt{\rho_w} \left(\frac{\dot{\gamma}_w^r}{\dot{\gamma}_0}\right)^{1/m} \tag{23.7}$$

Here G is the shear modulus, $\dot{\gamma}_0^r$ is a reference strain rate, $1/m$ is the strain rate sensitivity parameter, and α is a numerical constant. The strain hardening behavior of the "composite" is defined by a scalar quantity, τ^r, that is obtained by applying the rule of mixtures:

$$\tau^r = f\tau_w^r + (1 - f)\tau_c^r \tag{23.8}$$

A final "ingredient" that completes this dislocation density-related strain hardening model is the ansatz

$$f = f_\infty + (f_o - f_\infty)\exp(-\gamma/\tilde{\gamma}) \tag{23.9}$$

It describes the variation with strain γ of the volume fraction f of the cell wall "phase" from an initial value f_o to an asymptotic value f_∞ ($< f_o$), the rate of this variation being determined by the parameter $\tilde{\gamma}$. A decrease of f [2, 3] can be rationalized if the increase of the total cell wall area associated with the growth in ρ and the concomitant decrease in d is outstripped by the "sharpening" of the cell walls, i.e. a decrease of their thickness, with elimination of redundant dislocations with progressive straining. The model [2, 3] was shown to provide a unified description of all strain hardening stages (including stage IV), cf. Fig. 23.2. A similarly structured phase mixture model, albeit differing in detail from the one outlined above, was proposed by Zehetbauer [9, 10].

The constitutive model discussed above describes the variation of the density of dislocations and of the cell size as well as the strain hardening associated with it. Variation of the misorientations between neighboring cells, which is believed to lead eventually to grain refinement, can also be monitored, as described in Section 23.3. Combined with a crystal plasticity model, the above constitutive formulation can provide a tool for modeling texture evolution in a polycrystal as well [3].

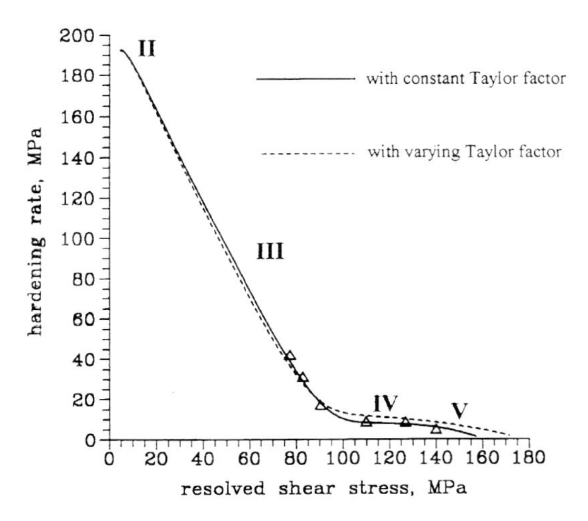

Fig. 23.2 A unified description of hardening stages II to V in a strain hardening versus stress diagram for copper tested in torsion (after [2]). The triangles represent the experimental data [10].

In the preliminary version of the texture part of the model used in the simulations presented below, the misorientations between the dislocation cells within a grain are disregarded, however, and the grain is characterized by a unique equivalent resolved shear strain rate that can be expressed in terms of the individual resolved shear strain rates $\dot{\gamma}_s^r$ on slip systems s as [3]

$$\dot{\gamma}^r = \left[\sum_{s=1}^{N} \dot{\gamma}_s^{r(m+1)/m} \right]^{m/(1+m)} \tag{23.10}$$

Here N is the number of active slip planes in the grain. In the simulation exercises conducted so far, the sum in Eq. (23.10) was calculated by randomly selecting 5 slip systems from among the 12 potentially active slip systems in each grain. In finite element calculations with ABAQUS a full constraints Taylor model with 300 grains per element (initially randomly oriented) was used. Pole figures were determined by aggregating the grain orientations. The results are shown in Section 23.3.

23.3 Application of the Model to the ECAP Process

The constitutive frame outlined above was employed to simulate the variation of the microstructure and the mechanical properties under ECAP, e.g. for Route Bc in which after each pass the workpiece is rotated about its axis by 90°, always in the same direction. Details of the finite element analysis used can be found in [11, 12]. Figure 23.3 shows the evolution of the dislocation cell size with the num-

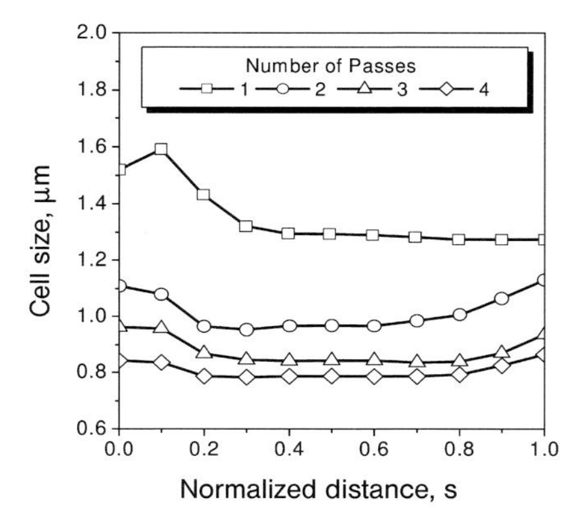

Fig. 23.3 Evolution of the cell size of aluminum with the number of ECAP passes (Route Bc) (after [11]). The nonuniformity of the microstructure, seen as a dependence of the cell size on the distance from the bottom of the workpiece (normalized with respect to the specimen thickness), is seen to become weaker with increasing n.

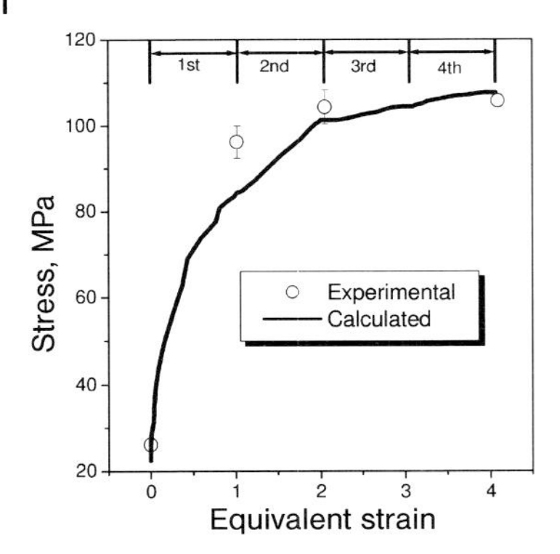

Fig. 23.4 Evolution of strength of Al with the number of ECAP passes (after [11]).

ber n of ECAP passes in pure aluminum obtained in this way. Increasing homogeneity of the microstructure with n is evident from this figure. The rise of the strength is presented in Fig. 23.4. It should be noted that, as in most known cases of ECAP processed materials, both the cell size and the strength tend to saturate after a few ECAP passes.

Similar results were obtained for pure copper, the difference being that Cu exhibits a more pronounced grain refinement, the average grain size reaching values as small as 200–250 nm [12]. As mentioned in Section 23.2, texture evolution can also be traced as part of the simulations based on the model presented. For the case of copper, this is illustrated by the calculated pole figures in Fig. 23.5, showing a reasonable predictive capability of the model.

It was also demonstrated [13] that ECAP of body-centered cubic (bcc) materials, such as IF steel, can be simulated using the same kind of model. The evolution of strength with strain, or the number of passes, for route A ECAP is illustrated in Fig. 23.6 [13].

In a separate exercise [14], the evolution with strain of the average misorientation angle between neighboring dislocation cells was considered. This is an

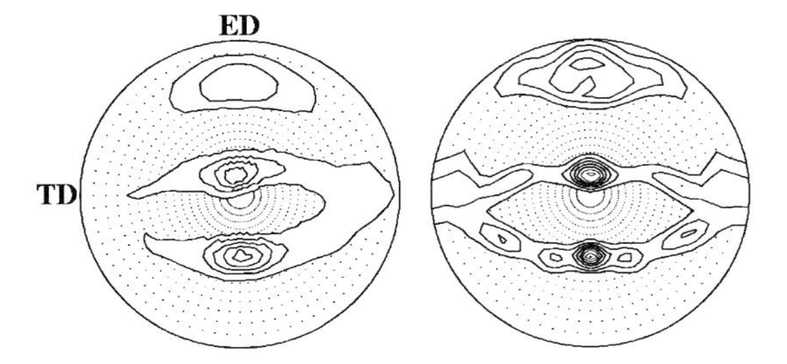

Fig. 23.5 Comparison of calculated (right) and measured (left) (111) pole figures for Cu after four ECAP passes (Route C_r) [12]. Contours are shown in intervals of 0.5 × random density.

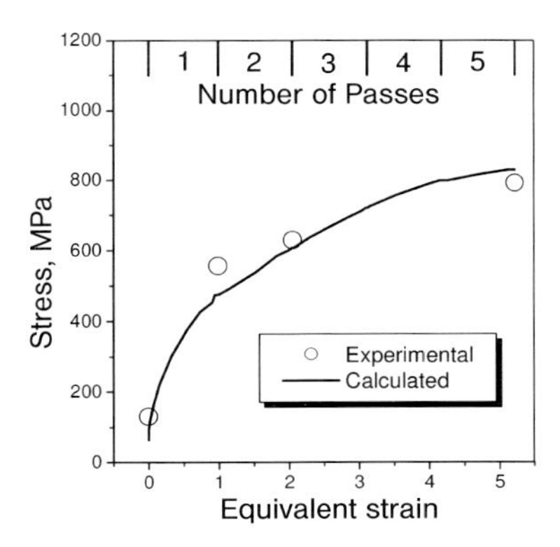

Fig. 23.6 Equivalent stress vs. equivalent strain/number of ECAP passes for IF steel: simulation (solid line) and experiment (open circles) (after [13]).

important issue in view of the underlying assumption that it is the cell structure that eventually, with accumulation of misorientations with strain, turns into a new grain structure. To account for misorientations, the cell wall dislocation density ρ_w is split in two separate populations: the geometrically necessary dislocations which contribute to Burgers vector imbalance in a cell wall and produce a misorientation across the wall, with density ρ_w^g, and the statistical dislocations of density ρ_w^s. The absolute value of the misorientation angle θ between the neighboring cells separated by a cell wall is given by

$$\theta = \arctan(b\sqrt{\rho_w^g}) \cong b\sqrt{\rho_w^g} \tag{23.11}$$

In [14] it was assumed that a certain fraction ξ of the dislocations incoming into cell walls from the cell interiors contribute to the mentioned imbalance in the Burgers vector thus leading to misorientation build-up. This implies that while dislocations with the Burgers vectors of opposite signs arrive at cell walls in equal numbers, locally there is some fluctuational bias that leads to build-up of misorientation. This is expressed by the equation

$$\dot{\rho}_w^g = \xi \frac{6\beta^* \dot{\gamma}(1-f)^{2/3}}{bdf} \tag{23.12}$$

No recovery of geometrically necessary dislocations was included. Furthermore, for simplicity the parameter ξ was considered to be constant. The evolution equation for the average misorientation angle following from Eqs. (23.11) and (23.12) reads

$$\dot{\theta} = \frac{\chi}{2}\frac{\dot{\gamma}}{\theta} \tag{23.13}$$

where

$$\chi = 6\beta^* \xi \frac{b}{d} \frac{(1-f)^{2/3}}{f} \tag{23.14}$$

The evolution equation for the density of statistical dislocations is written as

$$\dot{\rho}_w^s = (1-\xi)\frac{6\beta^* \dot{\gamma}(1-f)^{2/3}}{bdf} + \frac{\sqrt{3}\beta^* \dot{\gamma}(1-f)}{fb}\sqrt{\rho_w^s + \rho_w^g} - k_o\left(\frac{\dot{\gamma}}{\dot{\gamma}_o}\right)^{-1/n} \dot{\gamma}\rho_w^s \tag{23.15}$$

while that for the density of cell interior dislocations, Eq. (23.3), remains unchanged. (Note that ρ_w in the first term of Eq. (23.3) is then given by the sum of ρ_w^g and ρ_w^s.) When the average cell size d and the volume fraction of the cell walls f saturate at large strain, Eq. (23.3) yields basically a square root dependence of θ on strain. In a general case, the whole set of equations constituting the model need to be solved. A dependence which is steeper than the square root one then follows [14], but the average misorientation angles the model predicts are in the range of several degrees only, which does not support the notion of transformation of cell wall structure to a grain boundary structure proper in a quantitative way. Further work is needed to provide an adequate description of the misorientation angle evolution. A tool for describing the variation of the misorientation angle distribution with the ECAP pass number is provided by the Langevin approach based description proposed in [14, 15]. The evolution of the distribution function for the misorientation angle calculated for copper using the ensuing Fokker–Planck equation is presented in Fig. 23.7.

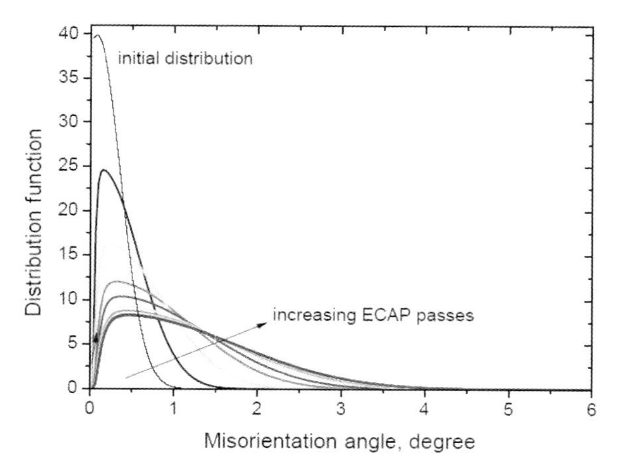

Fig. 23.7 Evolution of the distribution function for the misorientation angle with the number of ECAP passes (copper, [14]).

23.4 Conclusion

The above examples have demonstrated that the model for large strains [2, 3] provides a reliable platform for numerical simulations of the ECAP process. Such aspects as a gradual refinement of the microstructure, particularly a decrease in the dislocation cell size, with the concomitant strength growth, and texture evolution are covered by the model in a more than satisfactory way. Some problems still do exist with regard to an accurate description of the accumulation of dislocation cell misorientations with the number of ECAP passes. If one major goal to be pursued in the near future was to be named in the context of ECAP modeling, then it would certainly be a consistent description of the dislocation cell structure evolution and the implementation of the transformation of this "precursor structure" into a fully developed grain structure with a significant proportion of large angle grain boundaries. Texture simulation that is somewhat eclectic at this stage, as the mentioned transformation is disregarded, will also need to be put on a more reliable footing, in that gradual subdivision of the existing grains changing the grain population during ECAP processing will need to be considered.

However, even in its present form the model already provides a useful means for numerical simulations of ECAP, which permits optimization of the ECAP process parameters with regard to the degree of uniformity, microstructural features and texture and, as an ultimate goal, the mechanical performance of the material produced.

Acknowledgments

The author would like to acknowledge valuable contributions of S.C. Baik, Y. Brechet, R. Hellmig, H.S. Kim, A. Molinari, and L.S. Tóth to the modeling work reported in this paper. Financial support from DFG (Grant ES 74/12-2) is gratefully appreciated.

References

1 U.F. Kocks, H. Mecking, *Progr. Mater. Sci.* 48, 2003, 71.
2 Y. Estrin, L.S. Tóth, A. Molinari, Y. Brechet, *Acta Mater.* 46, 1998, 5509.
3 L.S. Tóth, A. Molinari, Y. Estrin, *J. Eng. Mater. Technol.* 124, 2002, 71.
4 P. Cugy, Y. Brechet, Y. Estrin, O. Bouaziz, to be published.
5 V.M. Segal, V. Reznikov, A. Drobyshevkiy, V. Kopylov, *Russian Metall.* 1, 1981, 99.
6 V.M. Segal, *Mater. Sci. Eng. A* 197, 1995, 157.
7 R.Z. Valiev, R.K. Islamgaliev, I.V. Alexandrov, *Progr. Mater. Sci.* 45, 2000, 103.
8 R. Valiev, Y. Estrin, Z. Horita, T.G. Langdon, M.J. Zehetbauer, Y.T. Zhu, *JOM* 58, 2006, 33.
9 M. Zehetbauer, H.P. Stüwe, A. Vorhauer, E. Schafler, J. Kohout, *Adv. Eng. Mater.* 5, 2005, 245.

10 M. Zehetbauer, *Acta Metall. Mater.* 41, 1993, 589.

11 S.C. Baik, Y. Estrin, H.S. Kim, R.J. Hellmig, *Mater. Sci. Eng. A* 351, 2003, 86.

12 S.C. Baik, R.J. Hellmig, Y. Estrin, H.S. Kim, *Z. Metallkd.* 94, 2003, 754.

13 S.C. Baik, Y. Estrin, H.S. Kim, H.T. Jeong, R.J. Hellmig, *Mater. Sci. Forum* 408, 2002, 697.

14 Y. Estrin, L.S. Tóth, Y. Brechet, H.S. Kim, *Mater. Sci. Forum* 503–504, 2002, 675.

15 W. Pantleon, *Mater. Sci. Eng. A* 400–401, 2005, 118.

Subject Index

Integral Materials Modeling: Towards Physics-Based Through-Process Models
Edited by Günter Gottstein
Copyright © 2007 WILEY-VCH Verlag GmbH & Co. KGaA, Weinheim
ISBN: 978-3-527-31711-0

Related Titles

Wiley-VCH (Ed.)

Ullmann's Modeling and Simulation

2006
ISBN 978-3-527-31605-2

W.D. Callister

Materials Science and Engineering
An Introduction

2006
ISBN 978-0-471-73696-7

W.F. Riley, L.D. Sturges, D.H. Morris

Mechanics of Materials

2006
ISBN 978-0-471-70511-6

G. de With

Structure, Deformation, and Integrity of Materials
Volume I: Fundamentals and Elasticity
Volume II: Plasticity, Visco-elasticity, and Fracture

2006
ISBN 978-3-527-31426-3

D. Raabe, F. Roters, F. Barlat, L.-Q. Chen (Eds.)

Continuum Scale Simulation of Engineering Materials
Fundamentals – Microstructures – Process Applications

2004
ISBN 978-3-527-30760-9